"This rich and diverse collection of essays takes the notion of research imagination to a new level. It challenges us to think about matters of crime, deviance and criminal justice in frames that they have so far rarely been thought within. It makes for a unique and fascinating read."

Katja Franko, *Professor of Criminology, University of Oslo*

"Criminology is a boundless discipline but this was never as obvious as it has become with this book that shoots criminologists and their critical thinking to outer space. This new edited book by Yarin Eski and Jack Lampkin is a mesmerising journey into the future of the discipline. Perceptions of harms and identification of crimes in outer space, and also ethical conundrums, will accompany the creative social scientists of today and tomorrow."

Anna Sergi, *Professor of Criminology, University of Essex*

"Interdisciplinary scholarship in crime, criminal justice, and ethics in outer space have become more and more important along with space exploration. The new excellent book edited by Yarin Eski and Jack Lampkin delves into unknown/strange diversities, complexities, and entanglements. An indispensable guiding star for a new Space Criminology Odyssey 2024."

Noriyoshi Takemura, *Emeritus Professor of Criminal Law and Criminology, Toin University of Yokohama, Japan*

"This book is a groundbreaking achievement, a first and a must read/ride. A truly global accomplishment, contributions hailing from six continents set out the terrain of the study of space crime, criminal justice and ethics, sitting within, yet considerably and tantalisingly extending, the boundaries of criminology."

Steve Tombs, *Emeritus Professor, Department of Social Policy and Criminology, The Open University*

"Crime will always exist despite all our best efforts, so space activities will not stay clear from it. But what is 'space crime'? Does space itself, even if non-living, deserve protection? This book gives a very good overview of, and anticipates the potential harmful acts and behaviour in, outer space by individuals, (big) corporations and nations. It also discusses legal issues and ethical dilemmas, philosophical questions, anthropology and, last but not least, (geo)politics. But foremost, it very clearly brings out, given the irreversible character of some space crime, the need to think about this NOW and see what prevention is possible. Definition, legislation, regulation or at least a code of conduct is urgent, even if enforcement and follow-up will remain problematic. Food for thought!"

Philippe Schoonejans, *Noordwijk Space Expo Board Member and former project manager in ESA's space exploration programme*

Crime, Criminal Justice and Ethics in Outer Space

Breaking new ground in criminology, this book reflects on the expansion of outer space endeavours, the new pathways this presents for crime, challenges to Earth-based conceptions of justice, and the ethical issues raised.

This book is the first edited collection of chapters focused on how to prepare for, address and respond to, instances of criminal and harmful behaviour in (and related to) outer space. It also considers what criminal justice might look like in outer space, and how the important arena of ethics might play a pivotal role in helping overcome problems related to crime and crime control. The book comprises 24 chapters from authors spanning six continents, giving a truly international dimension to the first anthology relating to the intersection of space criminology, space criminal justice and space ethics. It is this international dimension that is essential to the development of a holistic understanding of crime, criminal justice and ethics in outer space.

Exploring recent topics, including the dark origin of space exploration, expansion of satellite industries, space tourism, asteroid mining and human settlement on the Moon and Mars, the book will appeal to space professionals, and students and researchers working in criminology, critical security studies, law and ethics.

Yarin Eski is an Associate Professor in Public Administration, doing research on and at the intersection of criminology, governance and policing. He is co-director of the Resilience, Security and Civil Unrest (ReSCU) R&I Lab at the Vrije Universiteit (VU) Amsterdam. His research focuses on maritime crime and policing, as well as arms dealing, genocide and space crime (control).

Jack Lampkin is a Senior Lecturer in Criminology at York St John University, UK. He is a green criminologist and completed his PhD in 2018 at the University of Lincoln's Law School (UK) where he studied the environmental and social impacts of unconventional hydraulic fracturing for shale gas production purposes. Jack is interested in all things relating to human interactions with outer space and the impact this has on both terrestrial and extra-terrestrial environs.

Routledge Studies in Crime and Society

Radicalisation
A Conceptual Enquiry
Gilbert McLaughlin

Women's Criminalisation and Offending in Australia and New Zealand
Edited by Victoria M. Nagy and Georgina Rychner

Justice, Indigenous Peoples, and Canada
A History of Courage and Resilience
Kathryn M. Campbell and Stephanie Wellman

Fairness and Crime
A Theory
Mark S. Davis

Extremism and Radicalization in the Manosphere
Beta Uprising
Deniese Kennedy-Kollar

Neoliberalism and Punishment
Ignacio González-Sánchez

Victimisation in the Digital Age
An Online/Offline Continuum Approach
Edited by Tine Munk and Morag Kennedy

Disability Hate Crime
Perspectives for Change
Edited by Leah Burch and David Wilkin

Crime, Criminal Justice and Ethics in Outer Space
International Perspectives
Edited by Yarin Eski and Jack Lampkin

For more information about this series, please visit: www.routledge.com/Routledge-Studies-in-Crime-and-Society/book-series/RSCS

Crime, Criminal Justice and Ethics in Outer Space

International Perspectives

Edited by
Yarin Eski and Jack Lampkin

Routledge
Taylor & Francis Group
LONDON AND NEW YORK

First published 2025
by Routledge
4 Park Square, Milton Park, Abingdon, Oxon OX14 4RN

and by Routledge
605 Third Avenue, New York, NY 10158

Routledge is an imprint of the Taylor & Francis Group, an informa business

British Library Cataloguing-in-Publication Data
A catalogue record for this book is available from the British Library

Library of Congress Cataloging-in-Publication Data
Names: Eski, Yarin, editor. | Lampkin, Jack, editor.
Title: Crime, criminal justice, and ethics in outer space : international perspectives / edited by Yarin Eski and Jack Lampkin.
Description: Abingdon, Oxon [UK] ; New York, NY : Routledge, 2024. | Series: Routledge studies in crime and society | Includes bibliographical references and index. |
Identifiers: LCCN 2024041786 (print) | LCCN 2024041787 (ebook) | ISBN 9781032567907 (hardback) | ISBN 9781032567914 (paperback)1 | ISBN 9781003437178 (ebook)
Subjects: LCSH: Space law. | Outer space. | Criminal justice, Administration of. | Criminal liability (International law)
Classification: LCC KZD1145 .C75 2024 (print) | LCC KZD1145 (ebook)
LC record available at https://lccn.loc.gov/2024041786
LC ebook record available at https://lccn.loc.gov/2024041787

ISBN: 978-1-032-56790-7 (hbk)
ISBN: 978-1-032-56791-4 (pbk)
ISBN: 978-1-003-43717-8 (ebk)

DOI: 10.4324/9781003437178

Typeset in Times New Roman
by Taylor & Francis Books

Contents

Illustrations

Figure

Preface and Acknowledgements

Crime, Criminal Justice and Ethics in Outer Space: International Perspectives is an academic enterprise that explores strange new worlds of study, to seek out new lively insights and new collaborations between disciplines, boldly going where no other group of authors has gone before!

This book offers rich scholarship on space crime, criminal justice and ethics with interdisciplinary views on space criminal justice and ethical phenomena ranging from environmental crime on Mars, to space policing, to satellite ethics, to respecting non-living cosmic entity reclamation. On our mission, we aim to understand intersecting topics from different disciplines, resulting, we hope, in a holistic field of study on space crime and its criminal justice and ethical implications.

Having brought together nationally and internationally acclaimed scholars, we, the editors, gained a tremendous amount of insight for which we would like to thank you all greatly. Your chapters are inspiring. We are proud and are sure that the reading audiences will learn as much as we did from you. As important, we truly appreciate your patience throughout the writing process of this book.

We would also like to thank Tom Sutton, Euan Rice-Coates, and Jessica Phillips at Routledge/Taylor & Francis, for your very kind support and guidance. To the reviewers of the proposal for this book we also owe much debt. Your feedback provided the necessary critique to kick-start this book adequately! The faults that have remained are ours and ours alone.

Finally, we would like to thank our family and friends for being there for us. We love you to the Moon and back!

Foreword

For over half a century, astronauts have travelled into outer space and have stayed at space stations including the International Space Station (ISS) for varying periods of up to a year. While there, they have conducted diverse missions and spacewalks for external repairs to the ISS. Not all of the visiting astronauts, despite their defence background, regimented grooming, disciplined culture and prolonged training on handling stress and trauma, have displayed exemplary behaviour and best military composure. A few select instances of serious misconduct, lack of due diligence and apparently "uncondonable" indiscipline by some astronauts, narrated in succeeding paragraphs, will vindicate this assertion.

There is the infamous tale of defiance and insubordination by the three-member crew of Skylab-4 in 1973. The mission was bugged with glitches requiring the astronauts to put in extra work at irregular hours. Further, the crew were entitled to a holiday after every ten days of work, but this was denied thrice consecutively. The crew requested specifically for a day off for rest and recuperation from accumulated fatigue and mounting mental stress. This request was refused and, on the contrary, new experiments were ordered. This lack of appreciation of their depleted physical strength, slowing mental faculties and psychological stress led the crew to unilaterally decide to take a day off by shutting down their radio communication so as to remain incommunicado with ground control and the Capsule Commander. The day was well-enjoyed by the crew but with a tinge of guilt for wrong-doing and indiscipline.

Military law lexicon treats such acts of serious insubordination and defiance as forms of conspiracy and mutiny because the modus and actions were patently unauthorised and contrary to service discipline, regimented behaviour and in defiance of the command-control system. Many may not appreciate that military justice works on its own doctrines of strict obedience and hallowed concepts of command and authority. Despite condoning public postures by NASA authorities, the respective service commanders possibly did not take kindly to this rebellious conduct of self-assumed authority and the crew members concerned possibly would have been branded in their confidential dossiers. No wonder none of them got a space sortie again.

Take another instance of August 2018 when a Soviet habitation module, Soyuz MS-09 spacecraft was docked to the ISS. It was observed that pressure within this module and the ISS, in general, was falling. Ground control warnings led to detection of an unwanted and inexplicable hole drilled in the ceiling of the Soyuz vehicle that was causing the problem. This was filled with gauze and epoxy composites; the defect was rectified, and pressure stabilised. Nevertheless, the reason for the existence of the hole, and who caused it and why, had to be investigated as per military rules.

The American investigation generally regarded it as an erroneously drilled hole during servicing on the ground before launch. The Russians did not accept this conjecture and conducted their own inquiry, the results of which were made public to the embarrassment of NASA. Roscosmos squarely put the blame on the US astronaut Serena Aunon-Chancellor by alleging that she had psychological fits and bouts of depression while at the ISS due to a broken love affair, and circumstantially she could have possibly committed the crime. NASA refuted the allegation with dossiers of her fitness at the ISS and camera footage to show that she was nowhere near the site of the problem prior to detection of the hole. Recrimination apart, the factum of the occurrence cannot be denied; and whether the culprit, an astronaut or a ground technician, is found with evidence or not, an act with criminal implications was committed.

It seems pertinent to highlight another incident of neglect in space. Russia docked its module *Nauka* on the underside of the International Space Station in November 2021. This change was in order to upgrade its research laboratory at the ISS. After the module had latched successfully, the Soviet ground mission controllers commenced post-docking protocols and reconfiguration procedures. It was then that the thruster jets of *Nauka* suddenly restarted, and the space station spun one-and-a-half turns of its original position, i.e., about 540 degrees. The US mission director declared a space emergency and activated thrusters on another module to counter the spinning, thus bringing it back to its normal position. At that time, there were seven astronauts on-board and at risk. Roscosmos maintains that the mishap was caused due to a short-term software failure where a command was mistakenly implemented. This explanation seems too thin to be accepted and the real culprit, man or machine, may not be made public anytime soon.

There are other cases of indiscipline and misdemeanour despite rules of service conduct and functioning modalities. The American crew of Gemini-10 in a prankish act threw their dirty linen into outer space and this irresponsible act of senseless pollution of outer space was not even rebuked or reprimanded by NASA. Thereafter, the Soviet crew, also in a show of bravado and upmanship, did a similar act. Similarly, and recently, Chinese astronauts have done the same. The misconduct in each case was intentional and deliberate and deserving of the highest censure and international condemnation.

Again, in March 1965, John Young, an astronaut of Gemini-3, despite strict security checks, had smuggled a corn-beef sandwich into the space craft and offered it to the commander Grissom. As he took a bite, it started to disintegrate and crumbs started flying all over in the weightless environment of the cabin. To control damage, the sandwich was immediately put into the pocket but the already flying crumbs could possibly get behind the electrical panels to create an existential panic. Since then, NASA has banned all types of breads on space flights.

In November 2009, astronaut Heide Stefanyshyna-Piper, on a space-walk mission for external inspection of the ISS, lost her tool bag which was spotted orbiting by amateur space watchers. Apart from economic loss to the state, this reduction in available tool kits hampered efficiency in operations. Thus, even dereliction in duty or lack of due diligence in outer space can have serious repercussions and threatening consequences, and could be deemed as space crime.

And now the era of space travel has commenced, space tourism may follow soon and planetary residencies will be established as space-habitation technology is perfected. Given that disciplined soldiers have, at times, behaved so irresponsibly, one can hardly expect impeccable and orderly conduct from the space-travellers and tourist-guests who spend a life-time saving for the trip. No wonder, they would demand their pound of flesh in comfort, convenience, entertainment and hospitality for which lapses may not be condoned easily. And most of them hailing from different nationalities, ethnicities, cultures and social systems may not gel well, and behave differently or naively, resulting in arguments and disputes that may turn into acts of criminality.

Human perceptions, predilections and prejudices may be different, and an alleged streak of criminality may not be ruled out even in some assumedly innocuous or inadvertent acts. The human groupings, gatherings and congregations would become more heterogeneous with hospitality workers and long-term planetary residents escalating the scale of human interaction in the space domain. Public order may be at stake. Moreover, as the motivation and purpose of stay of each one would be different, the situation could provide a breeding ground for disputes, conflicts and crimes. The future holds ominous possibilities that clamour for remedies.

Apart from human crimes with physical force, from provocation or conspiracies, these can be caused from mental volition, prompted by *mens rea*, secretive attitude or collective aggression. And now the era of digital enablement and artificial intelligence has dawned; so, there can also be crimes committed with technology: robots, gadgets, triggers or exciters. In this digital age, it is often lamented that a keyboard is more powerful than a bomb and the criminal does not have to expose themself in proximity to the target. The same applies to other technologies like lasers, plasma blasting, micro-waves and other damaging techniques. In fact, crimes in outer space can be committed from any domain and with any platform, while technology is further perfecting itself.

Be that as it may, an old dictum maintains that there cannot be crime without a criminal. Nevertheless, crimes may be committed but the criminals may not be traced and apprehended and, as a result, justice for many space crimes may only hang in the balance. This may happen more often for crimes occurring with technological devices, which leave little trail or footprint for any viable investigation. This lacuna calls for new treaty law, an effective institutional system and organisational support infrastructure for handling and dispensing due justice in respect of space crimes.

This scenario adds to the hassles of judicial administration in cases where jurisdiction conflicts or extradition of the criminal is pleaded for. These are not straight equations but decisions of the state buffered with political influences and strategic nuances. Even where a criminal is arraigned, positive justice may not be 'just' because the ethnic background of the criminal and the applied codified law may turn out to be strange bedfellows resulting in a travesty of justice. Such simplistic acceptance of criminality and positivist application of laws in international predicaments may turn out to be iniquitous and unfair.

In a way, this may amount to a callous application of national law to an unmatched crime inducing social injustice. Hence, the need for care and caution to harmonise the social background, motivation and intention of the culprit with the definition of crime, typicality of the applicable legal system (adversarial, inquisitive, jury or Islamic) and its codified sanctions. This to me flags this pertinent point and advocates for veritable justice in a scenario of international human mixing with varying cultural norms, diverse perceptions of good behaviour and different legal wrongs.

The Editors of the book deserve congratulations for a very apt initiative for an oncoming future scenario. The title of the book *Crime, Criminal Justice and Ethics in Outer Space: International Perspectives*, seems highly appropriate for its visionary scope, dispersion of chapter-topics and the nuances expressed in the articulations. The importance of this subject also needs no justification as space crimes will be the newest and most pressing problem of governance in outer space. The successful handling of these crimes would need a new approach at policing, new expedients of handling criminal administration, experiments in technologies for judicial investigation and a new understanding of criminal jurisprudence for an international milieu.

The book, indeed, appears timely, topical and of contemporary relevance for a change of mindset from existing international space law and to put due systems and institutions in place for workable delivery of social justice, when the situation so confronts. It surely needs pro-active initiation of dialogue, discussion, research and negotiations. More attention is also wanted to fine-tune the proposals with modified or new solutions.

I wish this publication success in its avowed purpose and hope it excites scholarly minds towards research. I hope it motivates policy-makers to cogitate, and enthuses diplomatic corps for due negotiations to find better options and suitable initiatives on the subject. The book also seems to bear popular interest for space enthusiasts.

With my best wishes,

Dr Gurbachan Sachdeva
15 March, 2024
Professor Emeritus, Chandigarh University, Punjab, India
Adjunct Professor, NALSAR University of Law, Hyderabad, India
Former Adjunct Professor, S I S, Jawaharlal Nehru University, New Delhi, India

Contributors

Eman Aboelkhair is a Lecturer in Criminal Law at Menofia University, Egypt; she also serves as Sustainable Development Ambassador in Egypt. Eman obtained her PhD from Menofia Law School in 2024 and teaches primarily on Criminal Law and criminology programmes. Her research interests are in space law, international criminal law and sustainability.

Avi Brisman (MFA, JD, PhD) is a Professor in the School of Justice Studies at Eastern Kentucky University (Richmond, Kentucky, USA), an Adjunct Professor in the School of Justice at Queensland University of Technology (Brisbane, Queensland, Australia), an Honorary Professor at Newcastle Law School at the University of Newcastle (Newcastle, New South Wales, Australia), and a University Fellow at the Centre of Law and Social Justice at the University of Newcastle (Newcastle, New South Wales, Australia). He is also Associate Editor of *Crime Media Culture*, Book Review Editor of the *International Journal for Crime, Justice and Social Democracy*, and Immediate Past Editor-in-Chief of *Critical Criminology: An International Journal.*

Giulio Calcara works as a Lecturer in Law at the University of Plymouth, UK. Most of his research focuses on human rights, international police cooperation, transnational criminal law, and policing outer space. He has presented his research in several academic conferences. Furthermore, he has published numerous articles and papers and has acted as a reviewer and a peer reviewer for major academic publishing houses such as Brill, Oxford University Press, Routledge, Sage, Springer, and Wiley-Blackwell. He has given interviews for international media and acted as an expert on international police cooperation and human rights on several occasions.

José Luis Carpio-Domínguez is a Research Professor in Criminology at Universidad Autónoma de Tamaulipas (UAT), Mexico. He holds a PhD in Social Sciences with orientation in Sustainable Development and a PhD in Biomedical Sciences. He is a member of, among others, the Green Criminology Specialist Group of CEESP-IUCN and of the Research National System (SNII-1). He is the author of several academic publications including the book *Green Criminology for Mexico: Development of a Green Perspective on Mexico's Criminological Heritage*, published in 2023.

Kerry Clamp is an Associate Professor in Criminology at the University of Nottingham, UK. Her research focuses on the application of restorative justice within transitional settings, policing and the community. She is the author of *Restorative Justice in Transition* (2013), *Restorative Justice in Transitional Settings* (2016) and *Restorative Policing: Concepts, Theory and Practice* (2017, co-authored). She has always had a strong interest in practice and a desire to reduce the boundaries between academia, practice and policymaking. She currently sits on the International Advisory Board for *The International Journal of Restorative Justice.*

Yarin Eski is an Associate Professor in Public Administration, doing research on and at the intersection of criminology, governance and policing. He is co-director of the Resilience, Security and Civil Unrest (ReSCU) R&I Lab at the Vrije Universitcit (VU) Amsterdam. His research focuses on maritime crime and policing, as well as arms dealing, genocide and space crime (control). He is the author of several books, including *Maritime Crime and Policing* (2023, Routledge) and *A Criminology of the Human Species: Setting an Unsettling Tone* (2023, Palgrave).

Brian Patrick Green is the director of technology ethics at the Markkula Center for Applied Ethics at Santa Clara University, USA. His work is focused on the ethics of technology, including such topics as AI and ethics, the ethics of space exploration and use, the ethics of technological manipulation of humans, the ethics of mitigation of and adaptation towards risky emerging technologies (including ones with catastrophic risk potential), and various aspects of the impact of technology and engineering on human life and society including the relationship between technology and religion (particularly the Catholic Church). Green teaches AI ethics in Santa Clara University's Graduate School of Engineering and formerly taught several other graduate engineering ethics courses.

Chelsea Haramia is an Associate Professor of Philosophy and Director of Gender Studies at Spring Hill College, and is soon joining the team at the University of Bonn (Germany) working as a Senior Research Fellow on the Desirable Digitalisation: Rethinking AI for Just and Sustainable Futures project. Her current research focuses on AI ethics, the ethics of space exploration, and the ethics of science and technology – as well as environmental ethics, environmental justice, feminist philosophy, bioethics, and public philosophy. She is co-editor of the online journal *1000 Word Philosophy*, and she brings the benefits of philosophical inquiry to area children through her award-winning Philosophy for Children service-learning course.

Rebecca Kaiser is a PhD candidate in criminology at the University of Tasmania (Australia) with a Bachelor's degree in Psychology and an Honours in Criminology. Rebecca's primary focus is the behavioural analysis of conflict dynamics in isolated, confined and extreme environment populations with the aim of using an increased understanding of these dynamics to make future long-term space travel safer.

Karlijn Korpershoek is a PhD researcher at the Jagiellonian University, Poland. She is part of the ARIES (Anthropological Research into Imaginaries and Explorations of Outer Space) project group and specialises in the social and cultural consequences of large space infrastructures for communities living in their periphery. She currently lives in Kourou, French Guiana, and is exploring the historical, political and social dimensions of the European spaceport.

Anita Lam is an Associate Professor in the Department of Humanities at York University, Canada. Since her first monograph *Making Crime Television* (2014), she has focused on examining the intersection of crime, media and popular culture. More recently, she has been interested in reimagining crime scenes in the Anthropocene (*Criminal Anthroposcenes*, 2021). She currently sits on the International Advisory Boards for the academic journals *Crime Media Culture* and *Criminology & Criminal Justice*, and she is on the editorial board for the *Canadian Journal of Law and Society*.

Jack Lampkin is a Senior Lecturer in Criminology at York St John University, UK. He is a green criminologist and completed his PhD in 2018 at the University of Lincoln's Law School (UK) where he studied the environmental and social impacts of unconventional hydraulic fracturing for shale gas production purposes. More recently, Jack has expanded Professor Niroyoshi Takemura's seminal work in astro-green criminology, which involved 'Mapping the Terrain' of the discipline in his 2021 publication within *Astropolitics: The International Journal of Space Politics and Policy*. Jack is interested in all things relating to human interactions with outer space and the impact this has on both terrestrial and extra-terrestrial environs. In 2023, he published the book *Space Criminology: Analysing Human Relationships with Outer Space* alongside Professor Rob White (University of Tasmania, Australia).

Mika Launiala is an Associate Professor at the University of Eastern Finland and a docent in procedural law at the University of Turku, Finland. Prior to that, he worked for over a decade as a Rural Police Chief in the Finnish police service. Mika Launiala received his LLD at the University of Eastern Finland in 2013, where he currently teaches and researches criminal law, procedural law, and police law. In his areas of expertise, Launiala has published several academic works, including articles and papers.

Victoria Nagy is a Senior Lecturer in Criminology at the University of Tasmania, Australia. She was awarded her PhD from Monash University and completed her MA at Eötvös Loránd University in Budapest. Her research interests include historical criminology, gender and crime, corrections, and academic misconduct. Her most recent monograph, *Male Rape Victimisation on Screen*, was published in 2023 by Emerald Publishing.

Erika Nesvold has a PhD in Physics from the University of Maryland, Baltimore County, USA, and has performed astrophysics research at the NASA Goddard Space Flight Center, the Carnegie Institution for Science, NASA Ames, and SETI. She is currently an Astrophysics Engineer for the Universe Sandbox astronomy simulation software. In 2018, she co-founded a non-profit organisation, the JustSpace Alliance, to advocate for a more inclusive and ethical future in space and to harness visions of tomorrow for a more just and equitable world today. She is also the author of *Off-Earth: Ethical Questions and Quandaries for Living in Outer Space*.

Valérie Pijlman is a PhD candidate at the Netherlands Institute for the Study of Crime and Law Enforcement (NSCR) in Amsterdam. Her research is inspired by her educational background in Clinical Psychology (MSc) and Victimology (MSc), and focuses on the help-seeking behaviour of victims of sexual violence. She focuses on those victimised in cyberspace by forms of image-based sexual harassment and abuse, and barriers and facilitators victims encounter in seeking help from informal and/or formal sources, aiming to promote help-seeking by supporting changes in healthcare and policy.

Sarah Poss is a PhD candidate and junior lecturer in political science at Vrije Universiteit, Amsterdam, with a Master's degree specialising in political economy and a Bachelor's degree in international relations. Sarah's research focuses on US foreign policy in response to existential threats in geopolitically significant frontiers, namely the Arctic, outer space, and the deep sea.

Alla Pozdnakova holds a doctoral law degree from the University of Oslo (2007), LLM in International and European Law (2001) from the Riga Graduate School of Law and a law degree from the University of Latvia Law Faculty (1999). She is co-editor for the law journal *Oslo Law Review* published by the Scandinavian University Press, a board member of the Norwegian branch of the International Law Association, a member of the Northern Areas Committee and Chair of the Research Group International Law and Governance at the University of Oslo. Her research interests also include outer space law and she is a member of the Space Law committee tasked with the preparation of a draft proposal for the new Norwegian Outer Space Act.

Benjamin Segobaetso is a PhD candidate in the Ethics and Public Affairs programme at Carleton University, Canada. He holds a BA (Hons) in

Conflict Studies from Saint Paul University, Canada. He completed his MA in Ethics at Saint Paul University in 2018. Benjamin's current research dissertation examines the unfolding ethical implications of outer space developments in relation to the Moon and various planets' industrialisation as well as associated strategic security issues.

Iris van Sintemaartensdijk is an Assistant Professor at Universiteit Twente (The Netherlands) at the Faculty of Behavioral Management and Social Sciences. Her research and publications include books, chapters and scientific articles on criminal decision making, technology and behaviour change. She obtained her PhD at the VU Amsterdam University, for which she studied how to understand and tackle crime, in particular, burglary, through virtual reality. Her teaching consists of delivering courses that include the topics of Psychology & Crime, Systematic Intervention Design, Advanced Research Methods and Behaviour & Technology.

Bruno Reynaud de Sousa teaches at the School of Law (University of Minho, Portugal) and conducts research at the JusGov-Research Centre for Justice and Governance. He is an alumnus of the US Department of State, the International Space University, the National Defense Institute of Portugal, and the College of Europe. In 2023, as a Fulbright awardee, he held a visiting scholar position at the Space Policy Institute (Elliott School of International Affairs, George Washington University, USA). His affiliations include the International Institute of Space Law and EuroDefense-Portugal.

Nigel South PhD FAcSS is Emeritus Professor of Sociology and Criminology at the University of Essex, UK. He has made various contributions to the development of 'Green Criminology' (e.g., *Routledge International Handbook of Green Criminology*, 2020) and his recent work has explored the responses of the powerful and elites to threats of climate change, pandemics, and coming crises of global infrastructures. In 2022, he received the 'Outstanding Achievement Award' from the British Society of Criminology, and in 2013 he received a 'Lifetime Achievement Award' from the American Society of Criminology, Division on Critical Criminology.

Rob White is Emeritus Distinguished Professor of Criminology in the School of Social Sciences at the University of Tasmania and Adjunct Professor at the Centre for Rural Criminology, University of New England, Australia. He has written extensively in the areas of criminology, youth studies and eco-justice. Among his recent books are *Advanced Introduction to Applied Green Criminology* (2023) and *Space Criminology* (co-authored with Jack Lampkin in 2023). He is a Fellow of the Academy of Social Sciences in Australia, the Australian and New Zealand Society of Criminology, and the Academy of Social Sciences, United Kingdom.

Sabine Winters is a freelance science philosopher and lecturer in the fields of film and philosophy, philosophy and technology and future thinking. She creates interdisciplinary programmes where science, art and technology meet and she regularly gives workshops on the (role of the) imagination in science. She works together with scientists, artists, and designers, to discover where fiction touches science and where science flows over into fiction. Her professional contributions provide greater understanding in how fiction and science intersect, and what role the imagination plays in understanding, discovering, and developing scientific theory.

Emilia Ziosi is a Researcher at the Department of Sociology, University of Oxford (UK). She recently completed her PhD on illicit economies and extra-legal governance in contemporary Honduras at the Department of International, Legal, Historical and Political Studies, University of Milan in 2023. From February 2021 to June 2022, she was a Visiting PhD Student at the Centre for Latin American and Caribbean Studies (CLACS), University of London. Her research interests include the relations between organised crime, the state and civil society and the interconnections between legal and extra-legal governance. Emilia is also interested in environmental crime both on Earth and in outer space. She holds a BSc in Economics and Management from the University of Trento, Italy, and a MSc in Organised Crime, Terrorism and Security from the University of Essex, UK.

Introduction

Exploring the Final Frontier of Criminology, Criminal Justice and Ethics Scholars Together in Outer Space

Yarin Eski and Jack Lampkin

"The more we share the more we have", said Leonard Simon Nimoy, American actor and director, famed for playing Spock in *Star Trek*. His words touch upon something we took to heart for this book titled *Crime, Criminal Justice and Ethics in Outer Space: International Perspectives*, providing a unique and scholarly insight on unexplored subjects, and relations between space crime, criminal justice and ethics.

The intersection of space crime, criminal justice and ethics is a growth area, both in terms of academic theoretical progression and industrial practical development, but also as a subject for professional and academic study (Eski 2023; Green 2021; Lampkin 2021). The significant lack of analytical interest in space crime, criminal justice and ethics is problematic given the (relatively) recent expansion of outer space endeavours that present more pathways for criminal behaviour, such as the satellite industry and constellation expansion and outer space exploration advances (i.e. humans to physically return to the Moon imminently; plans for development of the Moon and Mars; mining of asteroids, etc.).

However, of all disciplines, the study of crime as a whole, securing against it and finding justice for it, criminology, has neither actively presented itself as relevant for, nor has it been paid attention to, by criminology, (critical) security studies and legal scholars. Exceptionally, there is pioneering work that explores space crime, working toward an astro-green criminology, focusing on human interactions with outer space that impact off-Earth environments, and attempting to understand such anthropogenic activities that may create exo-environmental harm (Lampkin & White 2023; Sachdeva 2023; Takemura 2019). Think of space tourism activities for instance, alongside space mining, and the problems associated with orbital space debris. But also think about how criminal justice can effectively address these and other types of crimes and harms in space. For example, how do we provide a fair trial in a spacecraft or on the Moon or Mars, whilst criminal justice systems are still Earth-based? What if the lack of space criminal justice would lead to impunity? And how do ethics play their part in such criminological and criminal justice questions?

All of these important questions – perhaps even existential questions given how space is there for all humankind to explore – remain difficult to ask and

DOI: 10.4324/9781003437178-1

answer. Meanwhile, the commercial growth of the space sector continues, and this has a big impact on global security and vice versa (Boschetti et al. 2022). Slowly, very slowly, a couple of pioneering academics, policy makers and other professional stakeholders are showing interest in the subject of space crime, criminal justice and ethics. They are trying to expand their knowledge.

However, research into space crime, criminal justice and ethics is rarely published as regularly as mainstream crime, criminal justice and ethics literature. Issues in space criminology are currently only being published in disparate venues, as shown herein; and articles are also frequently only published in non-archival publication outlets, notably exclusively online journals. Weightier papers usually appear in discipline-based journals. This perpetuates the pre-existing silos of discipline specialisation and threatens the collaboration and inter-disciplinary working required for effective, applied space crime, law and criminal justice, as well as policing practices, and security policies and (ethical) solutions. It can also mean that high-quality interdisciplinary research on space crime, criminal justice and ethics specifically fails to be published in academic texts and journals because the mind-set of the specialist reviewers, steeped in the editorial specific discipline, leads them to reject research which does not tick the traditional boxes of that discipline.

Therefore, this book is a critical response to both the taboo that exists around space crime, criminal justice and ethics questions in the public domain, but also in the academic domain. In this book we will address initial, *yet fundamental* criminological, criminal justice and ethical questions, which will add a valuable and currently absent addition to academic knowledge in the interrelated disciplines of criminology, criminal justice and ethics that engage with outer space issues. Also, and in many ways, this book, by merging the established disciplines of criminology, criminal justice and ethics, may be the beginning of a new (and hopefully lasting!) hybrid discipline that will study (the interaction and intersection of) space crime, criminal justice and ethics for many years to come.

The timing of the book is pertinent, given the imminent space expansion activities planned in the coming years that will unavoidably create many (interacting) criminological, criminal justice and ethical issues, as explored in the following pages.

These developments of space-related crimes, criminal justice and ethics issues will have their impact on space (crime control) in the future. Hence, space crime, criminal justice and ethics as a subject field demands (much more) analytical scrutiny and scholarship than it may have received thus far. Although authors in their respective fields – some only recently – have engaged with the themes addressed here (Green 2021; Lampkin 2021; Pozdnakova 2022; Takemura 2019), this book bridges the gap between key criminology, criminal justice and ethics thinkers and literature, starting to form what remains to this point a non-existing universal and well-founded academic area of study that specifically connects crime, criminal justice and ethics together in one place. Therefore, this book is an undeniably (and unashamedly) interdisciplinary endeavour that unifies previously connected, yet disparate, subjects.

This book is also one of the first texts devoted to the specific (intersecting) study of space crime, criminal justice and ethics, given how the studies of these topics have remained essentially Earth-based and -orientated until now. Although the 1967 *Outer Space Treaty*, the International Space Station (ISS) codes, and other regulations, are in place as existing legal agreements on what we – on Earth – have agreed on, these and other space (criminal) laws often fail to work, and lead to virtually no law enforcement follow-up. There is currently no universal space criminal justice act and there is no space police force, to put it differently. In fact, what space crime, justice and ethics comprise, and how they relate to one another, is a question that has not yet been asked, let alone one for which answers have been attempted. This book can be seen as an attempt to begin such vital discussions.

Hence, we will cover the following, broadly constructed and interpreted areas of crime, (criminal) justice and ethics, but also policing, security and restorative justice, including, but not limited to, (traditional) deviance, illicit activities and volume crime in space, cybercrime and security in space, space warfare, illegal space dumping, pollution and extra-terrestrial contamination, as well as space violations of human rights, space piracy, non-terrestrial terrorism and white collar space crime. The interplay and intersection of these and other space crimes, comprise an utmost important area that shall be explored. In relation to responses to space crime, relevant aspects of space criminalisation and securitisation, as well as the growing field of space criminology, criminal justice and ethics, will form a crucial part of this book. The focus will thus also lie on space criminal justice, policing and security dynamics, investigation, justice and the morality, ethics and philosophy that underpin them.

To emphasise, the book will provide a comprehensive single-volume book of readings in space crime, criminal justice and ethics. As stated, space crime – in comparison to space criminal law, justice, and ethics – remains largely under-researched and under-explored and the two have not been addressed as relating to one another within the context of a single book. To illustrate this, there is a lack of edited books on the subjects, nationally and internationally, which is understandable given the context of the Outer Space Treaty and the agreement on the International Space Station that point to state responsibility. Yet in practice, and in relation to the notion of "space warfare", understanding of space crime, criminal justice and ethics, is still at its very beginning, with only a few initiatives in the field. Particularly problematic is the virtually non-existing comprehension of advanced technologies (e.g. space-based quantum computing) in the context of space crime, criminal justice and ethics.

This book includes chapters accessibly written by internationally acclaimed scholars from the interconnecting and intersecting fields of criminology, policing studies and security studies, law, criminal justice, victimology and ethics. Given the very high level of niche and novelty involved regarding space crime, space criminal justice and space ethics, we have used a triple-approach in selecting relevant experts at the intersection of aforementioned fields. From across the globe, contributions have been made by:

- Hard-science space scholars (e.g. astrobiologists, space engineering scientists) not necessarily working directly on the themes of crime, law and ethics specifically but who do have soft-science perspectives on the matter;
- Criminologists, legal scholars, policing and critical studies scholars, victimologists and ethicists whose expertise lies within the fields of crime, criminal justice and ethics, not necessarily working on outer space issues, but who are interested in broadening their scope of practice and expertise to outer space; and
- A small group of scholars who are a hybrid of the previous two groups, focusing on crime, criminal justice and ethics as they pertain to outer space.

Not only do we guarantee an accumulation of different disciplines and international perspectives in this book, we also unite together scholars who have been scattered around the world and, in so doing, we are building an *international* interdisciplinary community of scholars committed to the study of crimes, criminal justice and ethics in outer space. Moreover, the sometimes opposing contributions in this book assure critical discussions and considerations for not only the criminology, criminal justice, policing and security studies, victimology, law and ethics domains, but also the professional audiences in space governance and the space economy, and the space educational sectors.

After this book's introduction, the first substantial chapter by Yarin Eski, Chapter 1, will take a step back into history, to the dark origin of space exploration and its connection with Nazi sciences, including Nazi aerospace engineering and Nazi criminology. Then more forward looking, Rob White will explore in Chapter 2 how ownership dynamics in the space industry on Earth foreshadow trends in outer space, focusing on privatisation, commercialisation, and their implications for resource exploitation and labour practices, painting a picture of a "Wild West" capitalism scenario where profit trumps rights and private interests overshadow public welfare. Trying to get closer to what space crime may look like, and thinking about how to overcome as much as possible the methodological difficulties of doing criminological research in outer space, Chapter 3 by Iris van Sintemaartensdijk makes the case for using virtual reality (VR) to provide people with the closest experience of outer space, while also exposing them to hyper-realistic VR scenarios in which they are in spacecraft. In Chapter 4, Emilia Ziosi explores what kind of criminological methods there are for researching and preventing environmental crime opportunities arising from private property rights in outer space, specifically "astro-mining". A niche form of space crime, that of illegal warfare in and via outer space, is going to be explored by Bruno Reynaud de Sousa in Chapter 5, by discussing how space commercialisation, in particular satellites, transformed Earth into a "glass house" where democracies and autocracies co-exist and whether transparent warfare will be the new norm. From there, Valérie Pijlman shows in Chapter 6 that satellite Internet provides connectivity even in the most remote and rural of areas, making satellite-enabled image-based sexual violence (e.g. sharing unsolicited nude or sexual images and the non-consensual sharing of others' nude or sexual images) a specific crime via space that could make anyone (simultaneously) a perpetrator, victim and spectator of sexual abuse, instantaneously and globally. Afterward, Jack Lampkin takes a small but first step in Chapter 7 on behalf of victimology, into the direction of developing a space victimology, in order to parallel space criminology by considering space victimhood with a precautionary approach, addressing space victims prior to the development of outer space locales, instead of after-the-fact. With a refreshing new perspective, Rebecca Kaiser in Chapter 8 delves into the world of space analogue missions, which are environments that simulate the isolation, confinement, and extreme conditions of space exploration, exploring opportunities for research into crime in outer space. Anita Lam, Nigel South and Avi Brisman critically examine in Chapter 9 to what extent "space criminology" should maybe not be seen as an entirely "new" field of exploration, showing how popular fiction and academic work have considered the idea of space as a domain of resource extraction and elite consumption before.

We then move on to the (criminal) justice perspectives this book has to offer, starting with Chapter 10 by Erika Nesvold on how science fiction could assist in delivering solutions for criminal justice systems in space, as well as possible alternatives to our current carceral system that would otherwise feel unfeasible in a terrestrial context. Sabine Winters explores in Chapter 11 how justice philosopher John Rawls and his veil of ignorance method could provide input on international space guidelines regarding space debris multi-actor governance. Criminal justice in outer space already starts with atmospheric justice, as Jack Lampkin and José Carpio-Domínguez elaborate on in their Chapter 12 on space exploration and the damage it causes to Earth's atmosphere through direct atmospheric pollution practices, including rocket launches. This is a bitter pill to take, given that launching platforms at Earth-bound and off-Earth spaceports, as environmentally polluting as they might be, fulfil a vital role in space exploration and in policing space crime, as Sarah Poss, Yarin Eski and Jack Lampkin argue in Chapter 13, by moving toward a critical geopolitical-criminological research agenda on spaceport crime (control) and policing. Eman Aboelkhair's Chapter 14 follows up, exploring space pollution by emphasising the gravity of the threat posed by orbital debris, as well as the sustainability challenges connected to space activities, answering the question of what the international criminalisation measures for orbital debris pollution may look like in the future. In Chapter 15, Giulio Calcara and Mika Launiala delve into the legal aspects of policing outer space, outlining what could be the legal framework for policing crime in outer space, and who is allowed to police crimes committed in outer space. In a similar vein, but with a legal dogmatic and socio-criminological approach, Alla Pozdnakova explores in Chapter 16 the extraterritorial criminal jurisdiction over individuals and companies committing space-related offences, delivering a holistic perspective. Kerry Clamp then focuses, in Chapter 17, on a restorative justice approach to tackle specifically sexual harassment and assault in outer space by identifying challenges and considerations that support the need for innovative responses beyond legalism and adapting a risk mentality when countering these forms of crime and aggression in outer space. In Chapter 18, Victoria Nagy questions what will happen to space offenders if we imprison them; and whether imprisonment can ethically be used as a punishment in outer space contexts?

Nagy's chapter bridges into where the book shall draw to a conclusion: ethics in space and their relationship to space crime and criminal justice. Similar to space crime and criminal justice, space ethics also start on Earth, as Karlijn Korpershoek reveals in her Chapter 19 on the ethics and politics surrounding the placement of space infrastructures, especially launch sites, across the world. Brian Green, in Chapter 20, will then show how the transcendent power of ethics allows law (enforcement) to work on Earth where there are borders, but how (once in outer space), ethics is confronted with its weakness. There, it is harder to delineate physical borders, but also for borders relating to the policing of crime and enforcement in space, which could trail off into infinity. Following Green's chapter, Benjamin Segobaetso will consider, in Chapter 21, the question of any moral limits to what intelligence officers may do in outer space during the industrialisation of the Moon and other celestial bodies. Afterward, in Chapter 22, Chelsea Haramia looks at the early stages of human space expansion, when outer space will be largely occupied not by humans but by non-living planetary bodies and environments, and by our abiotic robotic explorers. She raises questions about which obligations there may be for us in such a landscape, as these are not lifeless environments devoid of moral considerations, as she argues convincingly.

To conclude this opening chapter, it must be stated that if we want to go to space for all humankind, to find new hope and prosper, our nature to commit crime must be taken into account, as well as the criminal justice and ethical challenges awaiting us out there. We take into space our norms, cultures, behaviours, strengths and weaknesses. It is our hope that we can, through this book, critically inspire the reading audience and collaborating disciplines that are present throughout the chapters. We hope that they find each other, also beyond this book and beyond academia, perhaps one day in space.

References

Boschetti N, Gordon N, Sigholm J & Falco G (2022) *Commercial Space Risk Framework Assessing the Satellite Ground Station Security Landscape for NATO in the Arctic and High North*. IEEE MILCOM Conference. From: https://www.researchgate.net/publication/366050065.

Eski Y (2023) Space, the Final Frontier to Exploit and Annihilate: Space Exploration, Inhabitation and Settlement. In *A Criminology of the Human Species: Setting an Unsettling Tone* (pp. 71–98). Cham: Springer Nature Switzerland.

Green BP (2021) *Space Ethics*. Lanham, MD: Rowman & Littlefield.

Lampkin JA (2021) Mapping the Terrain of an Astro-Green Criminology: A Case for Extending the Green Criminological Lens Outside of Planet Earth. *Astropolitics: The International Journal of Space Politics & Policy*, 18 (3), 259–273. https://doi.org/10.1080/14777622.2020.1972542.

Lampkin JA & White R (2023) *Space Criminology: Analysing Human Relationships with Outer Space*. London: Palgrave Macmillan.

Pozdnakova A (2022) Pollution of the Marine Environment by Spaceflights. In Platjouw FM & Pozdnakova A (Eds), *The Environmental Rule of Law for Oceans -: Designing Legal Solutions*. Cambridge: Cambridge University Press. ISSN 9781009253741. pp. 76–92.

Sachdeva GS (2023) *Crimes in Outer Space: Perspectives from Law and Justice*. London: Springer Nature.

Takemura N (2019) Astro-Green Criminology: A New Perspective Against Space Capitalism. *Toin University of Yokohama Research Bulletin* (40),7–17.

1 Scientific Crimes Against Humanity for All Humankind

Accounting for the Space Legacy of Aryan Criminology and Nazi Aerospace Science

Yarin Eski

Introduction

The plans for outer space tourism are developing at a fast pace and are nearer than we may expect:

> Will you be planning a trip to an orbiting "space hotel" as early as 2025? The Gateway Foundation, a private company developing this "space hotel", thinks so. The organization plans to build what it describes on its website as "the first spaceport". This spaceport, the **Von Braun** Rotating Space Station, will orbit Earth and will accommodate not only scientific research but also visiting tourists looking to experience life away from our home planet. But, while any timeline for the creation of such a structure would be daunting, the Gateway Foundation plans to build the spaceport as early as 2025 (with the support of the space construction company Orbital Assembly).
>
> (Space.com 2019 – online source, bold emphasis added by author)

The Gateway Foundation (now Gateway Spaceport) a United States (US)-based space company, believes it is shaping the future of space travel with its space hotel concept. Originally called the "Von Braun Station" in 2019, this idea involves interconnected modules forming a rotating wheel in orbit around Earth. The company aims to launch that, and more space stations for tourists where up to 400 guests could stay in 2027 and also make use of office spaces, reflecting a shift towards a space "business park". The company believes that as space tourism gains momentum, the cost barrier will diminish. Moreover, as spaceports simulate gravity, they will provide familiar comforts and novel tourist experiences which combine luxurious interiors that have Earth-like ambiance with breath-taking space views.

Perhaps the most remarkable thing about the spaceports is not necessarily their ambitious aims, including simulated gravity. It is rather the initial name of the spaceport, the *Von Braun Station*, inspired by Wernher von Braun, a former German–US rocket scientist. But why is that so noteworthy?

DOI: 10.4324/9781003437178-2

This chapter addresses that question by unpacking the history of Von Braun the rocket "pioneer" and how space exploration "for all humankind" originates from a dark place; a place where the criminological and aerospace sciences, unitedly, profited from as well as assisted the Nazi regime and its crimes against humanity. First, this chapter will reveal how criminology – specifically Aryan biological criminology – can be considered an initiator, accelerator, and amplifier of Nazi crimes against humanity. Then it delves into the history of Wernher von Braun and his rocket team that developed the V-2 which weaponised the Nazi regime with a most destructive power. Afterward, the United States' poaching of Nazi scientists and knowledge is elaborated on, followed by an explanation of how from the Nazi (criminological) scientific history, space exploration originated. Finally, it will be discussed how, because of that dark origin of space exploration, the void Nazi ideology of anti-creation slithers through space travelling ambitions today and even manifests itself; a slithering that critical space criminology should account for before it can engage credibly in a holistic study of space crime.

Scientific Racial Conquest: Aryan Criminology

Whenever we think of Nazi science, we think of horrific experiments on Jewish prisoners in concentration and death camps, performed by scientists such as Dr Josef Mengele who, funded by a German research foundation, used Auschwitz for cruel medical experiments on prisoners, particularly twins and people with physical differences (Marwell 2020). He "earned" his alias "the angel of death" by sometimes being kind to the children that he would later send to their deaths in gas chambers, and for doing inhumanely terrible things to twins, including amputations and deadly infections, and attempting to experiment with changing their eye colours, removing eyes, teeth, and blood, and forcing inmates to undergo painful treatments (Halioua & Marmor 2020). However, Nazi science comprised more than these experiments. Criminologists were complicit, in such a way, that they provided the "evidence base" and justification for criminalising and dehumanising (a process that also enabled inhumane experiments on) Jewish people and many more groups of people:

> It will be most unfortunate for scientific progress if research, and theory too, must be tailored according to political patterns in order to produce an Aryan criminology.
>
> (Cantor 1936: 418)

In fact, following the same ideology of Nazi conquest and superiority, Nazi criminologists and aerospace scientists were driven by theory and science, saturated with ideas about superior–inferior dichotomies and delivering technological research that enabled the deadly and destructive Nazi racial and territorial conquest.

Apart from a few exceptions, the "dark hour" of criminology during the Nazi regime has never received serious critical criminological attention (see Rafter 2008). As a matter of fact, criminology, as well as victimology, have both been slow in addressing genocide, crimes against humanity, and war crimes, despite expectations and several criminologists and victimologists who have called for more attention to these issues (Brannigan 1998; Day & Vandiver 2000; Eski 2020; Kauzlarich et al. 2001; Letschert & Van Dijk 2011; Maier-Katkin et al. 2009; Morrison 2004; Pemberton 2015; Rafter & Walklate 2012; Woolford 2006; Yacoubian 2000). A significant gap remains in criminological theories and research on genocide, especially on the role of criminologists as (indirect) perpetrators (Rafter 2008; Spierenburg 2016).

Brannigan (1998) argued that criminological theories, such as self-control theory (Gottfredson & Hirschi 1990) about crime and delinquency, are ill-suited to study the Holocaust as ethnic genocide. He suggests considering a new theory; that of an evolutionary psychology of genocide, containing the presumption that "genocide thrives through a combination of cultural animosities, forces that are proximal, and our instinctual legacy for groupishness, forces that are distal" (Brannigan 1998: 272). Whether that may be the case or not, relevant for this chapter are his observations on how "theories of racial purity proposed in the leading schools of racial hygiene made the Jew a threat to Aryan purity and group survival" (Brannigan 1998: 269). It was these racist theories that were developed by Nazi criminology (Ambos 2020).

Prior to the rise of Nazi criminology, also referred to as Aryan criminology (Cantor 1936), the belief in the hereditary nature of crime was prevalent in conservative Bavaria, leading to the establishment of a Criminal Biological Agency in 1924 that collected prisoner dossiers to distinguish incorrigibles from those capable of resocialisation (Spierenburg 2016). Here, we can see clearly Cesare Lombroso's book *Criminal Man* (1876), inspired by Darwin's *On the Origin of Species* (1859), in which he suggested criminals were a unique group with physical and mental differences, like skull sizes and facial bone irregularities. He theorised they were like ancient humans and exhibited "atavism", having inherited certain ancestral traits that reappeared and influenced criminal behaviour. This made him classify them as a separate criminal subspecies, or, "born criminals" with distinct traits. Still, later on, he put less emphasis on biological explanations for crime (Spierenburg 2016: 383).[1]

With the rise of Hitler, biological criminology shifted its focus, condemning the supposed "softness" of the Weimar era and categorising criminals as subhuman. Criminologist Franz Exner theorised that Jews' behaviour aligned with their racial profile. Criminology, during the Nazi era, emphasised social causes of criminality and endorsing the legal system while advocating extreme measures like sterilisation and castration to combat crime (Spierenburg 2016).[2] Aryan criminology was something new and specifically

> [a] Nazi biocriminology [that] constituted an episode in the evolution of criminology, defined as the scientific study of crime and criminals [...]

> criminology was still an immature field when the Nazis became interested in it. In countries such as England, France, and the United States, criminology was turning in sociological directions; elsewhere it developed along biological lines. In Nazi Germany it fed into a political program of mass extermination. German criminal-biology was not directly responsible for that program, but it did provide one justification for it.
>
> (Rafter 2008: 288)

Rafter, who did a detailed political historical study of criminology's role in Nazi Germany (2008), showed that Aryan criminology used biological theories of crime to rationalise mass killings, associating inherent criminality with certain groups. She shows how they developed a science of criminal biology to identify hereditary criminals who needed indefinite incarceration to prevent them from reproducing. Nazis intensified their use of Aryan criminology to select allegedly mentally handicapped offenders for "euthanasia", and murder criminals and hereditary "asocials" in gas chambers or sent them to concentration camps for "extermination through labour"; Aryan criminology became a tool for political power and justification of the Holocaust.

As Wetzell (2006: 415) has argued, "Nazi-era criminology was characterised by a continuing tension between hereditarian biases and an ongoing process of increasing methodological sophistication. … [A] certain amount of 'normal science' continued under the Nazi regime." This implies there may have been a difference between mainstream criminology and Aryan criminology in Nazi Germany, a difference that has received critique, arguing that "even scientifically reputable 'mainstream' criminologists (to use Wetzell's term) became caught up in and partly even contributed to Nazi radicalisation" (Ambos 2020: 284). It is therefore not accurate to dismiss Aryan criminology as "bad science" created by "mad scientists", but as a type of criminology performed by criminologists who collaborated actively and willingly with the Nazi regime (Rafter 2008).

This obedient willingness to perform Nazi ideology-charged science by following orders (cf. Browning 1993) and following Lombroso-based biocriminological theory eagerly when doing their Aryan criminological research, is something Aryan criminologists have in common with Wernher von Braun and his aerospace engineering colleagues. Whereas Aryan criminology delivered science-based justification for racist, totalitarian, and totally annihilative Nazi racial domination, Von Braun certainly scientifically engineered machinery for racist, totalitarian and totally annihilative Nazi territorial domination.

Scientific Territorial Conquest: Werner von Braun and the V-2

Developed in Nazi Germany, the V-2 (in German: *Vergeltungswaffe 2*, transl.: Vengeance Weapon 2) was the world's first long-range guided ballistic missile that was fuelled by a liquid-propellant rocket engine. The intention behind the V-2 was to function as a "vengeance weapon", targeting Allied cities in retaliation for bombings (Neufeld 1995). On September 8, 1944, a V-2 rocket was launched from Wassenaar, a town not far from The Hague in the Netherlands (Campbell 2012). The rocket's journey had begun tentatively but gained momentum, ascending over the Dutch coast and quickly reaching the upper atmosphere. It had undergone several transformations in speed, breaking the sound barrier, and in altitude, shedding weight and shifting from rocket to projectile (Brzezinski 2007). At 6:41pm on that Friday evening, the V-2 eventually plunged toward the British capital, London, striking the neighbourhood of Chiswick and bringing death and destruction, as many more V-2s would deliver to British cities during World War 2 (Aygin et al. 2021).

In Operation Crossbow, which was spearheaded by Douglas Kendall, nearly 2000 British and American intelligence personnel used stereoscopic aerial photos to uncover German military secret testing sites for, amongst others, the V-2 rocket. They also discovered hidden advanced weaponry that eventually resulted in the destruction of V-weapon launch sites which altered the war's course (Williams 2013). It was then discovered that in developing the V-2 long-range missile, the Nazis utilised forced labour from prisoners of war, deported civilians and convicts in the Mittelbau-Dora concentration camp complex, treating them as expendable resources for their secret missile program (Wagner 2015). The building of the V-2 took the lives of tens of thousands of slave labourers, making the V-2 unique as it is the only Nazi weapon to have killed more people during its construction than its use (Crim 2017).

What is less well-known is that before the atrocities being brought upon British cites, on October 3, 1942, the V-2 was the first object launched from Earth that reached the edge of outer space (Kinstle & Kinstle 2018). The V-2 is the brainchild of Wernher von Braun, a German scientist during the Nazi regime and later a US rocket engineer that initiated US space travelling (Biddle 2009; Jacobsen 2015; Neufeld 2004; O'Brien & Sears 2011). Before his US space career, Von Braun worked with the German army before receiving his doctoral degree in physics at the University of Berlin in 1934, and this was before the Nazis came to power in 1933; he developed V-2 rockets. In 1936, he headed a military research centre in Peenemünde, a crucial base for Nazi Germany's rearming. Around 1937, he is thought to have joined the Nazi Party, most likely to secure his job, but it is unclear if he was part of Nazi crimes against people. There are indications though that he willingly assisted the Nazi regime (cf. Crowley & Trudeau 2011: 43).

Memoirs by Holocaust survivors who endured the brutal conditions of work camps responsible for mass-producing the V-2 rockets provide additional insights, including how Von Braun and his colleagues had made several visits to these facilities (NASA 2023). This means he was (and others were) aware of the suffering endured by those who toiled there, which is an indication of Von Braun's conviction that such suffering was an unavoidable means to a perceived utopian end (Bartels 2021). Therefore, as Neufeld (2002: 58) has concluded before on Von Braun,

> [despite] evidence of his lack of enthusiasm for the SS [*Schutzstafel*, transl.: Protection Squadron] and of his growing disillusionment with the regime in the last two years of the war [...] his involvement in concentration-camp labor [...] is central to any judgment of him, and the documents I and others have uncovered show him to be responsible, in some sense, for "crimes against humanity".

Altogether, this tells a different story than the one in which, for a long time, Von Braun and his team were thought of as only being interested in rockets that could go to outer space and that he had no choice in Nazi Germany but to design ballistic missiles (Scharmen 2021). To an extent, Von Braun was like a scientific Nazi general, in charge of a team of rocket enthusiasts, engineering college graduates and professors, comprising unrepentant Nazis who did their work in a culture of wartime science inside Nazi Germany and remained indifferent to the annihilation they wrought on Britain and the Netherlands (Crim 2017). Overall, they possessed an overriding ambition and amoral technocratic outlook, while being convinced they were on a mission, according to Goebbels, to save Germany (Crim 2017). Their work lives depended on being granted all the financial support, materials and personnel they required, meaning they were supported by and in return supported the German war machine: their continuation of the work was most important and heavily depended on continued Nazi conduct of the war (Crim 2017). Considerably, their work and the war were one. Like the previously described Aryan criminologists, Von Braun and his fellow researchers were wilful scientists, not just following orders out of fear but out of a generic sense of duty, actively theorising, designing and contributing to scientific and technological evil.

It was exactly this Nazi expertise and technology that, after the defeat of the Nazi regime, served as precious war bounty for the victors, in particular for the US and the Union of Soviet Socialist Republics (USSR). In fact, and once captured by the US military, Von Braun and his "rocket team" "were aware of their bargaining power with the Americans" (HQ Third US Army Intelligence Center 1945).

Securing "Malignant Ingenuity": Adlershof, Operation Paperclip and P.O. Box 1142

After the war, the allied American and Soviet armies were interested in not just destroying the enemy but even more so in seeking the Nazi expertise in various fields, such as armoured equipment, aviation, naval technology, chemistry, and rocketry. It led to the preserving of Nazi military research sites that harboured countless advanced technologies that they themselves could use, also potentially against one another (O'Reagan 2021), marking the first glimmers of what later would become the Space Race between the two global powers.

The Red Army and Adlershof

Whereas the US was mostly interested in retrieving Nazi scientists and their knowledge on the V-2 rocket (which this chapter will explore in-depth in the following section), the Soviets seemed more interested in the developed equipment itself. A few months after the Chiswick bombing, in April 1945, in post-war Berlin, the Soviet Army captured one of those secret Nazi military sites: Adlershof. As Joseph Stalin wanted the technological edge the Nazis had, Red Army Major Boris Chertok, a USSR electrical engineer, was tasked with the responsibility of documenting, packing, and shipping the advanced scientific equipment and technology discovered in these captured research facilities (Brzezinski 2007). The Germans developed precise engineering and cutting-edge scientific instruments from renowned brands like Siemens and Karl Zeiss. Despite the horrors committed by the Nazis, Chertok was impressed and amazed even by the Nazis' "malignant ingenuity": "Oh, this German love for details and this exactness, which has ingrained such top-notch work into the culture. [...] I am envious" (Brzezinski 2007: 8). The precision and quality of the German scientific equipment and instruments found at the facility remained largely unavailable in the Soviet Union during the war. It was the Soviet pressing need for advanced technology that overshadowed initial feelings of hatred and revenge, allowing for a large-scale process of collecting the scientific and technological treasures to be taken back to Moscow. In the end, it was this systematic appropriation of German innovations that helped set the stage for the Soviet space programme during the Cold War, enabling the launch of the World's first artificial satellite, Sputnik-1, into outer space in 1957 (Brzezinski 2007; Lampkin 2021).

US and Operation Paperclip

Whereas Soviet military interests focused on V-2 materials, US military interests targeted V-2 scientists and ambitions that significantly contributed to US Defence strategic advancement as well as to the progress of space exploration, including NASA missions. But how is it possible former Nazi scientists that were (assisting) war criminals managed to escape Germany and enjoy impunity?

Project Paperclip, also known as Operation Paperclip, is a significant post-World War II programme, bringing together German and Austrian scientists and technicians to the US to use their expertise for military and industrial purposes (Jacobsen 2015; Lasby 1966; Scharmen 2021; Wagner 2015). While Project Paperclip itself was short, similar programmes continued until 1962. Between 1,500–1,600 German and Austrian professionals, along with their families, were moved to the US, and most became US citizens. The programme began to gain knowledge from Germany after the war, because German and Austrian scientists had valuable technical knowledge, especially the team behind the advanced V-2 rocket system (Jacobsen 2015). They gave themselves up to the Western Allies in May 1945, after which the US brought them under Project Overcast to keep their research going, which was vital for US national interests. The Joint Intelligence Objectives Agency (JIOA), a group formed by the US Joint Chiefs of Staff, managed the capture and movement of German experts. As the programme became public, future recruitments used the name "Project Paperclip", referring to the paperclips holding documents about scientists with potential controversial pasts during the Nazi era. The final version of Project Paperclip was approved by President Harry S. Truman in September 1946, allowing the immigration of "chosen, rare minds" (Neufeld 2004: 198) under the sponsorship of US military departments. The programme had restrictions, and candidates went through a strict JIOA screening, with those having strong Nazi Party ties excluded.

One of the deeply problematic aspects of Operation Paperclip is that the large group of former Nazi scientists, were not ordinary Nazis following orders (cf. Browning 1993). For example, Otto Ambros, Theodor Benzinger, Kurt Blome, Walter Dornberger, Siegfried Knemeyer, Walter Schreiber, Walter Schieber, and Wernher von Braun worked closely together with Adolf Hitler, and many of them were Nazi Party members, also affiliated with the SA (*Sturmabteilung*, transl.: Storm Troopers) and the SS (Jacobsen 2015). Of these, only Ambros was tried during the Nuremberg Trials in 1948. He received an eight-year sentence in the IG Farben trial in Nuremberg, because he was the chief responsible for the Auschwitz-Monowitz camp and the use of forced labour. However, he only had to serve a couple of years of this sentence, as he was released in 1951 (Jacobsen 2015: 280–296). Still, out of fear the USSR would capture them before the US would or that they would escape to South America, as Nazi high official Adolf Eichmann did when fleeing to Argentina (Gil 2005), these Nazi scientists were welcomed in (even admired) by US military (Jacobsen 2015) and enjoyed a rather high standard of living in the (often secret) US military facilities they were taken to.

P.O. Box 1142

On November 2, 2021, Netflix released a documentary about Jewish WWII veterans who exposed a secret US military camp, Fort Hunt, where they as soldiers right after the war hosted and interrogated Nazi prisoners of war (POW) (Loushy et al. 2021). The veterans were refugees from Nazi Germany, willing to fight for the US government against Nazism back in Europe. However, they were stationed at "P.O. Box 1142", a secretive facility in Virginia that detained around 4,000 POWs, including Nazi scientists who would later play an important role in the US space programme. One of the Nazi POWs there was (in)famous rocket scientist Wernher von Braun. At Fort Hunt, the covert unit of Jewish US soldiers had to focus on interrogating especially prominent German scientists who willingly shared their knowledge in the hope of staying in the US.

After a long time of having to maintain silence about their work because the operations and interrogations at Fort Hunt did not fully conform to the Geneva Conventions, a group of veterans shared their experiences with interrogating the Nazi POWs at Fort Hunt. Peter Wiess and Arno Mayer, both Jewish refugee soldiers, explain in the Netflix documentary, they had to oversee, interrogate, and retrieve sensitive information and intelligence from Nazi POWs, by especially maintaining their high quality of life, leading to bizarre interactions (Loushy et al. 2021). Officially, Wiess and Mayer (Loushy et al. 2021) were selected for their language abilities and, in the cases of Fred Michel and H. George Mandel, also for their scientific expertise (Dvorak 2006).

Some of the interrogators tragically lost their entire families during the Holocaust, which they discovered while being stationed at Fort Hunt. Whether men like Mandel, Mayer, Michel, Pins and Wiess were indeed merely used for their language skills is doubtful. Given that their understandable anger toward the captured Nazi POWs could easily have turned into more hostile and aggressive, even torturous, interrogations, this was done purposefully by US military in order to retrieve (far) more information about highly advanced Nazi technology. Such instrumentalising of anger of Jews against the Nazi enemy, is similar to how in Nazi ideology the antisemitic anger among the German population was instrumentalised by German politicians against the Jewish minority (Arendt 1973). Whether such instrumentalisation of Jewish anger took place, will remain, however, unknown and made more difficult to explore given that Fort Hunt was bulldozed in 1946 and all documents were either destroyed or classified (Strickler 2014).

The moral complexity of these Jewish US soldiers' daily contrast of treatment and service, knowing about the inhumane Holocaust while having to treat the atrocious Nazi war criminals humanely, became extra complicated due to their wrestling "with the morality of bugging prisoners' cells with listening devices" (Dvorak 2007 – online source). Instead, they would prefer friendly games like chess and Ping-Pong to obtain information from the prisoners (Dvorak 2007, 2006; Strickler 2014). Or as Rudolph Pins, a 94-year-old former Jewish US soldier who was an interrogator at Fort Hunt as well, said: "You don't get people to talk by beating them or waterboarding or anything of that nature" (Pins in Strickler 2014 – online source).

Although Pins made clear that the strategy was put in place to make the prisoners feel comfortable so that they would talk, the interrogators did, however, use psychological techniques. For example, if the Nazi POWs would not cooperate, they were told to be turned over to the Red Army. In fact, the interrogators even went so far as to wear Soviet military uniforms so Nazi POWs would be convinced, they were to be transferred soon to a situation that involved much rougher treatment (Strickler 2014). Michel and Mandel, both of German-Jewish descent as well, explained they would use these techniques to gain valuable information about scientific advancements, including microwaves, atomic studies, weapons systems, submarine innovations like acoustic torpedoes, jet technology and rocketry, all to enable effective US countermeasures (Dvorak 2006), as well as enabling US space ambitions.

Discussion: Space Exploration Spawned by Total Annihilation

Once he was a naturalised US citizen, Von Braun used his V-2 design to become one of the founding fathers of US space travel. He led the team behind the development of crucial Saturn rockets, including the Saturn V, which facilitated the historic Moon landing in 1969. From 1960 until 1970, he was the NASA Marshall Space Flight Center Director, following President Eisenhower's decision to transfer his team to the agency. Both Presidents Eisenhower and Kennedy were instrumental about it, with Kennedy issuing the challenge to land on the Moon within the decade, a milestone accomplished with the Apollo-11 mission in 1969 (NASA 2023). Von Braun garnered widespread recognition in the US after contributing an article titled *Why I Chose America* (1952a), even earning him an award for "patriotic writing", as well as it being widely acknowledged as a crucial source by many journalists and authors (Jacobsen 2015). Von Braun became quite the celebrity and did not remain unnoticed by Walt Disney Studios, where he was invited to create space-related television shows for a new series, and eventually ended up working together with them (Newell 2013). His decision to collaborate with Disney paved the way for broader recognition, having produced "Man in Space" in 1955 that drew about 42 million viewers, becoming one of the highest-rated television shows in American history back then and popularising space travel. Moreover, in that same year, he and a couple of fellow German rocket scientists were granted US citizenship during a public ceremony. It altogether got him into NASA, where their achievements culminated in the successful launch of America's inaugural space satellite, Explorer-1, in 1958, a response to the Soviet launch of Sputnik the previous year. Lastly, Von Braun advocated for human Mars missions, received a National Medal of Science, and overall, was pivotal in the US–Soviet Space Race.

Despite appearing integrated religiously and politically, the Nazi ideology of conquest and politics of fear were still reflected in Von Braun's ideas for a future US space station, which should have become a new terror weapon to instil fear of random death from above; a similar fear-tactic used with the V-2 rocket (Bartels 2021 – online source):

> It's not just the German Nazi rocket science that von Braun brought over, but it's sort of Nazi methodologies for waging war that he brought over that were essential to his worldview.

Moreover, his post-war advocacy for space-based psychological warfare is also underscored by his contributions to science fiction literature, such as his book *Das Marsprojekt* (transl.: *The Mars Project*) (1952b):

> Von Braun's notion of a world united by mutual terror—by fear of seemingly random death raining down from above—is of a piece with the way in which Adolph Hitler's Third Reich utilized the inherent capabilities of the V-2 to terrorize England in its frantic attempt to win

> World War II. In his conception of a world government enforced from an orbital weapons platform, the whole Earth is London during the bombardment. The space station is just another weapon of fear and vengeance. [...] In *The Mars Project*, von Braun's protagonists encounter a similar predicament. In his future scenario—written in 1948 but set in 1980—von Braun's Lunetta space station had achieved what his V-2 rockets had failed to do in 1944: win a world war. But when the station's technicians turn Lunetta's space telescope away from Earth, and toward its nearest neighbors, they discover (like Bogdanov's hero in *Red Star*) that our planet is not the only one inhabited by potentially hostile intelligent life. The world war has been won, and war in general is potentially a thing of the past, on an Earth united into one world government through the occupation and weaponization of Earth's outside.
>
> (Hopf 2018: 84–85)

Perhaps the controversy is not merely the fact that Von Braun was a willing scientific executioner during Hitler's Nazi reign over Germany, but also how

> what he truly believed was unclear, though as long as he had the support to continue his work, he appeared to be content [...] Von Braun was loyal to his work, to space, and this led him to commit terrible atrocities and lead a country to international glory within one lifetime.
>
> (Hopf 2018: 83)

Without believing in the "for all humankind" space adage, he advertised and used a narrative of international cooperation and peace, preaching the inclusivity of space, which directly benefitted him and his former Nazi rocket science colleagues. They got their research funded (and work lives secured) while being portrayed as wanting to bring about world peace and feeling globally united (as was later realised in the International Space Station where astronauts from all over the world work together) (Hopf 2018). So, while publicly being a NASA employee, embodying good press for the US space programme, appealing to the American people to improve morale and raise public awareness about budget increases for US space travelling programmes, Von Braun benefitted from prestige and kept his dark past hidden, also that of other former Nazi rocket scientists of involvement in crimes against humanity (Hopf 2018). Perhaps the most perverse observation is how Von Braun and his Nazi scientific collaborators surreptitiously ingrained totalitarian Nazi ideology into US space science and exploration, *while* denying and benefiting from the total annihilation technologies he and his colleagues created. In that sense, he covertly used "his Nazi past to fit the narrative of the Cold War", reflected in his "calls to unite the [US] towards one common goal – outright defeating their enemy, the Soviets, in some, simply reaching the stars in others", revealing his rhetoric that "push[ed] for space exploration as a means to establish American greatness, American superiority" (Hopf 2018: 82).

Was Von Braun a genius or mad? Many great thinkers and artists, including Aristotle, Seneca, Molière, William Blake, Edgar Allen Poe, Friedrich Nietzsche, Marilyn Monroe, Steve Jobs, and Denis Diderot, have all shared their thoughts about how genius is tied to madness, and vice versa:

> Oh! How near are genius and madness! Men imprison them and chain them, or raise statues to them.
>
> (Denis Diderot n.d.)

In fact, it was Cesare Lombroso himself (whose ideas influenced the Nazis' Aryan criminology) who said, "Genius is one of the many forms of insanity" (Lombroso in Furlong & Fitzgerald 2020: 50). As much as these wisdoms about genius/madness dichotomies can be considered as stigmatising for brilliant extraverts, they are as neutralising for nefarious scientists; a neutralisation that should be dealt with critically.

Hence, whereas Nazi aerospace scientists and engineers may have been geniuses, they were first and foremost war criminals who escaped facing the Nuremberg trials and would most likely have been found guilty and imprisoned for their inhumane and banal madness, as happened with Joseph Mengele (Marwell 2020). Instead, symbolical "statues" were raised for them, reflected in how Von Braun (and his colleagues) received much American public appraisal, even having the first space hotel being named after him. Building a hotel named after a Nazi war criminal comprises the glorification of total annihilation and anti-creation.

It was existentialist Jean-Paul Sartre (1968, 2003) who emphasised the importance of intentionality in understanding genocide, focusing on the genocidal will, which is a will to achieve total annihilation, to create absolute nothingness, to anti-create. Anti-creating involves the existential destruction of a human collective, where the objective is to transform something into nothing (Huttenbach 1988: 174). Genocide, in its purest form, then, seeks complete nothingness, erasing all traces of a group's existence, including its historical memory and cultural continuity (Huttenbach 1988). This entails physical extermination and the removal of a group from its territorial existence (Butcher 2013: 255–256). The destruction of culture is integral to this process, as culture embodies collective memory and unique identities (Gauger 2012: 7). Therefore, in an ideal-type genocide, both perpetrators and victims are forgotten, and the act of genocide itself becomes a non-act, leaving no evidence of the group's existence, "as if the terminated group had never existed, consigning it to a time-less, formless, condition of precreation, pushed back to a nonexistence [...] the unformed void" (Huttenbach 2004: 87).

That philosophy of anti-creation slithered through Nazi superiority, justified by Aryan criminology that scientifically othered people into "*Lebensunwertes Leben*" (transl.: "life not worth living"), leading to totally annihilative practices of eugenic policies and eventually the Holocaust to destroy internal "enemies" of the Nazi regime. Anti-creation also slithered through Nazi engineers and aerospace scientists that weaponised the Nazi army with V-2 rockets for hostile military engagement to destroy the external enemies of the Nazi regime. To an extent, without criminology and aerospace engineers, the Nazi state would not have been scientifically justified to be so destructive, neither against its "enemy from within" (extermination of the Jewish population), nor would it have been able to attack its enemies abroad (including the V-2 bombing of the United Kingdom).

Conclusion

In examining the role of Nazi ideology in the development of both rocket engineering and criminology, we discover a peculiar connection between these seemingly disparate fields. Nazi ideology, driven by a quest for conquest and domination (*Lebensraum*) in the name of "the Aryan race", had far-reaching consequences. It intertwined the work of rocket engineers and criminologists in an almost absurd manner. Nazi criminologists sought to justify the dominance of a supposed superior race over others, while Nazi engineers enabled their conquest of space on Earth. In essence, it was the collaboration of scientists, including criminologists and engineers, that facilitated the realisation of the Nazi "one space for one race" geopolitics. This, in turn, indirectly contributed to the subsequent US–Soviet Space Race and, more recently, the Billionaire Space Race.

The intertwined domains of criminology and aerospace science were united in a destructive pursuit, providing both justification for and instrumentalisation of the Nazi regime's malevolent ambitions to conquer space for the Aryan race. It is the "malignant ingenuity" of these two fields that aided the Nazi cause. Even after their defeat, the knowledge derived from Nazi criminology and aerospace science continued to influence American and Soviet superiority thinking, obscuring its origins of evil.

The inhumane foundation of scientific racial conquest by Aryan criminology and territorial conquest by Nazi aerospace scientists, underpinning one of the most destructive episodes in human history, also forms the basis of US, USSR and eventually global space exploration. This legacy, tainted by crimes against humanity, still lingers within the realm of space exploration, a reminder of the darker chapters in our collective history. Therefore, criminology and the space (science) community must acknowledge and account for space genius that is rooted in historical totalitarian madness of racist conquest.

It is a type of insanity we see slithering in today's space billionaire moguls and their ambitions for human expansion into outer space, hence criminology and the space community must also anticipate and prevent NewSpace's totalitarian tendencies, and perhaps even more destructive outcomes than those of the Holocaust (e.g. Eski 2023). So, from earthly total annihilation to space exploration, what is as important to be critically aware about is that it was the "men" in Diderot's citation (read: the public) who raised the statues for

Nazi scientists. It was back then and still is us, the ordinary bystanders who made it possible for Nazi ingenuity to continue to exist to bring objects from Earth to space, allowing for scientific racial and territorial conquest thinking of the Nazi regime, to linger today as cosmic conquest thinking (Davis 2023; Spigel 2013; Stesner 2022). The US and USSR Space Race was "predicated on racist and sexist barriers that effectively grounded 'racially' marked Americans and women in general" (Spigel 2013: 47), extending "exceptionalism and global white supremacy [...] from Earth into the Heavens" (Davis 2023: 54). Also nowadays, in this time of the Billionaire NewSpace Race, racist territorialism manifests itself in the space exploration ambitions of some groups. Already in the 1990s, David Myatt put forth a speculative concept in his book *Western Imperium*, envisioning the creation of the Fourth Reich, an autocratic Aryan state where Aryan birth rates would increase significantly (Goodrick-Clarke 2003). In this future, a genetically enhanced super-race named "Homo Galactica" would emerge from the most perfected Aryans, possessing exceptional intelligence, heightened sensory capabilities, and superior dexterity, equipping them to venture into the Milky Way Galaxy for colonisation (Gardell 2003; Goodrick-Clarke 2003). More recently, within certain far right groups, such as the Order of Nine Angels, a similar (occult) neo-Nazi ideology persistently slithers, suggesting that "pursuing galactic exploration and the colonisation of outer space to establish a 'Galactic Empire'" is the end goal of "Aryan civilisation as the true vanguard of humanity, which needs to evolve into a superhuman species", because "the ultimate home of a 'pure' human species—homo galactica—is not on this Earth" (Shah et al. 2023: 9–11, 16).

Whether through Von Braun or nowadays through neo-Nazism, this chapter has shown that racism and space are socio-historically intertwined, and problematically so. A critical space criminology – one that is already addressing the state of denial criminology toward space crime in general (see Eski 2023; Lampkin & White 2023; Rothe & Collins 2023) – has as well the difficult task to acknowledge its own Aryan criminological roots and Nazi evil, from which space exploration originated. It asks for, perhaps obligates, space criminology to deliver an epistemological critique, becoming a criminology against criminology (Cohen 2017, 2013), accounting for its own fallacies (as that of aerospace science, and other sciences that provided malignant ingenuity). Only then can space criminology deliver credible criminologies about space crime, as much as enabling itself to broaden its scope convincingly for not just unlawful but also lawful space crimes and harms that would otherwise remain un(der)studied.

Notes

1 As the Nazi era unfolded, Dutch criminologist Willem Bonger criticised the rise of Nazi science, publishing *Race and Crime* in 1939, ridiculing Nazi authors who

glorified the fighting spirit of Nordic peoples while statistics showed higher rates of violent crime in southern Europe.

2 Spierenburg (2016: 386): "Mussolini's takeover [made] that Italian criminology turned to the service of Fascism."

References

Ambos K (2020) Nazi Criminology: Continuity and Radicalisation. *Israel Law Review* 53 (2): 259–287.

Arendt H (1973) *The Origins of Totalitarianism.* New York: Harcourt Brace Jovanovich.

Aygin DT, Cox LA, Faulkner SC, Stevens MC, Verity R & Le Comber SC (2021) Double Cross: Geographic Profiling of V-2 Impact Sites. *Journal of Spatial Science* 66 (2): 183–194.

Bartels M (2021) New Book 'Space Forces' Examines the Cultural Drivers of Space Exploration. From: https://www.space.com/space-forces-book-fred-scharmen-interview.

Biddle W (2009) *Dark Side of the Moon: Wernher von Braun, the Third Reich, and the Space Race.* New York: WW Norton & Company.

Brannigan A (1998) Criminology and the Holocaust: Xenophobia, Evolution, and Genocide. *Crime & Delinquency* 44 (2): 257–276.

Browning CR (1993) *Ordinary Men: Reserve Police Battalion 101 and the Final Solution in Poland.* London: Harper Collins.

Brzezinski M (2007) *Red Moon Rising: Sputnik and the Hidden Rivalries that Ignited the Space Age.* New York: Times Books.

Butcher TM (2013) A 'Synchronized Attack': On Raphael Lemkin's Holistic Conception of Genocide. *Journal of Genocide Research* 15 (3): 253–271.

Campbell C (2012) *Target London: Under Attack from the V-weapons during WWII.* London: Hachette.

Cantor N (1936) Recent Tendencies in Criminological Research in Germany. *American Sociological Review*1 (3): 407–418.

Cohen S (2013) *States of Denial: Knowing about Atrocities and Suffering.* Cambridge: Polity.

Cohen S (2017) *Against Criminology.* London: Routledge.

Crim B (2017) Wernher von Braun's 'Rocket Team' and America's Military-Industrial Complex. Library of Social Science. From: https://www.libraryofsocialscience.com/ideologies/resources/crim-von-braun/#fn:3.

Crowley IF & Trudeau JR (2011) *Wernher von Braun: An Ethical Analysis.* Retrieved from https://digital.wpi.edu/pdfviewer/f4752h50s.

Darwin C (1859) *The Origin of Species by Means of Natural Selection, or the Preservation of Favoured Races in the Struggle for Life.* London: John Murray.

Davis EC (2023) Cold Warrior Magic, Africana Science, and NASA Space Race Religion, Part One. *Reclaiming Space: Progressive and Multicultural Visions of Space Exploration*53.

Day LE & Vandiver M (2000) Criminology and genocide studies: Notes on what might have been and what still could be. *Crime, Law & Social Change*34: 43–59.

Dvorak P (2006) WWII Secret Interrogators Break their Silence: Fort Hunt Veterans Questioned Nazi Scientists, U-Boat Crews. From: https://web.archive.org/web/20060821231648/http://www.msnbc.msn.com/id/14431095.

Dvorak P (2007) Fort Hunt's Quiet Men Break Silence on WWII. From: https://www.washingtonpost.com/wp-dyn/content/article/2007/10/05/AR2007100502492.html.

Eski Y (2020) An Existentialist Victimology of Genocide? In *Genocide and Victimology* (6–22). London: Routledge.

Eski Y (2023) Space, the Final Frontier to Exploit and Annihilate: Space Exploration, Inhabitation and Settlement. In *A Criminology of the Human Species: Setting an Unsettling Tone* (71–98). Cham: Springer Nature Switzerland.

Furlong Y. & Fitzgerald M (2020). 'Diagnosing' Literary Genius: A Concise Look at the Life and Works of Alexander Pushkin. *Journal of Genius and Eminence* 5 (2): 37–52.

Gardell M (2003) *Gods of the Blood: The Pagan Revival and White Separatism*. Duke University Press.

Gil I (2005) Argentina, Israel, and the Jews: Peron, the Eichmann Capture and After. *Israel Studies* 10 (1): 216–222.

Goodrick-Clarke N (2003) *Black Sun: Aryan Cults, Esoteric Nazism, and the Politics of Identity*. New York: NYU Press.

Gottfredson M & Hirschi T (1990) *A General Theory of Crime*. Stanford, CA: Stanford University Press.

Gauger A (2012) Ecocide is the Missing 5th Crime against Peace. Retrieved from http://sas-space.sas.ac.uk/4686/1/Ecocide_is_the_missing_5th_Crime_Against_Peace.pdf.

Halioua B & Marmor MF (2020) The Eyes of the Angel of Death: Ophthalmic Experiments of Josef Mengele. *Survey of Ophthalmology* 65 (6): 744–748.

Hopf KM (2018) *America's Favorite Nazi: How Wernher von Braun Shaped His Memory*. Florida: FSU Libraries.

HQ Third US Army Intelligence Center (1945) *Interrogation Center: Special Screening Report*, June 12, 1945, RG 260, FIAT, Box 8, NARA.

Huttenbach HR (1988) Locating the Holocaust on the Genocide Spectrum: Towards a Methodology of Definition and Categorization. *Holocaust and Genocide Studies* 3 (3): 289–303.

Huttenbach HR (2004) Defining Genocide. In S Totten (ed.) *Teaching about Genocide: Issues, Approaches, and Resources* (75–93). Greenwich, CT: Information Age Publishing.

Jacobsen AM (2015) *Operation Paperclip: The Secret Intelligence Program that Brought Nazi Scientists to America*. New York: Little, Brown and Company.

Kinstle MF & Kinstle RM (2018) *Flight Trajectory of the German V-2 Rocket: Investigation through a STEM Project*. In 2018 AIAA SPACE and Astronautics Forum and Exposition (p. 5158).

Kauzlarich D, Matthews RA & Miller WJ (2001) Toward a Victimology of State Crime. *Critical Criminology* 10: 173–194.

Lampkin JA (2021) Mapping the Terrain of an Astro-Green Criminology: A Case for Extending the Green Criminological Lens Outside of Planet Earth. *Astropolitics: The International Journal of Space Politics & Policy* 18 (3): 238–259.

Lampkin JA & White R (2023) *Space Criminology: Analysing Human Relationships with Outer Space*. London: Palgrave.

Lasby CG (1966) Project Paperclip: German Scientists Come to America. *The Virginia Quarterly Review* 42 (3): 366–377.

Letschert RM & Van Dijk JJM (2011) *The New Faces of Victimhood: Globalization, Transnational Crimes and Victim Rights*. Dordrecht: Springer.

Lombroso C (1876) *L'uomo delinquent. Studiato in rapporto alla antropologia, alla medicina legale ed alle discipline carcerarie*. Milan: Ulrico Hoepli, Libraio-editore.

Loushy M, Sivan D, Bergmann B & Bergmann J (2021) *Camp Confidential: America's Secret Nazis*. Netflix.

Maier-Katkin D, Mears DP & Bernard TJ (2009) Towards a Criminology of Crimes Against Humanity. *Theoretical Criminology* 13 (2): 227–255.

Marwell DG (2020) *Mengele: Unmasking the 'Angel of Death'*. WW Norton & Company.

Morrison W (2004) Criminology, Genocide, and Modernity: Remarks on the Companion that Criminology Ignored. *The Blackwell Companion to Criminology*, 68–88.

NASA (2023) Wernher von Braun. Former Marshall Space Flight Center Director. From: https://www.nasa.gov/people/wernher-von-braun/.

Neufeld MJ (1995) *The Rocket and the Reich: Peenemünde and the Coming of the Ballistic Missile Era*. New York: The Free Press.

Neufeld MJ (2002) Wernher von Braun, the SS, and Concentration Camp Labor: Questions of Moral, Political, and Criminal Responsibility. *German Studies Review* 25 (1): 57–78.

Neufeld MJ (2004) Overcast, Paperclip, Osoaviakhim: Looting and the Transfer of German Military Technology. In D Junker (ed.) *The United States and Germany in the Era of the Cold War, 1945–2000*. Cambridge: Cambridge University Press.

Newell CL (2013) The Strange Case of Dr. von Braun and Mr. Disney: Frontierland, Tomorrowland, and America's Final Frontier. *The Journal of Religion and Popular Culture* 25 (3): 416–429.

O'Brien JL & Scars CE (2011) Victor or Villain? Wernher von Braun and the Space Race. *The Social Studies* 102 (2): 59–64.

O'Reagan DM (2021) *Taking Nazi Technology: Allied Exploitation of German Science after the Second World War*. Baltimore: Johns Hopkins University Press.

Pemberton A (2015) Victimology with a Hammer: The Challenge of Victimology (Inaugural speech). Tilburg: Tilburg University.

Rafter N (2008) Criminology's Darkest Hour: Biocriminology in Nazi Germany. *Australian & New Zealand Journal of Criminology* 41 (2): 287–306.

Rafter N and Walklate S (2012) Genocide and the Dynamics of Victimization: Some Observations on Armenia. *European Journal of Criminology* 9 (5): 514–526.

Rothe DL & Collins VE (2023) Planetary Geopolitics, Space Weaponization, and Environmental Harms. *British Journal of Criminology*. From: https://doi.org/10.1093/bjc/azad003.

Sartre JP (1968) Genocide. *New Left Review* 48: 13–25.

Sartre JP (2003[1943]) *Being and Nothingness: An Essay on Phenomenological Ontology*. Abingdon, Oxon: Routledge.

Scharmen F (2021) *Space Forces: A Critical History of Life in Outer Space*. London: Verso Books.

Shah S, Cooper J & Newcombe S (2023) Occult Beliefs and the Far Right: The Case of the Order of Nine Angles. *Studies in Conflict & Terrorism*, 1–21.

Space.com (2019) Yes, the 'Von Braun' Space Hotel Idea Is Wild. But Could We Build It by 2025? From: https://www.space.com/gateway-foundation-von-braun-space-station.html.

Spierenburg P (2016) The Rise of Criminology in its Historical Context. In Knepper P & Johansen A (eds), *The Oxford Handbook on the History of Crime and Criminal Justice* (373–395). Oxford: Oxford University Press.

Spigel L (2013) White Flight. In *The Revolution Wasn't Televised: Sixties Television and Social Conflict* (47–68). Taylor and Francis. https://doi.org/10.4324/9780203700426-8.

Stesner S (2022). Ask What Your Country Can Do for You: The US–Russia Space Race in the Context of Race Relations. Available at SSRN 4149850.

Strickler L (2014) Secret WWII Camp Interrogators Say Torture Wasn't Needed. From: https://www.cbsnews.com/news/secret-wwii-camp-interrogators-say-torture-wasnt-needed/.

Von Braun W (1952a) Why I Chose America. *American Magazine* 2: 111–112.

Von Braun W (1952b) *Das Marsprojekt*. Frankfurt: Umschau Verlag.

Wagner JC (2015) *Produktion des Todes: Das KZ Mittelbau-Dora*. Göttingen: Wallstein Verlag.

Wetzell RF (2006) *Criminology in Weimar and Nazi Germany*. In P Becker & RF Wetzell (eds), *Criminal and Their Scientists: The History of Criminology in International Perspective* (pp. 401–424). New York: Cambridge University Press.

Williams A (2013) *Operation Crossbow: The Untold Story of the Search for Hitler's Secret Weapons*. London: Random House.

Woolford A (2006) Making Genocide Unthinkable: Three Guidelines for a Critical Criminology of Genocide. *Critical Criminology* 14 (1): 87–106.

Yacoubian GS (2000). The (In)Significance of Genocidal Behavior to the Discipline of Criminology. *Crime, Law and Social Change* 34: 7–19.

2 What on Earth is Happening in Outer Space? Questioning the Space Oligarchy

Rob White

Introduction

Outer space ethics pertaining to crime and criminality only make sense in the context of definitions of 'harm' and 'morality' and who gets to define what. That is, criminal justice is fundamentally oriented to evaluations and sanctions linked to both harm (e.g., assessments of damage and destruction) and morality (e.g., conceptions of right and wrong). Part of the work of a criminal court, therefore, is to ascertain the degree or seriousness of the harm and to weigh up the moral elements pertaining to the offence and its aftermath. This process of assessment has objective components, for example, close consideration of the offence from the point of view of overall harm caused. It also has subjective dimensions, for example, examination of the offender with respect to matters of culpability as well as situational factors affecting the potential effectiveness of sanctions (e.g., deterrent effect of fines). All of this assumes, of course, that there is a suitable court to weigh up such matters, that the nature of the harm and the morality of acts (and omissions) are universally agreed to and embedded in law, and that effective enforcement and compliance measures are in place.

Space criminology, involving the study of serious harms and criminal justice systems associated with outer space, is hamstrung, however, by the fact that the formal legal infrastructure guiding behaviour in space is underdeveloped (Lampkin & White, 2023; Sachdeva, 2023). More importantly, though, is the phenomenon of the privatisation of space travel, exploration, resource extraction and colonisation. This puts extraordinary power into the hands of a small number of companies and individuals while simultaneously signalling the limitations of the nation-state in space-related activities. The power of the global billionaires, built on Earth, is extending into the realms beyond the planet. Each domain is thereby impacted, and not necessarily for the good.

From an ethical point of view, for instance, the role of individuals such as Elon Musk in crowding space with his satellites, stopping Starlink connections when it suits him (e.g., against the Ukrainian government versus Russia), and expanding his rocket-led empire, harks back to the great imperialist (and usually monopoly) companies of the 18th century, such as

DOI: 10.4324/9781003437178-3

Hudson's Bay and East India companies (ABC News, 2023a; Alamalhodaei, 2023; Beven & Parry, 2023). The biggest of these companies also had their own armies to repress resistance and counter competition. Indeed, several hundred years of capitalism have been driven and underpinned by powerful forces (nation-states, companies, armies) pursuing sectional interests. Economic gains for the few have been achieved through global imperialism, colonialism, and militarism that have served to entrench a dominant world-view and the material basis for certain types of production, consumption, and reproduction. This historical process is underscored by the ethics of no-holds-barred capitalism that tramples rights and ignores the normal rules of morality (right and wrong) in favour of private commercial and ego-based interests. This chapter argues that similar processes of domination and subjugation continue apace in activities and initiatives associated with outer space.

The Space Entrepreneurs

The transformation of outer space from ostensibly a 'commons' for Earth-bound humanity to that of the personal property of and market venue for the billionaire begins with the mass insertion of private money into outer space projects (ABC News, 2023b; Chambers, 2023; Eski, 2023; Purtill, 2024; Williams, 2024). This privatisation of space ventures has been driven by several interrelated tendencies. These include:

- a reluctance by some national governments to invest heavily in space activities, even when government policy and reputation is tied up with such activities, because of budgetary constraints;
- the enthusiasm of some nation-states to embrace space ventures precisely because of the international kudos of such activities, but which can be accelerated through contributions from private partners;
- the evident fiscal benefits of having private partner investors alongside government provided funding, so that each can claim credit and acclaim for furthering the space agenda;
- the shift in taxation away from the rich and the corporate sector, further putting pressure on nation-states to make up savings in expenditure in areas such as social security and healthcare, and space initiatives;
- the sheer amount of money at the disposal of companies and individuals as corporations merge, grow and consolidate market share and move toward a monopoly position;
- the institutionalised evasion of taxes, combined with government tax cuts, provision of subsidies, and long term service and supply contracts, which frees up capital for space investment;
- the ego of billionaire capitalists who desire engagement in projects that paint them, at a personal level, as adventurers and pathfinders, leading humanity into a new era of space exploration and freedom from the shackles and potential dooms of the Earth.

The strategies that nation-states use to navigate outer space matters are contingent upon the class interests associated with political power. The power of transnational corporations finds purchase in the interface between the interests and preferred activities of the corporation and the specific protections and supports proffered by the nation-state (Tombs & Whyte, 2015). The latter can be reliant upon or intimidated by specific industries and companies. Tax revenue and job creation, as well as media support and political donations, hinge upon state-corporate synergies. This undermines the basic tenants of democracy and collective deliberation over how best to interpret the public or national interest. Basically, it has led to the private sector having a much freer hand to do what they wish in the reaches of outer space.

On Earth, the evolution of capitalism has been associated with trends such as privatisation and commodification. Common property has been made into private property, there have been concentrations of land ownership and management globally, and under neoliberalism, reliance on market mechanisms rather than government controls to fulfill use-value needs (e.g., food, housing, water). Commodification has seen the transformation of these use-values into exchange-values insofar as more and more aspects of social life and environment are commercialised and inserted into the buyer-seller nexus (e.g., freshwater provision). As a global phenomenon, the penetration of the transnational corporation into local markets and the dominance of capitalist production/consumption practices are evident worldwide.

Space Oligarchy

The changes have been profound. So much so, that today we know, by name, the key capitalist entrepreneurs of the planet, people such as Mark Zuckerberg, Bill Gates, and Rupert Murdoch. The new space oligarchs – those very rich business leaders with a great deal of political influence and outer space presence – also include very familiar names. The short list includes Richard Branson and *Virgin Galactic*, Jeff Bezos and *Blue Origin*, and Elon Musk and *SpaceX*. It is a core tendency within capitalism for private capital to be concentrated in ever fewer hands and centralised in ever larger corporations, as less competitive enterprises crumble and smaller enterprises get bought out and absorbed to create monopoly market conditions. The contours of unfettered and concentrated capitalist market power over our daily lives are manifest in precisely these personages and companies.

One consequence of this concentration of money and power is the diminishment of democratic say over key decisions in our lives. This is a form of structural authoritarianism that is intrinsic to monopoly capitalism (van der Velden & White, 2021). In essence, capitalism is an authoritarian form of social production and appropriation of the social surplus in which the privatised control and appropriation of socialised production is imposed from above. Specifically, what, how, and to what purpose social production is carried out, and how it is distributed, remains in the hands of a small elite minority. This power is held and wielded by the same cohort of capitalist billionaires with respect to developments and ventures in outer space. Importantly, the decision-making is arbitrary – decisions made are not accountable to public officials, nor do they have to reflect majority public sentiment. What the billionaire says, goes. At least as long as they have the economic and political leverage to do so (*Guardian*, 2024).

And 'what goes' is frequently informed by and rhetorically defended in the name of the market. Philosophically, for example, neoliberalism tends to reduce the human condition to one of market exchanges, within which instrumental use of people and environments is acceptable, and where personal wealth and capital accumulation are seen as the highest goods that define human success. It is a rough and tumble, and unforgiving, moral universe. In effect, however, neoliberalism is built upon a falsehood. Powerful sectional interests already substantially own and control the bulk of the world's resources. There is no 'free market' as such. Land, water, food and energy are under the control of a small and shrinking number of private firms, and the community outside is both growing and increasingly powerless in the face of this concentrated ownership and control. The same tendencies apply to outer space, and the consequences are significant. This is evident in how harms and crimes are socially constructed, now and potentially into the future.

Harms and Crimes on Earth and in Outer Space

Amongst diverse motivations, a contemporary driving force for human movement into space is financial advantage. On Earth, this is accomplished by effective control of distinctive markets, development and deployment of innovative technologies, ownership of land and intellectual property, financial astuteness, and insertion into the political economy (both global and domestic) in ways that firm up state-corporate collusion.

Underpinning wealth in the global capitalist system is commodity production (Mandel, 1975; White, 2002). Built into the logic and dynamics of capitalism is the imperative to expand. Increasing labour's production of surplus value – the source of profit – demands constant changes in the ways labour is exploited and in the things that can be transformed from simple use-values (objects of need) into exchange-values (commodities produced purely for exchange). This can be achieved via changes in the organisation of work, by manipulation of the conditions within specific workplaces, and by transforming previously unproductive or non-capitalist forms of activity into sites of productive labour (e.g., family farming into agribusiness). The extended reproduction of the system rests upon the expansion of commodity production, since only labour hired by commodity-producing capital is directly productive of surplus value.

Forty years of neoliberalism from the late 1970s onwards saw the commodification of a wide range of services and products (e.g., water, waste, and energy utilities) which were formerly state-owned and operated ostensibly for public benefit. In addition, commodification was extended to new areas of human need and endeavour. Under capitalism, every aspect of human existence is subject to transformation insofar as capital seeks to create new forms of consumption (e.g., fads, fashions) and to convert existing or potential use-values into exchange values by commodifying human activity and human requirements (e.g., water, food, entertainment, recreation). For example, formerly 'free' services are now sold to consumers (e.g., paying for private music lessons or paying levies to private security firms or garbage collectors, when previously such services were offered as tax-funded, socialised services available at school and local neighbourhood levels). The process involves the transformation of the production of goods and services according to the dictates of exchange, rather than immediate use and need.

The privatisation of state services and goods is largely explainable in terms of the dynamic of capitalism to incorporate and subsume all parts of daily life into the web of accumulation. It is evident in the corporatisation of public institutions and in the direct transfer of such institutions into private hands (e.g., telephone, utilities, banks). Similarly, the introduction of user-pays as a form of cost-recovery (e.g., medical and educational fees) represents steps toward the commodification of goods and services, which in turn lend themselves to further privatisation of provision over time. The privatisation of space ventures largely follows this logic and this pattern. That is, the pay-off for business is the profitability stemming from engagement in the exploration process. This simultaneously involves the relative detachment of the state from regulation of the corporate enterprise or control over its productive (i.e., profitable) activities.

Nature is subject to continual processes of commodification under capitalism; namely, the transformation of use-values (e.g., water is essential for life) into exchange-values (e.g., water sold for profit). This extends into outer space as well insofar as the intrinsic value of celestial objects such as the Moon is ignored or downplayed in favour of how best to exploit it in commodity markets (e.g., mining, tourism). The space entrepreneur views and exploits celestial objects primarily through the lens of natural resources. Natural resources and landscape features in space, such as craters on the Moon, are subject to varying processes of commodification involving the transformation of existing or potential use-values (e.g., aesthetic appreciation or spiritual status) into exchange-values (e.g., profit-oriented tourist expeditions). The appropriation of nature in outer space may also, in future, involve the re-making of nature and its products biologically and physically. Examples of this include potential attempts to terraform Mars and/or to genetically alter humans to be more space-worthy (e.g., bone mass and weightlessness, responsivity to solar radiation exposure).

Natural resource extraction on Earth involves many different industries and includes the exploitation of oceans, mountains, rivers, trees, and lands for profit, not to mention the ways in which non-human animals are treated depending on how they contribute to or interfere with commodity production. The commodification of nature (and of humans, who sell their labour power) is evident in areas such as mining, logging, fishing, agribusiness, and fossil fuel production where the intent is to maximise profit by minimising or externalising costs (e.g., not including environmental damage in cost-benefit analysis). One result is the exploitation of the environment with little restraint, the exploitation of domesticated animals without due regard for welfare or rights, and the exploitation of humans without empathy. Profit-oriented practices combined with consumerist ideologies ultimately lead to human-induced scarcity in many needed resources (such as freshwater) and ramped up competition among corporations for what is left. This also means expanding the horizons of what can be exploited beyond the perimeters of Earth itself. Hence, the attractions of outer space as a potential source of additional resources to exploit.

Crimes of the Powerful

Many harms are linked to capitalistic mass production methods of natural resource exploitation. Indeed, this is the bread and butter of much of the green criminology research and commentary (Brisman & South, 2021). Transformations of the planet are evident across various indicators (White, 2022). Some of these include:

- **Resource depletion** – extraction of non-renewable minerals and energy without development of proper alternatives, over-harvesting of renewable resources such as fish and forest timbers
- **Disposal problems** – waste generated in production, distribution and consumption processes, pollution associated with transformations of nature, burning of fossil fuels and using up of consumables
- **Corporate colonisation of nature** – genetic changes in food crops, use of plantation forestry that diminish biodiversity, preference for large-scale methods that degrade land and oceans and affect species' development and wellbeing
- **Species decline** – destruction of habitats, privileging of certain species of grain and vegetable growing over others for market purposes, super-exploitation of specific plants and animals due to presumed consumer taste and mass markets
- **Global warming** – the rapid escalation in global temperatures due to atmospheric pollution, leading to climate change, the reconfiguration of the planet's ecosystem, and extreme climate and weather disruption.

Critical discussions of the impact of space junk, and the dilemmas raised about the prospects and outcomes of space mining, mirror some of the concerns about destructive social processes already evident on Earth (Eski, 2023; Lampkin & McClanahan, 2023; Lampkin & Wyatt, 2023). While mining, growing, harvesting, and extracting are not intrinsically 'bad' activities, the question is whether these are being done ethically and ecologically sustainably. Rising temperatures (due to carbon emissions), diminishment of biodiversity (and extinction of millions of species) and widespread pollution (such as plastics in oceans) indicate that the answer is in the negative. How can we trust companies that have denigrated the Earth to not continue doing so in outer space?

The key culprits in this destruction and degradation are, of course, transnational corporations operating within the pressures and limits of global capitalism. Crimes by the powerful involve actions (or omissions, failures to act) that are socially harmful and carried out by elites and/or those who wield significant political and social authority in the sectors or domains of their influence. One of the hallmarks of the harms and crimes perpetrated by the powerful is the taken-for-granted nature of their harms and crimes. Anything which impedes business is deemed to be unreasonable, faulty, bad for the economy, not the rightful domain of the state, to undermine private property rights, and so on.

> The 'naturalness' of corporate crime – the way in which social harms, economic exploitation and environmental destruction is built into the fabric of everyday, ordinary life as a 'normal' feature of how we produce and consume – makes it that much more difficult to challenge. This is compounded by the fact that much of what occurs does so in a fully 'legal' way, regardless of actual harm. Moreover, where external controls (materially and ideologically) on profit maximisation are weakened, then we can expect to see an increased incidence in illegal corporate activity and, more generally, greater propensity for social harm regardless of legal definition.
>
> (White, 2008: 39)

The capacity and propensity of corporations and their billionaire executives to do harm is immense. Yet, they are invariably shielded from criminal prosecution and status denigration by a complex protective web of media spin, deep pockets, lawyers-for-hire, and nation-state largesse.

Ultimately, when it comes to the harms of the powerful on Earth, state-corporate collusion virtually guarantees a free hand in natural resource exploitation (Whyte, 2020). Around the world, state regulation has generally proved to be woefully inadequate and/or subject to capricious political decisions. The trajectory of fossil fuels and carbon emissions since their deadly effects were widely acknowledged in the 1990s (see especially the Rio Conference on climate change) is a profound reminder of just how powerful the corporate denial and will to profit are, even in the face of planet-threatening repercussions. It is little wonder then that people may be sceptical as to whether the public interest will be served in outer space, especially when key decisions are made by the space oligarchy, the main concerns of which are self-interest, private profit, and competitive advantage.

Securitisation and the Private Interest

To secure their wealth-making capacity, private corporations are also in the best position to enforce a specific kind of social order, whether this is on Earth or in outer space. The history of the European corporation internationally, from the mercantile period onwards, is one that included direct conquest, including the mobilisation of private armies to subjugate local populations, enforce labour discipline, and maximise productive efficiency. This, too, involved varying forms of state-corporate collusion.

The appropriation of natural resources involves very specific forms of securitisation, that is, methods by which access to and control over these resources are secured to the benefit of the company. Securitisation is constructed around two key allied forces: natural resource extraction industries and specific companies; and nation-states that exert hegemonic power and control over natural resources. Historically, national security has and continues to be defined in terms of (nationally linked) company security with respect to the access to and exploitation of natural resources. As the saying goes, what is good for General Motors is good for the United States. In other words, specific corporations, while transnational in scope and activities, are identified with specific nation-states in ways that reflect synergistic interests.

Securing exploitation of natural resources via state-corporate collusion reflects not only shared elite interests and ideological agreeance. It is also intertwined with varying forms of corruption (White, 2017). This can include moral corruption, involving the undermining of trust and respect for established governmental processes and institutional practices that ostensibly should be guided by democratic oversight. It also includes direct corruption, that takes the form of direct breaches of criminal laws, facilitated by government officials and non-government actors, as well as indirect exploitation via informal networks and practices. For example, some harmful activities are perfectly allowable (i.e., legal), and receive the approval of state authorities regardless of the actual harms caused by the activities (e.g., clearfell logging, intensive salmon farming). Other activities may be deemed illegal, but without adequate state resources, are in effect allowed to occur simply as a matter of course (e.g., illegal land clearance, land grabs). State intervention and non-intervention, positive and negative, shapes how and where exploitation of natural resources occurs.

The increasing reliance by publicly funded space agencies to work in partnership with private companies, entails greater latitude in the regulation and even criminalisation of space-related activity to the advantage of the space oligarchy. This is because the weight of investment by the private sector is enough to reinforce the private interest over the public. Otherwise, the space endeavour is threatened with retarded progress and/or potential collapse on the part of public authorities. In the context of cross-national rivalry (economic, military, reputational), the interface between nation-state and corporation becomes even more close and complex. Fundamentally, it implies that decision-making about investment flows, technological uses, and next step exploration resides predominantly in the hands of the emergent space oligarchy. This represents a shift in decision-making that is one-step-removed from democratic accountability.

To protect their investments, transnational corporations will draw upon nationalist rhetoric (e.g., US achievements), market narratives (e.g., freedom to operate), and space adventure romance (e.g., the excitement of exploration) to consolidate their hegemony within the space industry expansion. But protection is also materially supported via market share dominance (such as satellite networks) and militarised space technology (that can be utilised as weapons) (ABC News, 2023c, 2024; Rothe & Collins, 2023). Moreover, what happens in space is vital to life on Earth in several respects (e.g., satellite use related to surveillance and telecommunications systems) and thus those who produce space-related infrastructure, especially satellites, also create reliance by governments on their expertise and technologies. This dependency is not necessarily mutual insofar as the wealth of the largest corporations dwarf that of many sovereign nations.

Exploitation in Space

Corporate interests in outer space are informed by how specific activities can be plugged into the general commodification production process. Travel to the Moon, for example, will be sold as a tourist package, a money-spinner for the travel entrepreneur. At the moment it is wealthy people who can afford to fly into space as tourists, but this is only the beginnings of a much larger influx of 'customers' into these domains. Health, medical and pharmacological research will be backed by investors, not because it is intrinsically good or helpful for humanity, but because it has huge profit margins. Resource extraction, especially space mining but also solar power, will generate profit for those companies with sufficient finances to manage the initial outlay. Production efficiency and cost cutting will be enhanced by further developments in artificial intelligence and robotics. Those who own and control transport to and from outer space will wield enormous power in these transactional processes.

Mining operations are also potentially very lucrative from an investment point of view. This will be corporate driven and private profit based, supported by specific nation-states who will continue to conflate the national interest with private interests. Space exploration and exploitation will be based on money and resources back on Earth, and these, in turn, hinge upon political economy (which nation-states put money into space ventures and how) and private investment (which companies spend money on which commodities, such as satellites, minerals, and medicines). The ownership and control of outer space will initially consist of a dog-eat-dog contest, with large corporate conglomerates competing to seize and consolidate advantage now, in order to secure large profits later. Monopolies will be created through both hegemonic state intervention (e.g., countries such as China, the US, and Russia supporting specific enterprises) and market pressures (e.g., the costs of establishment and maintenance of space infrastructure versus how this translates into purchased consumables and profits back on Earth).

The race to exploit natural resources and natural settings in outer space is far from equal and currently dominated by the space oligarchs. Likewise, the status of outer space and of celestial objects such as the Moon as 'commons' for humanity is less than clear, given contemporary trends and developments. While it may be the case that discussion is ongoing in United Nations forums regarding potential legal models for activities in exploration, exploitation, and utilisation of space resources, laws and regulations are well behind what is occurring *in situ*. The proliferation of new satellites, new Moon vessels and accumulated space junk all point to a rapidly changing space landscape. For now, there are no enforceable laws in place, something that is reinforced by the distance and technology involved.

The fundamentally undemocratic nature of space expansionism basically excludes the non-elite majority of Earth. It also is reflected in the position of non-hegemonic nation-states which occupy an unequal position relative to both other nation-states and private corporations. To put it simply, the present spacefaring nations have the resources and capacity to send expeditions into space as do billionaire-backed private initiatives. Many countries do not have the resources to be part of this potential exercise nor to give weight to alternative ecologically informed visions of celestial bodies and outer space. It has been noted that while rhetorically there is affirmation that 'space is for everyone', and this is embedded in documents such as the *Outer Space Treaty* and *Moon Agreement*, the reality is far from this (Lampkin & White, 2023). Accordingly, UN delegations from the excluded nations have argued that 'the exploration, exploitation and utilization of space resources should be coordinated at the international, multilateral level in order to ensure the continued peaceful and sustainable use of outer space and to ensure that such activities were carried out in accordance with international law and for the benefit and in the interests of all States' (United Nations, 2021: 25). How to guarantee inclusiveness whether through international law or otherwise is a matter of deep and continuing concern.

Back on Earth, big questions need to be asked regarding how and why, historically, the world has been riven by deep social inequalities, and about the continuing legacy of imperialism and colonialism and the role of capitalist institutions in this process. Similar inequalities are being replicated in relation to space exploration and with respect to what corporations and governments do with space resources. As it stands, access to space depends on nation-state largesse, corporate profit requirements, and billionaire interest. The commodification of space – space travel, minerals in space, health innovations related to space laboratories – represents a further intensification of the privatisation of space, its activities, and thereby its *de facto* private ownership (reinforced by state subsidy and support).

Ownership and Control

Current plans for outer space expansion thus signal a large and growing interest by private individuals and corporations in space colonisation and exploration. This comes with a substantial price tag and hence there is great emphasis on profit-making activities and industries. Because of costs, the private ownership of spacecraft will also likely be replicated when it comes to space stations, settlement colonies, and mining outposts. Most governments will be unable or unwilling to engage in such gross expenditure. Thus, the worlds being created beyond Earth will largely be owned by large corporations and controlled by the space oligarchs.

This means that the apparatuses of social control will also largely be in the hands of those spending their money on space-related ventures. For instance, the physical infrastructure in outer space that is built by and for private companies – the satellites, space stations, transportation vehicles, supply rockets, and extra-terrestrial outposts – will be regulated and policed by these same companies. This will likely consist of different, possibly overlapping, personnel with specific skill sets. For example, cyber experts will be hired to protect data banks and communication systems, security guards employed to guard access to spacecraft, stations and settlements, and armed militia deployed to enforce order in workplaces and to protect workers as well as the products produced in space operations (such as solar energy and mineral extractions) from competitors and organised criminal groups (Lampkin & White, 2023). If this scenario comes close to reality, it will very much hark back to the imperialist era of global capitalism which saw private companies literally directly rule vast regions of the planet, for example, in places such as India.

Private police, company security guards and mercenaries are beholden to their employer. The use of military and paramilitary forces in space requires examination of a series of interrelated matters including training, activities, powers (especially lethal force), and accountability. By contrast, government funded police and armies are ostensibly subject to democratic control or at the least control by the public institutions of the state and meant to act in the public interest.

But in outer space, who will pay for police, and to whom will they be accountable? The United Nations police service has not proved particularly effective on Earth in resolving conflicts or maintaining social peace. Policing by national police services in international contexts needs to be coordinated by agencies such as Interpol and Europol as otherwise they might be considered partisan, biased, or culturally inappropriate. Whether this collaborative model would work in outer space, given the smaller number of parties involved, is questionable.

Meanwhile, where the activities and infrastructure provision of private corporations and governments overlap – for example, in jointly constructed space stations, collaborative efforts relating to rocket launches and state-corporate mining enterprises – big questions remain regarding jurisdiction, authority structures, and financial contribution. Where there is continuing silence and inaction around these questions, power will simply relegate to the private corporation.

Conclusion

What is happening on Earth is indeed what is unfolding in outer space. A small handful of individuals are making momentous decisions that will have ramifications for many years to come. The rich are being made even more rich (and famous), while the masses beneath fall prey to romanticised tales and exclusionary processes that promise much but deliver little for them. From the perspective of harm and morality, this is not a social good.

The fight for climate justice on Earth illustrates the struggles and the enormous challenges that lie ahead in regards to outer space. Fundamentally, the fight for climate justice involves assertion of democratic control over land, air, water and energy and advocacy of radical egalitarianism (involving eco-justice for humans, ecosystems, and non-human species). This means divesting present corporate owners of their private property and re-asserting communal control (van der Velden & White, 2021; Whyte, 2020). Just a handful of transnational corporations contribute a significant proportion of the world's carbon emissions. These companies are responsible for more greenhouse gases than most countries. They are an integral part of the originating problem. This will most likely continue to be the case in off-world contexts.

Juxtaposed and contrary to these specific interests are the collectivist ideals of 'universal human interests' and 'ecological citizenship'. The appeal of ecological citizenship as a concept stems from recognition of the universal interests that underpin the relationship of humans with their environment. Ecological citizenship allows for stepping outside prescriptive patriotism (e.g., Australia first, America first, China first) when global ecological health and well-being demands a planetary response (e.g., Earth first). Ecological citizenship is also tied to the notion of 'Earth Rights' and the survival needs of all species and biospheres on the planet. Similar demands are required in relation to the exploration and exploitation of outer space.

First, there is need to reaffirm the notion of 'commons' and putting the needs and interests of humanity over and above those of the private corporation and billionaire boss. This requires collective action by citizens and governments around the world and the pressuring of organisations such as the United Nations and European Union. Words must be accompanied by action that confirms positive social rights and negates the emphasis on private interests.

Second, universal interests linked to the idea of the commons should include the notion of celestial justice – that is, that 'justice for all' must be inclusive of the non-human as well as the human. Demands in relation to the Rights of the Moon tap into this vital recognition of intrinsic value and against the reduction of activity to commodified exchange-value. Eco-justice demands nothing less than justice for humans, specific environments, non-human animal and plant species, and celestial objects.

Third, financing space ventures can be achieved by taxing the rich on Earth. This can be further advanced by divesting the billionaires of their ownership and control of vast financial empires in favour of democratic public institutions and decision-making. The money for space is certainly there. But the naturalness of Earthly exploitation of natural resources and human labour obfuscates and legitimates the ruthless destruction and degradation underpinning contemporary global capitalism. Yet, the six pillars of social life – water, air, food, energy, shelter, and security – are too important for decisions to be left in private hands (the billionaires) or secretive state (bureaucrats and self-serving politicians): governance must be public, open, cooperative, and democratic. The same must be said about governance in outer space.

References

ABC (Australian Broadcasting Corporation) News (2023a) 'Elon Musk refused Ukraine's request to activate Starlink network, fearing "major" act of war', 9 September 2023.

ABC News (2023b) 'Virgin Galactic flies first tourists to edge of space, Russia takes off to the Moon', 11 August 2023.

ABC News (2023c) 'Kim Jong Un celebrates North Korea's new "space power" after spy satellite launch', 25 November 2023.

ABC News (2024) 'Russia has obtained a "troubling" emerging anti-satellite weapon, the White House says', 16 February 2024.

Alamalhodaei A (2023) 'Musk: SpaceX's Starlink "has achieved breakeven cash flow"', Yahoo!finance, 3 November 2023.

Beven M & Parry Y (2023) 'Elon Musk wouldn't give Ukraine's military access to his Starlink satellites at a crucial moment. Did he make the right call?', ABC News, 7 October 2023.

Brisman A & South N (eds) (2021) *The Routledge Handbook of Green Criminology*. London: Routledge.

Chambers G (2023) 'Space edict: keep US in dark', *The Weekend Australian*, 9–10 September, p.1.

Eski Y (2023) *A Criminology of the Human Species: Setting an Unsettling Tone*. London: Palgrave Macmillan.

Guardian, The (2024) 'Elon Musk moves SpaceX incorporation to Texas after Delaware judge axed $56bn Tesla pay', *The Guardian*, 15 February 2024.

Lampkin J & McClanahan B (2023) 'Astronomical withdrawals: a green criminological examination of extreme energy mining on extraterrestrial objects', *Crime, Law and Social Change*, https://doi.org/10.1007/s10611-023-10123-9.

Lampkin J & White R (2023) *Space Criminology: Analysing Human Relationships with Outer Space*. London: Palgrave Macmillan.

Lampkin J & Wyatt T (2023) 'An Astro-Green criminological examination of orbital space debris', *Criminology and Criminal Justice*, 23(3): 1–18.

Mandel E (1975) *Late Capitalism*. London: Verso Books.

Purtill J (2024) 'Why did America return to the Moon? What does Odysseus mean for future lunar missions?', *ABC News*, Science, 24 February 2024.

Rothe D & Collins V (2023) 'Planetary geopolitics, space weaponization and environmental harms', *The British Journal of Criminology*, 63: 1523–1538; https://doi.org/10.1093/bjc/azad003.

Sachdeva G (2023) *Crimes in Outer Space: Perspectives from Law and Justice.* Singapore: Springer Nature.

Tombs S & Whyte D (2015) *The Corporate Criminal: Why Corporations Must Be Abolished.* London: Routledge.

United Nations (2021) *Report of the Committee on the Peaceful Uses of Outer Space.* Sixty-Fourth Session (25 August to 3 September 2021). General Assembly, Official Records, Seventieth-sixth Session, Supplement No.20.

van der Velden J & White R (2021) *The Extinction Curve: Growth and Globalisation in the Climate Endgame.* London: Emerald Publishing Limited.

White R (2002) 'Environmental harm and the political economy of consumption', *Social Justice*, 29(1–2):82–102.

White, R (2008) 'Class analysis and the crime problem', in T Anthony & C Cunneen (eds), *The Critical Criminology Companion.* Sydney: Hawkins Press.

White R (2017) Corruption and the securitisation of nature. *International Journal for Crime, Justice and Social Democracy*, 6(4): 55–70. https://doi.org/10.5204/ijcjsd.v6i4.449.

White R (2022) *Theorising Green Criminology: Selected Essays.* London: Routledge.

Whyte D (2020) *Ecocide: Kill the Corporation Before It Kills Us.* Manchester: Manchester University Press.

Williams T (2024) 'First US Moon landing in decades appears doomed after "critical" fuel leak in Peregrine lander. Here's how it has played out', *ABC News*, 9 January 2024.

3 Imagining Space Crime

Using Virtual Reality to Advance our Understanding of Space Crime

Iris van Sintemaartensdijk

Introduction

Outer space is something in our minds that we relate to adventure, technology, exploration (Tutton 2021). We think about our future and how we could potentially live on Mars. What does not come to mind are the negative aspects of space and space exploration. In particular, it seems that the topic of crimes in outer space or "space crime" is one that is wilfully ignored – in particular by the spacefaring community.

Space crimes might however have already happened. In 2019, there was reporting on what was called the first-ever crime in space (Sachdeva 2021). Astronaut Anna McClain was accused of having accessed the bank account of her former spouse while serving on the International Space Station in order to gain information about the financial situation of her former spouse. Eventually, this accusation was classified as false, and no crimes had been committed by McClain. The widespread attention for the case does signal an interest for the fact that crimes can happen in outer space, and what the legal ramifications are of such crimes.

The new race to the Moon is happening as we speak. India has recently landed a spacecraft on the south pole of the Moon (Parvaiz 2023). Meanwhile, space exploration is not only a governmental affair anymore, as commercial companies like Virgin Galactic and Blue Origin transported passengers to zero gravity and SpaceX is planning to transport people and cargo soon (Davidian 2024). With all of these international players in the space market, and the geopolitical implications of their space explorations, we need to pro-actively study space crime.

Our Inability to Imagine Crimes in Space

Understanding space crime poses a challenge given the limited physical access to space itself, access to the spacefaring community and the dark figure of space crime. This means relying on the imagination rather than on situations that have taken place. Imagining such situations, however, is no easy feat, let alone imagining our responses. In order to study situations and responses to

DOI: 10.4324/9781003437178-4

situations that have not necessarily happened to people (yet), researchers use vignettes or written scenarios. In these scenarios readers are asked to imagine a certain scenario taking place, "Imagine you are....". Subsequently, participants respond to these scenarios by answering a number of questions. Although such scenarios have the benefit of isolating specific factors embedded in a situation, the contextual scenario is limited in these scenarios, and relies heavily on the participants' ability to imagine themselves in the described situation (van Gelder et al. 2019). Written scenarios place participants in an artificial situation, where there is a high chance of eliciting artificial responses (Kozlov & Johansen 2010).

In particular for understanding space crime, such written scenarios can be troublesome. First, there is the question of whether the scenario can capture the distinct feeling of being in outer space. Feelings of isolation are likely important for our behaviour in space, as research on places where isolation is frequent, such as submarines (Brasher et al. 2010) or penitentiary institutions (Schliehe et al. 2022), has shown. Only a limited number of people have been to space, and imagining the feelings of isolation and responses to situations in this extraordinary place are likely harder for people to imagine than the same situation taking place on Earth. Second, there is the "giggle" factor to take into account. When reading a scenario about outer space and crimes taking place, participants might imagine themselves in situations that are far from reality, such as the involvement of aliens.

Creating scenarios however is the best opportunity to study space crime scenarios. This will allow for proactively understanding scenarios that will take place in the near future. We need methods that can bring space crime scenarios to Earth. Analogue space missions in which a crew on Earth works in environments to mimic circumstances in space, such as recreating long-distance space exploration missions in Antarctica (Bell et al. 2019) or Mars missions in Morocco (Groemer et al. 2014) have proven invaluable for researchers studying these missions. However, analogue research suffers from small sample sizes (Bell et al. 2019) and is costly and time-intensive to set up. Our best option to create space crime scenarios that go beyond written scenarios, and make use of the ability to experience space scenarios like in analogy space missions, is to use virtual reality.

The Benefits of Using VR in (Space Crime) Research

Virtual reality (VR) is a computer-generated environment (Brey 2020). This virtual world is experienced usually through stereoscopic VR-glasses that allow for viewing a virtual world as you would in real life (Fox et al. 2009). The recipients of this VR are transformed into another world in which nearly everything is possible: walking around in a haunted house, riding on a roller-coaster, or even standing on a plank on the edge of the roof of an enormous building.

Slater and Sanchez-Vives (2016) reviewed the most common uses of VR as a research method. VR has first found its place in psychological research and neuroscience to understand virtual embodiment (Yee & Bailenson 2007) and to study how people navigate spaces (Gould et al. 2007). For scientific and data visualization VR proved to be optimal to visualize complex data, such as for example interaction patterns of molecules (Leinen et al. 2015). In education, VR has provided ways to practice safely before proceeding to real-life situations, such as for example allowing doctors to practice on virtual patients (Alaraj et al. 2011). Physical training has also benefited from VR through the opportunities for repeated practice with varying levels of difficulty (Miles et al. 2012). Lastly, with VR the first steps were taken to understand social and cultural experiences, and moral behaviour. Bailenson et al. (2003) were able to show people keeping a greater distance from virtual others when approaching them from the front. Slater et al. (2006) were even able to virtually replicate the famous Milgram experiment. In the original paper (Milgram 1965), participants were found to deliver a deadly electric shock to a fellow participant when pressured by authority; they later discovered the electric shocks were not administered and the fellow participant was a confederate. Slater et al. (2006) found that pressure from authority still increased the likelihood of participants delivering a deadly shock, only now to a virtual avatar.

VR allows for the creation of two types of scenarios. First is the opportunity to create training scenarios, in which recipients are expected to train certain aspects of a task in which feedback can be given quickly. For example, with VR-environments new crime scene investigators can be trained in how to conduct their work at crime scenes (Kanable 2012). The second type of scenarios are those to study behaviour, for which Blascovich et al. (2002) named VR as useful to observe behaviour as it unfolds.

In recent years, several criminological researchers have made use of VR as a method for criminological research and understanding the behaviour of those in the criminal justice system (see van Sintemaartensdijk & Meenaghan 2024 for a recent review). Researchers in the field of burglary have made the most use of this method, finding burglars taking the same approach to burgling houses in real life and in virtual reality (Nee et al. 2015). Subsequent studies reveal that VR provided an excellent tool to understand the expertise of burglars (Meenaghan et al. 2023) and even to understand how burglars choose their targets in neighbourhoods (van Sintemaartensdijk et al. 2021). Another prominent field within criminology using VR is the study of the bystander effect. With VR there was the possibility to expose participants to various bystander scenarios, such as for example a bar where a group member is harassed (Slater et al. 2013). These virtual scenarios outperform the use of the same scenarios written out (Xue et al. 2021). Lastly, due to the fact that VR can create any scene possible, researchers on Crime Prevention Through Environmental Design (CPTED) have made use of VR to link features of the environment to criminal activity. One of the first criminological studies using VR assessed the safety of train stations (Cozens et al. 2002), and subsequent reviews show that VR provides the adaptability for scenarios necessary for CPTED research (Noh & Lee 2019).

Based on the existing literature, there seem to be two main benefits of using VR as a research method for understanding space crime. The first main benefit is maximizing (experimental) control. With VR there is the opportunity to simulate any scene that we can potentially think of in a realistic way (Mania & Chalmers 2001; van Gelder et al. 2017). Within these scenes, we are able to control exactly what is going on (Blascovich et al. 2002; Fox et al. 2009). Any recipient of the VR system will thus be able to see the same scene as the previous or next recipient. In addition, there is a stark contrast with written scenarios, which depend on the imagination (van Gelder et al. 2019). To study space crime scenarios, this means people would likely be less inclined to imagine wild scenarios and potentially have a better understanding of how crimes in space could take place.

The second main benefit of VR is the opportunity to study scenarios that are otherwise unobtainable. VR allows researchers to study phenomena that are hard to study, whether this is due to ethical reasons, practical reasons or even financial reasons (van Gelder et al. 2014). Some situations cannot be studied by a researcher due to the lack of access. Going to outer space is not yet affordable nor achievable for the average researcher. However, even when some situations would be practically and financially achievable, there are ethical considerations that could warrant against studying the situation in real life. The aforementioned replication of the Milgram experiment (Slater et al. 2006) would never have been approved by an ethical review board in contemporary society with the exact same set-up of deceiving participants into thinking they are giving electrical shocks to another person. In order to understand crime and prevent crime from happening, we need to be able to understand situations that are hard or near-impossible to study in the real world as the behaviour unfolds. With VR, we have the possibility to study these situations without compromising the safety of potential victims, the (former) offenders or the researchers.

Applying VR to Understand Space Crime

VR is not unknown to those interested in space and teaching others about space. Cao et al. (2019) used a VR game to see if people playing such a game would be more interested in learning about lunar missions. They compared people playing a VR game to those playing the same game on a desktop computer. In the game, participants had to go through several phases of a lunar mission: planning, preparing and driving. Although the VR-game took participants longer to play, the engagement of those in the VR-condition was higher and the motivation to learn more about lunar missions increased. VR thus provided an excellent option for teaching others about space.

At the moment, VR provides the closest approximation to people experiencing crimes in space. With VR, you can immerse someone completely into another world. Recipients of the VR-environment experience it through a head-mounted display that impacts the engagement of the users in making them feel as if they are somewhere else (Mania & Chalmers 2001). This feeling of being somewhere else is also called "presence" and is essential to foster when creating VR-environments (von der Pütten et al. 2012). With a VR-simulation the goal is to achieve a high presence, in order to make people feel as if they are truly in the virtual world, rather than feeling as if they are playing a game.

A good example of the power that VR-scenarios hold over written scenarios, and how we can learn from situations that have not happened to the participants, is the work by van Gelder et al. (2019). In their study, van Gelder et al. (2019) exposed participants to either reading a bar fight scenario or to experiencing the scenario using VR. The scenario was written in the second person and describes an altercation in a bar between the participant and another male who tries to make a pass at the girlfriend of the participant. In the written scenario participants read the text on a laptop, whereas in the VR condition there was a 360° video filmed from the perspective of the participant. The participant heard a voice-over explaining the context of the scenario, in the same way participants in the written scenario read the same information. Following exposure to the scenario, participants were asked questions on perceived risk, intention to aggress, perceived realism, anger and anticipated shame and guilt. Within the VR condition, participants rated a higher presence and higher realism and even showed a higher intention to aggress toward the man hitting on their girlfriend.

Engaging people in scenarios on space crime in which they feel truly involved in the virtual world could elicit a better understanding of what could potentially happen in these scenarios, and perhaps even show how the confinement of space impacts our decision-making process. In the study by Cao et al. (2019) where VR was used to get people interested in lunar missions, VR provided a higher engagement and motivation to learn than the desktop game. When it comes to transforming people into another world to understand space crime, VR scenarios will provide more immersion over written scenarios that depend on the imagination of the recipient (Kozlov & Johansen 2010).

With VR we can also create scenarios varying in complexity and realism. Any scenario that we can potentially think of is possible within VR. This means firstly that we can vary the realism of the scenario, choosing a scenario close to home such as an act of violence on a space station. Some VR environments are more life-like than others (Brey 2020). With the design of every VR-environment, the researcher has to make choices regarding what to include or exclude and to what degree detail and realism play a role. A list of rules for using VR in (psychological) research by Vasser and Aru (2020) highlights the use of beautification of the environment, with participants experiencing the highest potential graphics. With VR, it would be possible to create visually pleasing scenarios that are hyper-realistic. For example, images from onboard the International Space Station could be recreated and serve as a basis for scenarios of crimes that consider the international relationship between the participating nations.

Interestingly, it is now possible also to study interactions between recipients of a VR-environment. Traditionally, VR is used to study the decision-making process and the behaviour of one participant. Interactions within the VR-environment were mainly between the participant and an avatar: a computer character that interacted with the recipient through pre-programmed responses. This of course is a limited interaction, as the avatar cannot react spontaneously. Now it is however possible to place multiple recipients within an VR-environment (Yaremych & Persky 2019). This opens up a realm of new possibilities for understanding space crimes. First, we can study the responses towards actual others, and see how they interact. Collaborating on tasks in VR and collaborating face-to-face have been shown to be strikingly similar (Moser et al. 2020), and with increasingly realistic environments we are able to better observe collaborative behaviour (Oksanen & Hämäläinen 2012). Second, researchers can participate in the virtual environment and replace the pre-programmed avatars currently used. With pre-programmed avatars, there is a script the avatars follow (e.g. speak to a participant when they are at a specified distance). There is no possibility to respond spontaneously to the participant in the VR-environment. For example, if the participant in a space crime scenario responds to a verbal harassment by a fellow astronaut, a pre-programmed avatar can provide a fixed number of responses whereas a researcher would be able to tailor their responses in real-time, if they were acting as another person in the scenario.

Important to note is that immersion can happen irrespective of the likelihood of the scenario. In the work of Seinfeld et al. (2018, 2023), people get transformed into the body of another person using VR. Violent offenders took on the perspective of a child exposed to domestic violence, and were found to be able to be more sensitive in recognizing anger in faces of both men and women (Seinfeld et al. 2023). The same paradigm also worked successfully for placing offenders into the shoes of a woman (Seinfeld et al. 2018). Several studies have even successfully had participants act with a digital realistic version of their older self (Ganschow et al. 2021; van Gelder et al. 2022), which impacts the willingness to make decisions that benefit the future self rather than only the present self. With these studies serving as examples we can see that even with varying degrees of realism participants are still immersed into the VR-environments. For understanding space crime, we thus can also explore scenarios that are more likely to happen (e.g. conflicts in a spaceship between two people of different nationalities) to scenarios far into the future or more akin to sci-fi (e.g. the possibilities of other than the human race being present in outer space).

First Steps to Take in Understanding Space Crime Scenarios

With numerous opportunities to create space crime scenarios in VR, the question is how to proceed to capitalize on the use of this method and which steps need to be taken in order for the space community and researchers interested in space crime to be able to use VR-scenarios to understand and study space crimes.

The first step that needs to be taken is the prototyping of the first VR space crime scenario. Here, decisions need to be made about what the first scenario should be, and to what degree this scenario needs to be realistic. With the first scenario being a proof-of-concept that is aimed to show the use of these scenarios over written scenarios, it is necessary to keep the scenario close to home. A balance might need to be made between creating a scenario with the highest experimental control versus a scenario that has higher ecological validity, a trade-off that is always present when designing new VR-environments (Blascovich et al. 2002; Kozlov & Johansen 2010). Adjacently, the best practices of the VR industry must be followed to reduce the chances of cybersickness, and to pick which type of VR setup would be the best option for this scenario (Vasser & Aru 2020). For example, trying to realistically display zero gravity by having the participants float in the VR-environment will only increase the chances of cybersickness.

When the scenario is completed, the next step is to compare the responses of those immersed within the VR-scenario to the responses of those who have experienced the written scenario. Here, it is essential to figure out whether presence and immersion within the VR-scenario exceeds that of the written scenario, and whether the willingness to take crimes in space seriously or to learn more about them, has increased as well. This will highlight why it is essential to invest more time and effort into developing space crime scenarios.

When the initial scenario is completed, it is time to engage with experts in the fields, such as space industry professionals or space agencies like NASA. These experts will be able to shed light on the likelihood of the scenario happening in space and could provide input on which scenarios are vital to study with the next iterations. In addition, they could also comment on visual aspects of the scenario and highlight aspects they find (un)realistic.

When the initial scenario has shown its merit, it is time to expand to new scenarios. This would first mean exploring crime scenarios that would happen on Earth, such as murder, assault or theft. The complexity of the scenarios can also be changed by allowing multiple agents into the scenarios in order to study the interactions between people. Lastly, we could also think about dialling down the realism of the scenarios and adding more sci-fi elements.

These scenarios could help in understanding how crimes in space could take place, and open up the study of a whole new field. There is also a great potential for these VR-scenarios to act as a conversation starter. Thinking about crimes in space seems uncomfortable, and not something that experts seem willing to do. If we cannot even talk about crimes in space in the abstract, we will not proceed to take the necessary action. Exposing the aforementioned experts to VR-scenarios could make them more willing to initially start talking. Exposure to VR-environments has been shown to increase engagement with the topic, such as for example burglars who share more about their burglar experiences (Meenaghan et al. 2018). Showing what space crimes can look like and the devastating personal, and potentially international, consequences these can have will hopefully instigate uncomfortable conversations that would not otherwise have taken place.

The Dark Side of the Moon: Disadvantages of Using VR for Space Crime Analysis

The dark side of the Moon when it comes to VR does need to be addressed. Although VR has numerous benefits when it comes to exposing people to new scenarios, there are some downsides. One of the most prominent is cybersickness. Not every person immersed in VR will experience this as pleasant (Hartmann et al. 2016). Cybersickness unfortunately can decrease cognitive performance, as well as confound physiological data (Vasser & Aru 2020). Recent research however indicates that cybersickness can be reduced and largely avoided without compromising the necessary level of presence (Vasser & Aru 2020) due to advancements of developer knowledge on, for example, being aware participants remain stationary while their virtual body moves around and how to limit this from happening.

The second downside to VR is the time and money that it can take to create VR-environments (van Gelder et al. 2014). A skilled programmer is necessary who can devote enough time to create a good environment. Additionally, researchers have to be willing to devote their time to translate their research ideas into requirements for the program. This means quite a substantial time investment. Devoting this time and attention however will result in a VR-environment that meets the researcher's needs, and could be used by others to foster replication (Vasser & Aru 2020). In addition to this, there are costs associated with VR-equipment and the development of VR-environments (Vasser & Aru 2020). Fortunately, VR has become more accessible over time, and the costs of VR-equipment have slowly been decreasing (Ratcliffe et al. 2021). The time and money involved in creating such VR-environments will likely be a fraction of the costs involved in analogue space missions, and rapid development and testing of multiple scenarios could also provide a valuable addition to studying these more intense analogue missions.

The third and final downside of VR is the inability to fully capture the experience of being in outer space and experiencing a criminal situation there. No matter how realistic a VR-environment is, there will never be a VR-environment that will show people what being in the confinement of space is actually like. It is the question to what degree people can let go of the awareness that the virtual environment is not real (van Gelder et al. 2014), in particular for being in outer space where they will never have been. Here it is important to make use of knowledge on feelings of isolation from research on, for example, submarines or penitentiary institutions, to see how feelings of isolation can impact behaviour and potentially how to (partially) invoke such feelings in a VR-environment. However, researchers should also never claim that what they are capturing is real-world behaviour when employing VR to understand behaviour. Rather, VR should be seen as the closest approximation to behaviour possible currently and researchers should be realistic about the applicability of the studied behaviour to the real world (Vasser & Aru 2020).

An example is a study where participants were exposed to a nine-minute VR-simulation of solitary confinement. The study found that empathy towards people who experience solitary confinement increased (Clifford & White 2020). Real-life solitary confinement will undoubtedly provoke more extreme reactions than VR solitary confinement, but the VR-experience can provide people with a little more insight into what solitary confinement could look like. VR can thus not expose behaviour that is identical to real-world behaviour, but can provide more insight into this real-world behaviour than was previously possible. With the developments in VR such as holographic images, motion sensors or the contributions of AI (Ludlow 2015), the opportunities to imagine space crime will only increase.

Conclusion

Since the introduction of VR for criminological research, we have seen many fields use this method to further understand behaviour when it comes to burglary, bystander effect and aggression. Now, we should use this method to take the next step and study space crime. We can use the ability to create any scenario possible for normally hard-to-study scenarios and expose people to various new scenarios. We can not only learn what people think of crimes in outer space but also encourage people to be interested in learning about this type of crime. It can serve as a conversation starter and thus as a starting point for further discussion in the spacefaring community. Imagining crimes in space is hard; using VR we can start to see the first glimpse of how these crimes might take place and why they are essential to study.

References

Alaraj A, Lemole M, Finkle J, Yudkowsky R, Wallace A, Luciano C, Banerjee, P, Rizzi S, & Charbel F (2011) Virtual Reality Training in Neurosurgery: Review of Current Status and Future Applications. *Surgical Neurology International* 2(1): 52.

Bailenson JN, Blascovich J, Beall AC, & Loomis JM (2003) Interpersonal Distance in Immersive Virtual Environments. *Personality and Social Psychology Bulletin* 29(7): 819–833.

Bell ST, Brown SG, & Mitchell T (2019) What We Know About Team Dynamics for Long-Distance Space Missions: A Systematic Review of Analog Research. *Frontiers in Psychology* 10: 811.

Blascovich J, Loomis J, Beall AC, Swinth KR, Hoyt CL, & Bailenson JN (2002) Immersive Virtual Environment Technology as a Methodological Tool for Social Psychology. *Psychological Inquiry* 13(2): 103–124.

Brasher KS, Dew ABC, Kilminster SG, & Bridger RS (2010) Occupational Stress in Submariners: The Impact of Isolated and Confined Work on Psychological Well-being. *Ergonomics* 53(3): 305–313.

Brey P (2020) The Ethics of Representation and Action in Virtual Reality. *Ethics and Information Technology*1: 5–14.

Cao L, Peng, C, & Hansberger JT (2019) Usability and Engagement Study for a Serious Virtual Reality Game of Lunar Exploration Missions. *Informatics* 6(4): 44.

Clifford K, & White R (2020) Mediated Representations of Prisoner Experience and Public Empathy. In *The Palgrave Handbook of Incarceration in Popular Culture.* London: Palgrave Macmillan, pp. 265–288.

Cozens P, Waters J, & Neale R (2002) A Virtual Reality Approach to Personal Safety and the Design of Built Environment Facilities. *Conference Proceedings of the 18th ACROM Annual Conference* 2: 461–473.

Davidian K (2024) *Key Space Highlights from 2023.* New Space.

Fox J, Arena D, & Bailenson JN (2009) Virtual Reality: A Survival Guide for the Social Scientist. *Journal of Media Psychology* 21(3): 95–113.

Ganschow B, Cornet L, Zebel S, & van Gelder J-L (2021) Looking Back From The Future: Perspective Taking in Virtual Reality Increases Future Self-Continuity. *Frontiers in Psychology* 12: 664687.

Gould NF, Holmes MK, Fantie BD, Luckenbaugh DA, Pine DS, Gould TD, Burgess N, Manji HK, & Zarate CA (2007) Performance on a Virtual Reality Spatial Memory Navigation Task in Depressed Patients. *American Journal of Psychiatry* 164(3): 516–519.

Groemer G, Soucek A, Frischauf N, Stumptner W, Ragonig C, Sams S, Bartenstein T, Häuplik-Meusburger S, Petrova P, Evetts S, Sivenesan C, Bothe C, Boyd A, Dinkelaker A, Dissertori M, Fasching D, Fischer M, Föger D, Foresta L… Zanella-KuxK (2014) The MARS2013 Mars Analog Mission. *Astrobiology* 14(5): 360–376.

Hartmann T, Wirth W, Schramm H, Klimmt C, Vorderer P, Gysbers A, Böcking S, Ravaja N, Laarni J, Saar, T, Gouveia F, & Maria Sacau A (2016) The Spatial Presence Experience Scale (SPES): A Short Self-Report Measure for Diverse Media Settings. *Journal of Media Psychology* 28(1): 1–15.

Kanable R (2012) Virtual Reality: A Reality for Crime Scene Training. *Law Enforcement Technology* 39(11): 8–16.

Kozlov MD, & Johansen MK (2010) Real Real Behavior in Virtual Environments: Psychology Experiments In a Simple Virtual-Reality Paradigm Using Video Games. *Cyberpsychology, Behavior, and Social Networking* 13(6): 711–714.

Leinen P, Green MFB, Esat T, Wagner C, Tautz FS, & Temirov R (2015) Virtual Reality Visual Feedback for Hand-Controlled Scanning Probe Microscopy Manipulation Of Single Molecules. *Beilstein Journal of Nanotechnology* 6: 2148–2153.

Ludlow BL (2015) Virtual Reality: Emerging Applications and Future Directions. *Rural Special Education Quarterly* 34(3): 3–10.

Mania K, & Chalmers A (2001) The Effects of Levels of Immersion on Memory and Presence in Virtual Environments: A Reality Centered Approach. *CyberPsychology & Behavior* 4(2): 247–264.

Meenaghan A, Nee C, van Gelder J-L, Otte M, & Vernham Z (2018) Getting Closer to the Action: Using the Virtual Enactment Method to Understand Burglary. *Deviant Behavior* 39(4): 437–460.

Meenaghan A, Nee C, Vernham Z, & Otto M (2023) A Comparison of Younger and Older Burglars Undertaking Virtual Burglaries: The Development of Skill and Automaticity. *Journal of Experimental Criminology*: 1–24.

Miles HC, Pop SR, Watt SJ, Lawrence GP, & John NW (2012) A Review of Virtual Environments for Training in Ball Sports. *Computers & Graphics* 36(6): 714–726.

Milgram S (1965) Some Conditions of Obedience and Disobedience to Authority. *Human Relations* 18(1): 57–76.

Moser I, Chiquet S, Strahm SK, Mast FW, & Bergamin P (2020) Group Decision-Making in Multi-User Immersive Virtual Reality. *Cyberpsychology Behavior, and Social Networking* 23(12): 846–853.

Nee C, White M, Woolford K, Pascu T, Barker L, & Wainwright L (2015) New Methods for Examining Expertise in Burglars in Natural and Simulated Environments: Preliminary Findings. *Psychology, Crime & Law* 21(5): 507–513.

Noh S, & Lee Y (2019) How Useful Are Virtual Reality Simulations to the Field of Crime Prevention Through Environmental Design? A Case Study. *Journal of Digital Landscape Architecture* 4:158–176.

Oksanen K, & Hämäläinen R (2012) Collaborative Serious Game as a Sociable Computer-Supported Collaborative Learning. *Proceedings of the 6th European Conference on Game Based Learning*: 364–371.

Parvaiz A (2023) India's Space Missions Spark Science Funding Debate. From: https://www.proquest.com/docview/2870617590?pq-origsite=gscholar&fromopenview=true&sourcetype=Newspapers.

Ratcliffe J, Soave F, Bryan-Kinns N, Tokarchuk L, & Farkhatdinov I (2021) Extended Reality (XR) Remote Research: A Survey of Drawbacks and Oportunities. *Proceedings of the 2021 CHI Conference on Human Factors in Computing Systems* 1–13.

Sachdeva GS (2021) Lex Specialis for Space Crimes. *Astropolitics* 19(3): 223–229.

Schliehe A, Laursen J, & Crewe B (2022) Loneliness in Prison. *European Journal of Criminology* 19(6): 1595–1614.

Seinfeld S, Arroyo-Palacios J, Iruretagoyena G, Hortensius R, Zapata E, Borland D, de Gelder B, Slater M, & Sanchez-Vives MV (2018) Offenders Become the Victim in Virtual Reality: Impact of Changing Perspective in Domestic Violence. *Scientific Reports* 8(1): 2692.

Seinfeld S, Hortensius R, Arroyo-Palacios J, Iruretagoyena G, Zapata LE, de Gelder B, Slater M, & Sanchez-Vives MV (2023) Domestic Violence from a Child Perspective: Impact of an Immersive Virtual Reality Experience on Men with a History of Intimate Partner Violent Behavior. *Journal of Interpersonal Violence* 38(3–4): 2654–2682.

Slater M, & Sanchez-Vives, MV (2016) Enhancing Our Lives With Immersive Virtual Reality. *Frontiers in Robotics and AI*3:74.

Slater M, Antley A, Davison A, Swapp D, Guger C, Barker C, Pistrang N, & Sanchez-Vives MV (2006) A Virtual Reprise of the Stanley Milgram Obedience Experiments. *PLoS ONE* 1(1): e39.

Slater M, Rovira A, Southern, R, Swapp, D, Zhang, JJ, Campbell, C, & Levine, M (2013) Bystander Responses to a Violent Incident in an Immersive Virtual Environment. *PLoS ONE* 8(1): e52766.

Tutton R (2021) Sociotechnical Imaginaries and Techno-Optimism: Examining Outer Space Utopias of Silicon Valley. *Science as Culture* 30(3): 416–439.

van Gelder JL, Otte M, & Luciano EC (2014) Using Virtual Reality in Criminological Research. *Crime Science* 3(1): 10.

van Gelder J-L, Cornet LJM, Zwalua NP, Mertens ECA, & van der Schalk J (2022) Interaction With the Future Self in Virtual Reality Reduces Self-Defeating Behavior in a Sample of Convicted Offenders. *Scientific Reports* 12(1): 2254.

van Gelder J-L, de Vries RE, Demetriou A, van Sintemaartensdijk I, & Donker, T (2019) The Virtual Reality Scenario Method: Moving from Imagination to Immersion in Criminal Decision-Making Research. *Journal of Research in Crime and Delinquency* 56(3): 451–480.

van Gelder J-L, Nee C, Otte M, Demetriou A, van Sintemaartensdijk I, & van Prooijen J-W (2017) Virtual Burglary: Exploring the Potential of Virtual Reality to Study Burglary in Action. *Journal of Research in Crime and Delinquency* 54(1): 29–62.

van Sintemaartensdijk I, & Meenaghan, A (2024) The Use of XR Technology in Criminological Research. *Journal of Experimental Criminology*: 1–25.

van Sintemaartensdijk I, van Gelder J-L, van Prooijen J-W, Nee C, Otte M, & van Lange P (2021) Mere Presence of Informal Guardians Deters Burglars: A Virtual Reality Study. *Journal of Experimental Criminology* 17(4): 657–676.

Vasser M, & Aru J (2020) Guidelines for Immersive Virtual Reality in Psychological Research. *Current Opinion in Psychology* 36: 71–76.

von der Pütten AM, Klatt J, Ten Broeke S, McCall R, Krämer NC, Wetzel R, Blum L, Oppermann L, & Klatt J (2012) Subjective and Behavioral Presence Measurement and Interactivity in the Collaborative Augmented Reality Game TimeWarp. *Interacting with Computers* 24(4): 317–325.

Xue J, Hu R, Zhang W, Zhao Y, Zhang B, Liu N, Li S-C, & Logan J (2021) Virtual Reality or Augmented Reality as a Tool for Studying Bystander Behaviors in Interpersonal Violence: Scoping Review. *Journal of Medical Internet Research* 23(2): e25322.

Yaremych HE, & Persky S (2019) Tracing Physical Behavior In Virtual Reality: A Narrative Review of Applications to Social Psychology. *Journal of Experimental Social Psychology* 85: 103845.

Yee N, & Bailenson J (2007) The Proteus Effect: The Effect of Transformed Self-Representation on Behavior. *Human Communication Research* 33(3): 271–290.

4 Resource Exploitation in Outer Space

The Potential of Crime Scripting as a Prevention Tool for Environmental Space Crime

Emilia Ziosi

Introduction

Territorial claims in space are prohibited under international law. Article II of the Outer Space Treaty (OST) (1967) states that outer space, including the Moon and other celestial bodies, is not subject to national appropriation by claim of sovereignty, by means of use or occupation, or by any other means; a principle which is reaffirmed in Article 11(2) of the Moon Agreement (1979). Nevertheless, with a surging international interest in space resources and exploration, accompanied by unprecedented advancements in the space industry, different countries have been trying to find ways to gain access to outer space resources. For example, recent legislative developments in the United States (US) and Luxembourg, among others, have officially recognised private property rights to outer space resources (Peterson 2021).

As access to space becomes easier due to technological advancements, legislations that favour 'exploration', and the increasing ability of a series of private companies to develop space capabilities with limited support or funding from governments (Miller 2019), serious concerns are raised on the environmental consequences of our actions in outer space (Dunnett et al. 2017). This is even more true in the absence of a comprehensive protocol stipulated by individual countries or supranational institutions to assess the potential impacts of human actions on these extra-terrestrial environments (Kramer 2014), especially in terms of harm.

While criminology has yet to fully engage with space-related crime and harm discourses (for exceptions, see Eski 2023; Lampkin and White 2023), scholars have recently started to explore the potential of the subdiscipline of green criminology in understanding extra-terrestrial environmental harm (Lampkin 2021; Rothe and Collins 2023; Takemura 2019). There is, however, still a noticeable absence of discussion on the possible environmental crimes and harms caused by private actors' property rights to space resources. Even more so, there is a lack of research surrounding the ways in which we can study and hopefully prevent such issues, especially within the criminological literature.

Given the increasing steps taken by individual countries to permit private mining and ownership of outer space resources, research on the matter

DOI: 10.4324/9781003437178-5

appears to be timely and of utmost importance. Yet, how can we study potential opportunities for crime in a realm that is largely inaccessible to the general population, particularly when the crimes have not yet occurred?

Over the past 30 years, specific techniques, namely Situational Crime Prevention Theory (SCPT) and Crime Script Analysis (CSA), have emerged in the study of crime prevention within the criminological debate (see Cornish 1994). These tools have been used to explore the crime-commission processes of a vast array of crimes and develop more informed prevention strategies (Huisman and van Erp 2013). Given the prevention-oriented nature of these techniques, alongside their capability to explore the underlying motivations behind potential offences, these techniques hold significant promise for assessing and mitigating opportunities for criminal activity in extra-terrestrial environments. Until now, however, these have solely been applied to crimes committed in the earthly realm, leaving outer space unexplored.

With this in mind, this book chapter pioneers the conceptual application of SCPT and CSA in the outer space realm, with the aim of exploring their potential in researching and preventing opportunities for environmental crimes resulting from private property rights to outer space resources. In doing so, it highlights the importance of criminology and its tools in researching crime in outer space, contributing to the existing literature on the criminology of the outer space (Eski 2023; Lampkin 2021; Lampkin and White 2023) and on crime scripting (Brayley et al. 2011; Chiu et al. 2011; Huisman and van Erp 2013).

To do so, this book chapter first introduces SCPT and CSA while showcasing their applications within criminology. It then delves conceptually into how these crime prevention techniques can be instrumental in researching and preventing specific opportunities for environmental crime in outer space, applying them to the case of 'astro-mining', that is to say, the extraction of natural resources from celestial bodies. In conclusion, this effort provides a preliminary exploration of opportunities for environmental crime in outer space within criminology, while also emphasising the crucial role of prevention in addressing environmental crimes and harms in underexplored territories.

Criminology and Crime Prevention Techniques

Crime Script Analysis (CSA) and Situational Crime Prevention Theory (SCPT) have been widely adopted by scholars to predict human behaviour and for crime and harm prevention purposes. Both approaches, departing from an opportunity-based perspective, have their roots in environmental criminology and rational choice theory and seek to understand crime as a process formed by different stages with the aim of preventing it (Weirich 2019). The Situational Crime Prevention criminological framework, which has grown to become fundamental in crime prevention theory, aims to identify the opportunities that enable the commission of a crime and determine possible means to reduce them (Clarke 1980; Clarke 2009). As Huisman and van Erp (2013: 1181) point out, 'SCPT starts from the assumption that potential offenders decide whether or not to engage in crime, based on the attractiveness of criminal opportunities, in terms of costs and benefits of crime'.

Within this framework, scholars have increasingly adopted Crime Script Analysis as an analytical tool to identify the actions required for a crime to occur. A script is 'simply a sequence of actions which make up an event' (Brayley et al. 2011); and CSA is described as 'a framework for breaking down and examining the crime-commission process and, second, create potential for situational prevention at every step of the crime' (Chiu et al. 2011: 359). This analytical tool, essentially, enables the creation of a step-by-step breakdown of the crime-commission process by deconstructing the actions taken by the actors involved, with the aim of informing better policy decisions (Weirich 2009).

Criminologist Cornish (1994) first introduced the script-theoretic approach to crime analysis two decades ago, and, since then, this practice has been widely adopted by scholars to explore different crime types. These include, among others, cybercrime, corruption and fraud offences, drugs offences, sexual offences (Dehghanniri and Borrion 2019), and green, environmental, and corporate crimes (Huisman and van Erp 2013). More recently, scholars have also delved into the financial aspects of profit-driven crime, introducing financial crime scripting (see Snaphaan and van Ruitenburg 2024; van Santvoord and van Ruitenburg 2023).

Despite the wide use of CSA, the existing applications have only considered the Earth realm until now, leaving outer space unexplored. Especially when it comes to environmental crime and harm, however, the extra-terrestrial realm is not less vulnerable. Given the disruptive impacts that human activities and exploration in outer space could have both on terrestrial and extra-terrestrial environments, there is a need to explore and hopefully prevent potential environmental crime opportunities in the complex realm of outer space.

Scripts, as can be evidenced by the different applications of CSA outlined above, can be used to analyse a broad range of crimes, proving especially valuable for understanding new or complex criminal activities (Brayley et al. 2011). This is especially true as CSA does not rely on a single scripting method, but rather, specific methods can be developed and tailored to the specific crimes under analysis. This implies that, within the context of outer space, where potential environmental crimes are both novel and intricate, this technique holds significant promise. It would enable the development of a specific model customised for the crime under analysis, the extra-terrestrial environment in which it occurs, and the intricate interplay of legal and atmospheric factors that make these opportunities for crime exceptionally complex.

This becomes even more evident in the exploration of contexts where specific data are not only missing, but also are inevitably incomplete. CSA is promising in this scenario, as it can be built using different types of data, developed from incomplete data, and updated as additional information becomes accessible (Brayley et al. 2011). As emphasised by Huisman and van Erp (2013), the key lies in focusing on the opportunity structure of specific crime categories. This means that generalised descriptors like 'environmental crime' prove insufficient when applying these frameworks. Therefore, in exploring how these tools can help identifying and preventing potential opportunities for crime in outer space, the focus must narrow down to a specific environmental crime opportunity that could be resulting from private property rights to outer space resources, like illicit astro-mining.

In the following section, the activity of 'astro-mining' and the opportunities for illicit behaviours around it will be presented. These will be used to think conceptually about how crime prevention techniques can be helpful at the time of researching and preventing specific opportunities for environmental crime in outer space. Despite this, the following sections do not aim to develop a crime scripting model of such activities, which would exceed the aim of this chapter. Instead, they provide a case to explore how these techniques could be applied to opportunities for crimes in the extra-terrestrial realm.

The Case of (Illicit) Astro-Mining: Towards a Conceptual Application of Crime Prevention Techniques to Environmental Space Crime

Environmental Crime and Harm in Outer Space

With an increasing number of states, individuals, businesses, and organisations aiming to establish their presence in space (Johnson-Freese 2017), the space-related environment is facing crucial challenges (Dalledonne 2021), and environmental harm happening in space is increasingly becoming a global societal concern. These concerns are amplified by the growing global interest in the exploitation of valuable extra-terrestrial resources, including, among others, rare earth minerals, precious metals, iron, nickel found in near-Earth asteroids, and helium-3 on the Moon (Na 2021).

The existing research on the environmental impacts of space exploration has primarily approached the subject from the perspectives of natural sciences and international law (Lampkin 2021), focusing especially on the available legal instruments for regulating environmentally harmful space activities. Law scholars have pointed out the inadequacy of international law to deal with the environmental challenges arising in outer space (Dalledonne 2021) and, together with space-security analysts and criminologists, have argued for the need to develop an international code of conduct for managing the space environment sustainably (Dalledonne 2021; Johnson-Freese 2017; Lampkin 2021).

Criminologists have recently started acknowledging the value of green criminology as a subdiscipline to explore space-related environmental harm and crime. Takemura (2019) and Lampkin (2021) have worked on an 'astro-green criminology' to acknowledge outer space-related environmental harm caused by humankind, while Rothe and Collins (2023) have delved into the environmental harms of space weaponisation. In their latest work, Lampkin and McClanahan (2023) have also adopted a green criminological approach to explain the emergence of 'astro- or space mining' and delve into its social and environmental impacts on both terrestrial and extra-terrestrial environments.

Astro-mining or space mining can be understood as a form of energy extraction that involves 'extracting natural resources that exist in outer space and returning them to Earth for human consumption' (Lampkin and McClanahan 2023: 2). With a surge in interest about outer space and the launch of satellites during the latter part of the 20th century (Takemura 2019), the idea of mining in outer space commenced as a series of extremely valuable resources were identified in the extra-terrestrial realm. This meant that both nations and private companies started increasingly paying attention to the profit opportunities of outer space mining, especially as countries such as the United States and Luxembourg, among others, started putting their efforts into approving legislations that allow private actors to own, sell, and transport outer space materials (Takemura 2019).

In the US especially, there are a series of private companies that are working towards space-mining, such as Moon Express; Planetary Resources Inc.; Deep Space Industries; and Shackleton Energy (Takemura 2019). However, as Lampkin and McClanahan (2023) point out, there are a series of limitations on space travel capabilities that currently prevent the exploration of mining opportunities on every celestial body. The extra-terrestrial bodies that are being targeted, considering the natural resources they are rich of and the possibility to reach them or set foot on them thanks to logistical developments are the 'Earth's Moon; asteroids and meteoroids; Mars; and the two Martian Moons (Phobos and Diemos)' (Lampkin and McClanahan 2023: 4). The general interest lies especially on the following activities: extracting water from the lunar poles, extracting water, volatiles and metals from near-Earth objects, and the manufacture of equipment from space resources, as well as the production of oxygen, among other things (Cilliers et al. 2023).

Current extraction efforts primarily centre around scientific experiments, however, the systems and technologies necessary for enabling large-scale commercial mining in space could be developed within our lifetimes (Denny et al. 2023). Research missions focused on returning samples from asteroids, exemplified by missions like Hayabusa, Hayabusa-2, and OSIRIS-REx, have highlighted the difficulties associated with extracting minerals from space using existing technology (Zhang 2018). As of 2023, in fact, the amount of asteroid material successfully brought back to Earth from space is less than 7 grams (Zhang 2018). Furthermore, a pioneering demonstration of space resource utilisation is showcased through the Mars Oxygen In Situ Resource Utilization (ISRU) Experiment (MOXIE) (Hoffman et al. 2022). This ground-breaking experiment was conducted within NASA's Mars 2020 Perseverance rover, successfully generating oxygen from the carbon dioxide that constitutes approximately 95% of the Martian atmosphere (Hoffman et al. 2022).

In the United States specifically, these advancements have been made possible through the enactment of the US Commercial Space Launch Competitiveness Act of 2015, also known as the Space Act of 2015, that the US congress passed into law on May 21 of the same year. The legislation allows US industries to 'engage in the commercial exploration and exploitation of space resources' (Space Act 2015), that is to say, it gives US companies the rights to own and sell natural resources they mine from celestial bodies in space (Lampkin and Wyatt 2022). The enactment of the law has, however, provoked significant debates around its feasibility and around its potential repercussions on the outer space environment.

Although the act, designed around the existing treaties regulating outer space, specifically recites that 'the United States does not [by this Act] assert sovereignty, or sovereign or exclusive rights or jurisdiction over, or the ownership of, any celestial body' (Space Act 2015), it has been widely criticised for potentially exploiting loopholes within the existing international regulations on the matter. Some scholars have argued that the act violates the 1967 Outer Space Treaty as the ownership of resources can be understood, in a way, as an act of sovereignty (Davies 2016). Others, such as Lampkin (2021: 241), have pointed out that the Space Act 2015 promotes the use of space resources for financial gain, potentially conflicting with Articles I and II of the 1967 OST. Notwithstanding these discussions, there is presently no international treaty or law explicitly forbidding the commercial utilisation and extraction of space resources.

Despite the actual legality or illegality of the activity, particularly considering, as highlighted by Lampkin (2021: 241), that 'the interplay between national and international laws regarding space exploration and resources is complex and often contradictory', these legislation advancements increase the likelihood that private companies will exploit such loopholes for profit, and as a result, potentially engage in environmental crime opportunities. Additionally, while these actions may not strictly constitute crimes, their potential for harm both on and beyond Earth is too significant to overlook, therefore necessitating careful consideration and proactive prevention measures. In this milieu, crime prevention techniques such as CSA and SCPT, as is showcased in the following sections of this chapter, can both help identify potential environmental crime opportunities arising from this legislation, or its absence, and suggest preventive recommendations and strategies to address these issues.

Identifying Opportunities for Crime Connected to Astro-Mining

Illicit astro-mining is particularly relevant at the time of thinking conceptually about the applications of crime prevention techniques for two reasons. On the one hand, the criminal opportunity has extensive analogies with the Earth realm (Dunnet et al. 2019; Lampkin and McClanahan 2023). Since we can only predict space-related environmental crimes and harms stemming from private property rights, analysing the opportunities for environmental offences considering how SCPT and CSA have been applied to crimes committed in the Earth realm can serve as a valuable initial step towards prevention.

On the other hand, illegal mining is, most of the time, a type of crime based on a rational opportunity, which makes it extremely applicable to crime prevention techniques. Opportunity-based theories, in fact, view offenders as intentional actors who engage in a decision-making process where they assess the costs and benefits of committing a crime in a particular circumstance (Huisman and van Erp 2013). This cost–benefit calculation depends extensively on the legal framework regulating the potential offence under analysis. Thus, it is evident that while this chapter does not delve into the legal intricacies of recent legislative advancements concerning extractive activities, their legal status undeniably shapes the potential opportunities for environmental crime associated with them.

On the one hand, in the scenario where astro-mining is considered illegal, the opportunities for environmental crime become more apparent. In fact, companies involved in this activity could already be seen as participating in a deliberate series of acts preceding the unlawful extraction of natural resources in outer space. In this case, it is evident how illicit astro-mining could be regarded as a crime of commission, as it would involve a decision-making process in which actors weigh the costs and benefits of engaging in such activities to determine their attractiveness. This makes SCPT, that is primarily designed to address 'predatory crimes of commission' (Huisman and van Erp 2013), a particularly useful technique to explore the actions and opportunities standing at the basis of the potential crime, which could start with the identification of a loophole in the existing legislation.

This is particularly true given the potential for substantial profitability in astro-mining, which due to high entry barriers that limit accessibility to a few companies, makes it an appealing proposition that private companies may increasingly be willing to pursue. Based on the striking similarities with illegal mining on Earth, crime prevention techniques could thus help identify first and foremost the opportunities for crime commission connected to the exploration and exploitation of valuable resources in outer space without appropriate rights to do so, and the costs and benefits private companies might take into consideration while deciding to commit an offence. This is especially beneficial at the time of designing tailored policies that consider the different motivations standing at the basis of the offence and the actions that private companies might be willing to undertake.

Recently, for example, media outlets have highlighted how AstroForge, a US asteroid-mining firm that is planning to launch a spacecraft to an undisclosed near-Earth asteroid in 2024 is keeping its target asteroid secret (*New York Times* 2023). Although the company's actions are not illegal, this case has sparked concerns in media outlets about private companies' motives and secrecy in commercial space exploration (see *New York Times* 2023). Additionally, it has highlighted the need to preventively address the existing gaps in spaceflight and resource ownership regulation, and the consequent potential exploitation of these gaps by those companies with the ability to go to space and engage in resource extraction.

As a matter of fact, if astro-mining is conversely deemed legal, opportunities for environmental crimes may involve exploiting existing, and often incomplete, treaties and legislations regulating outer space. This has striking similarities with the Earth realm, where the identification of 'new' territories with seemingly limitless resources and the subsequent mass exploitation by private companies, often in collaboration with global-north governments (McQuade 2019) has frequently resulted in extractive firms leveraging state-delegated authorities to exploit loopholes and facilitate extraction (Szablowski 2019). As discussed in the preceding section, the potential for private companies to engage in astro-mining has also arisen from a regulatory grey area, and the legality of such activities is often a subject of contention.

Moreover, besides finding loopholes to engage in resource extraction, private companies with advanced technologies for space exploration could potentially also restrain resources they are not authorised to extract. Alternatively, they could resort to employing unauthorised methods or chemicals to enable such activities – a strategy commonly employed in fracking on Earth (see Lampkin 2018). Especially in the absence of effective control and enforcement of laws in outer space, and in the lack of a single jurisdiction, this becomes a feasible scenario, resembling historical resource extraction practices on Earth. Furthermore, in both the legal and illegal scenarios, the potential profitability of astro-mining activities – coupled with a scarcity of oversight and regulation – introduces the risk of companies delving into illegal trade and smuggling of space-derived resources to maximise their gains, thereby adding an illicit dimension to the evolving dynamics of outer space exploration.

Regardless of the legality of the activity, however, it is evident that the opportunities for engaging in different types of environmental crimes connected to astro-mining are several. Especially through Earth-analogies, we are able to think about potential scenarios offering opportunities for exploitation to private companies that enjoy private property rights privileges over outer space resources. In their recent work, Rothe and Collins (2023: 2) have advocated for adopting a 'prospective approach' to examine in a critical way the potential environmental harms stemming from the space weaponisation race. Similarly, by drawing parallels with environmental offences on Earth, one can critically and prospectively assess not only specific opportunities for environmental crimes in the outer space, but also the various stages and actors implicated in the commission of the crimes.

This approach would allow for the development of customised crime script models tailored to specific environmental space crimes. Since we are unable to research crimes that have not been committed yet, crime prevention techniques offer an invaluable resource to identify and map these opportunities out. Additionally, as we shall see in the next section, identifying these opportunities and what stands at their basis is fundamental to recognise the potential harm they might cause and develop effective measures to mitigate their social and environmental impacts within the complex legal frameworks of outer space.

Environmental Space Harm: Showcasing the Need of Devising Preventive Interventions Connected to Astro-Mining

As highlighted earlier, the crime prevention techniques of focus in this chapter serve a dual role: first, to identify opportunities for environmental crime in outer space, and second, to formulate preventive interventions aimed at reducing their social harm. Going back to our Earth analogy, it is well-known that extraction processes of natural resources on Earth have often led to significant environmental harm. While Earth and outer space differ remarkably, especially in terms of available natural resources and atmospheres, drawing parallels with the harm caused by resource extraction, exploitation, and commercialisation on Earth can be extremely useful also at the time of considering the environmental consequences of private property rights over resources in outer space and designing tailored policies.

According to Lampkin and McClanahan (2023), much of the academic literature concerning potential environmental harms addresses the increasing dangers associated with orbital space debris (Migaud 2020). Space-related activities can however impact the environment both on Earth, for example through harmful emissions and pollution, and beyond Earth, potentially contributing to the deterioration and depletion of often non-renewable terrestrial and extra-terrestrial resources and environments (Lampkin and McClanahan 2023).

As addressed by Oduntan (2015), the pristine conditions of outer space could undergo irreversible changes as a result of mining activities, which could lead to a series of potential harms, such as the contamination of celestial bodies with Earth's microbes, known as forward contamination (Qizhi 1988). This could pose a threat to extra-terrestrial life and damage the environment surrounding Earth (Oduntan 2015). Simultaneously, there is also the risk of backward contamination, that is to say, the risk that extra-terrestrial micro-organisms might instead contaminate the Earth (Qizhi 1988).

Additionally, the unparalleled prospects for profit and influence may not only fuel economic competition between private companies engaged in space mining, but also give rise to complex conflict-related issues. Heightened competition for access to and control over valuable extra-terrestrial resources could potentially escalate, leading to conflicts with far-reaching consequences, posing risks to both the Earth and the outer space realms and their respective environments. The competition for valuable resources in outer space may trigger disputes over territorial claims, extraction rights, and resource allocation, and these conflicts could manifest in various forms, from legal battles to geopolitical tensions and, in the worst-case scenario, even physical confrontations.

To address these challenges, there is an urgent need for designing policies tailored to the outer space realm. As highlighted throughout this chapter, potential opportunities for environmental crime could primarily arise from the existence of loopholes within the existing national and international legislations around private property rights to space resources. First and foremost, this demonstrates the need for international cooperation on the matter and the establishment of robust regulatory frameworks that address the different ways in which private companies might engage in crime commission and environmental harm as a result of the existence – or absence of – legal frameworks regulating the ownership of space resources.

Moreover, the opportunities outlined above demonstrate the need to start looking forward in policymaking, and work towards prevention rather than tackling problems as they come along. As addressed by Oduntan (2015), the number of states with access to outer space is likely to increase, together with the number of states developing their own mining programs. This accelerates the urgency of discussing and implementing preventive international measures to address the environmental and social risks associated with the exploitation of outer space resources and demonstrates the importance of prevention in addressing these pressing issues.

Especially in light of the similarities of astro-mining with illegal mining on Earth, it is clear that prevention techniques such as SCPT and CSA could prove invaluable in designing effective policies to ensure responsible and sustainable practices in the exploration and utilisation of outer space resources. These techniques, as addressed throughout the chapter, could have the ability to prospectively identify opportunities for crime commissions and the different stages and actions that could form and facilitate a crime, as well as the related preventive measures to tackle both the crimes and the harms stemming from them.

Conclusions

The rise in investment, cost reduction, and notable technological advancements has rendered space-based activities that were previously economically non-viable, feasible. This is the case, for example, with outer space resource mining, which is becoming a closer reality than we think. Scholars have thoroughly engaged in the discussion around the benefits of such possibilities. As pointed out by de Zwart and colleagues (2023), for example, utilising resources extracted in outer space could have positive outcomes for the Earth's environment, potentially decreasing the energy required for transporting additional resources to space for local utilisation. However, these new opportunities also come with unprecedented challenges, risks, and important consequences, particularly concerning the safeguard of both the Earth and the extra-terrestrial environments.

This chapter has explored the case of astro-mining, focusing on the potential opportunities for environmental crime that might arise as a result of legislative advancements favouring private property rights to outer space resources. In doing so, it aimed to showcase the importance of a preventative approach in researching space-related matters, especially within criminology. Drawing parallels with the Earth realm, this chapter has focused on specific crime prevention techniques like SCPT and CSA to conceptualise their application to emerging challenges related to opportunities for environmental crime within the space realm, proposing a first conceptual exploration of their potential in addressing these issues.

Usually, when crafting specific prevention strategies using techniques such as SCPT and CSA, the process involves breaking down a particular crime commission into stages and proposing preventive recommendations for each stage. However, given that the primary objective of this chapter was conceptual, this chapter has presented general considerations associated with potential harms resulting from astro-mining and the corresponding reflections on preventive measures. In doing so, however, it also wishes to advocate for the need to develop actual crime scripting models to tackle these opportunities at every stage of the potential crime commission process and inform the public policy agenda.

In presenting the possible environmental harms arising from the crime opportunities connected to astro-mining in outer space, this chapter has also highlighted the importance of developing comprehensive and anticipatory regulatory frameworks to mitigate the risks associated with potential illegal activities connected to outer space resources' exploration and exploitation. This urgency is magnified by the absence of clear international regulation of the issue, as noted by Lampkin (2021), which showcases the vulnerability of international law on the matter. Another example, in this sense, could be represented by the increasing population of space debris, or space junk, in the outer space, and the lack of coordination and regulation surrounding the matter (Kerrest 2001; NASA 2021, 2022).

In conclusion, there are also limitations connected to the use of crime prevention techniques to explore environmental crime in outer space that need to be addressed. These are particularly associated with the inherent difficulty in carrying out research in this underexplored domain. One of the main limitations lies in studying something lacking a clear definition. To date, there is a dearth of consensus regarding the definitions of space-related crime and harm, which contributes significantly to the existing scarcity of research in this domain. The absence of a single jurisdiction in outer space, governed by a series of existing international treaties, further complicates research and prevention efforts. With this in mind, while this chapter does not aim to provide an exhaustive methodological framework for exploring outer space-related environmental crime, it hopes to serve as a conceptual starting point for the future development of international preventive policies and guidelines related to outer space environmental protection in the use and occupation of space.

References

Article 11(2), Agreement Governing the Activities of States on the Moon and Other Celestial Bodies (1979), 1363 UNTS 3.

Brayley H, Cockbain E and Laycock G (2011) The Value of Crime Scripting: Deconstructing Internal Child Sex Trafficking. *Policing* 5 (2): 132–143.

Chiu YN, Leclerc B and Townsley M (2011) Crime Script Analysis of Drug Manufacturing in Clandestine Laboratories: Implications for Prevention. *The British Journal of Criminology* 51 (2): 355–374.

Cilliers J, Hadler K and Rasera J (2023) Toward the Utilisation of Resources in Space: Knowledge Gaps, Open Questions, and Priorities. *Microgravity* 9 (22): 1–5.

Clarke RV (1980) Situational Crime Prevention: Theory and Practice. *British Journal of Criminology* 20 (2): 136–147.

Clarke RV (2009) Situational Crime Prevention: Theoretical Background and Current Practice. In MD Krohn, AJ Lizotte and GP Hall (eds). *Handbook on Crime and Deviance*. Dordrecht: Springer, 259–276.

Cornish DB (1994) The Procedural Analysis of Offending and its Relevance for Situational Prevention. *Crime Prevention Studies* 3 (1): 151–196.

Dalledonne S (2021) International Environmental Law and Environmentally Harmful Space Activities: Learning from the Past for a More Sustainable Future. *Journal of Property, Planning and Environmental Law* 13 (2): 139–151. https://doi.org/10.1108/JPPEL-09-2020-0040.

Davies R (2016) Asteroid Mining Could Be Space's New Frontier: The Problem is Doing it Legally. *The Guardian*. 6 February 2016. Last accessed 5 November 2023. https://www.theguardian.com/business/2016/feb/06/asteroid-mining-space-minerals-legal-issues.

de Zwart M, Henderson S and Neumann M (2023) Space Resource Activities and the Evolution of International Space Law. *Acta Astronautica*211: 155–162.

Dehghanniri H and Borrion H (2021) Crime Scripting: A Systematic Review. *European Journal of Criminology* 18 (4): 504–525.

Denny A, Sauzay A, Stoyanov M, Birch K and Busby J (2023) Mining and Manufacturing in Outer Space: The New Frontier in Disputes Risks. *Allen & Overy*. 03 October 2023. Last accessed 6 November 2023. https://www.allenovery.com/en-gb/global/news-and-insights/publications/mining-and-manufacturing-in-outer-space-the-new-frontier-in-disputes-risks.

Dunnett O, Maclaren A, Klinger J, Maria K, Lane D and Sage D (2019) Geographies of Outer Space: Progress and New Opportunities. *Progress in Human Geography* 43 (2): 314–336.

Eski Y (2023) *A Criminology of the Human Species: Setting an Unsettling Tone*. Palgrave McMillan.

Hoffman JA et al. (2022) Mars Oxygen ISRU Experiment (MOXIE)-Preparing for Human Mars Exploration. *Science Advances*8.

Huisman W and van Erp J (2013) Opportunities for Environmental Crime. A Test of Situational Crime Prevention Theory. *The British Journal of Criminology* 53: 1178–1200.

Johnson-Freese J (2017) Build on the Outer Space Treaty. *Nature*550: 182–184. https://doi.org/10.1038/550182a.

Kerrest A (2001) *Space Debris, Remarks on Current Legal Issues*. Proceedings of the 3rd European Conference on Space Debris. ESOC, Darmstadt, Germany, 19–21 March 2001.

Kramer WR (2014) Extraterrestrial Environmental Impact Assessments: A Foreseeable Prerequisite for Wise Decisions Regarding Outer Space Exploration, Research and Development. *Space Policy* 30: 215–222.

Lampkin JA (2018) *Will Unconventional, Horizontal, Hydraulic Fracturing for Shale Gas Production Purposes Create Environmental Harm in the United Kingdom?* Unpublished PhD Dissertation, University of Lincoln.

Lampkin JA (2021) Mapping the Terrain of an Astro-Green Criminology: A Case for Extending the Green Criminological Lens Outside of Planet Earth. *Astropolitics: The International Journal of Space Politics and Policy* 18 (3): 238–259.

Lampkin JA and McClanahan B (2023) Astronomical Withdrawals: A Green Criminological Examination of Extreme Energy Mining on Extraterrestrial Objects. *Crime, Law and Social Change*. https://doi.org/10.1007/s10611-023-10123-9.

Lampkin JA and White R (2023) *Space Criminology. Analysing Human Relationships with Outer Space*. Palgrave Studies in Green Criminology Series.

Lampkin JA and Wyatt T (2022) Widening the Scope of 'Earth' Jurisprudence and 'Green' Criminology? Towards Preserving Extra-Terrestrial Heritage Sites on Celestial Bodies. In: Gacek J and Jochelson R (eds) *Green Criminology and the Law*. London: Palgrave Macmillan.

McQuade J (2019) Earth Day: Colonialism's Role in the Overexploitation of Natural Resources. *The Conversation*. 18 April 2019. Last accessed 2 November 2023. https://theconversation.com/earth-day-colonialisms-role-in-the-overexploitation-of-natural-resources-113995.

Migaud MR (2020) Protecting Earth's Orbital Environment: Policy Tools for Combating Space Debris. *Space Policy* 52 (101361): 1–9. https://doi.org/10.1016/j.spacepol.2020.101361.

Miller GD (2019) Space Pirates, Geosynchronous Guerrillas, and Nonterrestrial Terrorists: Nonstate Threats In Space. *Air & Space Power Journal* 33 (3):33–51.

Na C (2021) Space Mining for Minerals Including Rare Earths and Tungsten. *Tungusten*. 24 August 2021. Last accessed 21 October 2023. https://tungusten.com/space-mining-for-minerals-including-rare-earths-and-tungsten/

NASA (2021) Space Debris and Human Spacecraft. 26 May 2021. Last accessed 21 October 2023. https://www.nasa.gov/mission_pages/station/news/orbital_debris.html.

NASA (2022) Astromaterials Research & Exploration Science, Nasa Orbital Debris Program Office. Last accessed 21 October 2023. https://orbitaldebris.jsc.nasa.gov/faq/#:~:text=3.-,How%20much%20orbital%20debris%20is%20currently%20in%20Earth%20orbit%3F,1%20mm%20exceeds%20100%20million.

New York Times (2023) The First Secret Asteroid Mission Won't Be the Last. *New York Times*. 27 December 2023. Last accessed 15 January 2024. https://www.nytimes.com/2023/12/27/science/secret-asteroid-mission-astroforge.html.

Oduntan G (2015) Who Owns Space? US Asteroid-Mining Act is Dangerous and Potentially Illegal. *The Conversation*. 25 November 2015. Last accessed 21 October 2023. https://phys.org/news/2015-11-space-asteroid-mining-dangerous-potentially-illegal.html.

Outer Space Treaty, 1967. United Nations Office for Outer Space Affairs. Last accessed 21 October 2023. https://www.unoosa.org/oosa/en/ourwork/spacelaw/treaties/outerspacetreaty.html.

Peterson L (2021) Governing the Unknown: How the Development of Intellectual Property Law in Space Will Shape the Next Great Era of Exploration, Exploitation, and Invention. *Northwestern Journal of Technology and Intellectual Property* 18 (3): 335–353.

Qizhi HE (1988) Environmental Impact of Space Activities and Measures for International Protection. *Journal of Space Law* 16 (2): 117–127.

Rothe DL and Collins VE (2023) Planetary Geopolitics, Space Weaponization and Environmental Harms. *The British Journal of Criminology* 63 (6): 1523–1538. https://doi.org/10.1093/bjc/azad003.

Snaphaan T and van Ruitenburg T (2024) Financial Crime Scripting: An Analytical Method to Generate, Organise and Systematise Knowledge on the Financial Aspects Of Profit-Driven Crime. *European Journal on Criminal Policy and Research*. https://doi.org/10.1007/s10610-023-09571-9.

Space Act 2015. H.R.2262–114th Congress (2015–2016): U.S. Commercial Space Launch Competitiveness Act. 25 November 2015.

Szablowski D (2019) "Legal Enclosure" and Resource Extraction: Territorial Transformation Through the Enclosure of Local and Indigenous Law. *The Extractive Industries and Society* 6: 722–732.

Takemura N (2019) Astro-Green Criminology: A New Perspective Against Space Capitalism Outer Space Mining May Make the Same Mistakes in Space as We Have on Earth. *Toin University of Yokohama Research Bulletin* 40: 7–17.

van Santvoord VD and van Ruitenburg T (2023) Financial Crime Scripting: Introducing a Financial Perspective to the Dutch Cocaine Trade. *The Police Journal* 96 (3), 374–389. https://doi.org/10.1177/0032258X221083449.

Weirich C (2019) *Situational Crime Prevention of Antiquities Trafficking: A Crime Script Analysis.* PhD thesis. Univeristy of Glasgow.

Zhang C (2018) The Tale of 2 Asteroid Sample-Return Missions. *Chemical and Engineering News.* 30 September 2018. Last accessed 23 October 2023. https://cen.acs.org/physical-chemistry/astrochemistry/tale-2-asteroid-sample-return/96/i39.

5 Cynical Actors in the Age of Transparency

Illegal Warfare and State Propaganda in a Glass House

Bruno Reynaud de Sousa

Introduction

The philosopher Byung-Chul Han's view of post-modernism equates living in the 21st century to living in the age of transparency (Han and Butler 2015). The telecommunications revolution, characterised by the smartphone and ubiquitous connectivity, created the ideal conditions for the progressive erosion of individual privacy powered by social media (Han and Steuer 2022). With "Spaceship Earth" (Fuller 1969) reaching eight billion inhabitants around 2023, statistics place the total number of smartphone mobile network subscriptions at 6.9 billion, with predictions for 2028 putting the same number above 7.7 billion (Statista 2024).

At the same time, thanks to advances in satellite technology, companies like Planet or Maxar harness imagery of the entire world every 24 hours, with a few commercial space companies capable of photographing specific areas of the planet several times daily (Planet 2024). As satellite imagery is continuously generated and kept as a digital record, comparative analysis is no longer a privilege of the most advanced and resourceful intelligence agencies (Planet 2024). Interestingly, the media has taken notice and has leveraged the possibility of gathering news by comparing images of a location spanning back days, weeks, months or even more than a year (Corcoran 2018).

Overall, US-based Earth Observation (EO) companies now possess a higher number of satellites than state actors, being capable of providing better capabilities to the public sector, some of which are out of reach for even the most powerful countries (Harrison 2023). Crucially, the development of competitive launch technologies came hand-in-hand with the demand to launch small-sized satellites essential for the operation of large constellations of satellites, such as Starlink (Harrison 2023). Remarkably, space commercialisation is set to revolutionise internet connectivity, making it ubiquitous by introducing direct-to-satellite cellular service (Alamalhodaei 2023).

Over the last five years, the world has witnessed two unprecedented events that have forever changed how we live and perceive the world around us. The first was the SARS-CoV-2 outbreak, which spread rapidly across the globe, bringing with it death, economic instability, and a new appreciation for the

DOI: 10.4324/9781003437178-6

importance of public health. The second is the conflict in Ukraine, where technology and misinformation gained prominence as nations and individuals alike struggled to navigate a rapidly shifting geopolitical landscape. Both events challenged our assumptions about safety, security, and the resilience of our societies and will undoubtedly shape our collective future in profound ways.

The Russian invasion of Ukraine highlighted the growing importance of US-based private sector innovation in space technologies for national security and defence. Advances in space technologies concurred to make the SARS-CoV-2 outbreak history's first transparent pandemic and the conflict in Ukraine humanity's first transparent war, laying bare the power the private space sector was able to yield. As the degree of dependency of Ukraine's Armed Forces on SpaceX technology became apparent, it was clear that the company transformed into a geopolitical actor with a vital role in the conflict.

Both the SARS-CoV-2 pandemic and the war in Ukraine illustrate how propaganda is deployed in contemporary international relations as part of a cynical actor toolkit to amplify foreign policy objectives. As explained in the present chapter, bad actors employ various tactics to construct and control narratives with multiple objectives. Remarkably, privately owned space technologies are already contributing to higher public scrutiny regarding actions by known cynical actors of international relations: China and Russia. For the present chapter, although recognising the underlying conceptual complexities (Baade et al. 2020), cynicism is equated to any behaviour that undermines the rules-based international order by interpreting International Law abusively to justify actions, seeking protection from international or external interference and obfuscating facts.

Space Commercialisation

Whereas commercialisation has made space more "congested, competitive and contested" (Harrison 2013), it has cemented the perception that this domain is the "future frontier of the global economy and society" (G7 2021: par. 31). Commercial space is one of the fastest-growing and most promising economic sectors, encompassing industry, science, and financial services (EU 2017). With existing concepts such as "space economy", "space sector", and "space company" insufficient to describe the spectacular dynamic of the last few years, the term "New Space" was coined to specifically refer to the "radical new commercial space initiatives" (Pelton 2017: 2) the scale of which led to predictions of a "New Space Economy" or a "New Commercial Space" (Pelton 2017).

As space commercialisation revolutionised launching, Earth observation and telecommunications, the space sector has entered the "third space age" (Harrison 2023). With an epicentre in the United States, the space sector is experiencing shockwaves of innovation unleashed by the 21st-century's "space barons" (Davenport 2018), such as Elon Musk, Jeff Bezos, and Richard Branson. The flexibility in the design and implementation of space missions has improved significantly, with the US-based private sector leading the way in driving innovation in space technologies and proposing new business models that are heavily data-driven, aiming to reduce barriers to accessing space and space-based technology. Regarding launch services, US-led innovation, especially SpaceX's reusability model, has lowered space launch costs while enabling more frequent launches and increasing the available launch capacity (Harrison 2023). Demonstrating that the private sector has achieved a remarkable level of proficiency and expertise in launching various types of equipment and spacecraft into Low Earth Orbit is the fact that the success rate of launches exceeds that of most G7 economies and established space-faring nations like India (BryceTech 2020; Harrison 2023). For the next five years, the major trend to watch is the increase in global launch capacity, which presently stands at more than "1800 metric tons annually in 2022" (Harrison 2023: 4). Regarding EO and telecommunications, advances in satellite technology enabled large constellations of smaller and more capable satellites. Furthermore, satellite communication companies provide internet connectivity worldwide and will soon offer direct-to-device 5G network access (Deloitte 2024).

On the one hand, this means that law enforcement at the state level now has access to satellite imagery at unprecedented levels regarding availability and quality. Space technologies can add value in the fight against environmental crime, certain types of tax evasion, and administrative law violations pertaining to specific economic activities subject to licensing, such as resource extraction or construction. On the other hand, at the international level, space technologies can deter bad behaviour by state and non-state actors in contravention of International Law, for example, as it relates to the environment or Human Rights.

The use of the smartphone coupled with ubiquitous connectivity in the third space age has significantly diminished the ability of a state to control the flow of information while simultaneously bringing unprecedented transparency into life on Earth. Whether through smartphone cameras or satellite optical and radar sensors, there are virtually no boundaries to the ability of EO companies and governments to document life on Earth. Crucially, advances in EO technology, artificial intelligence and information management will soon allow companies to provide near-real-time imagery (Planet 2024).

In the ever-evolving landscape of journalism, space commercialisation became central to integrating satellite images into news reporting as the costs associated with obtaining satellite imagery significantly decreased. Different from two decades ago, acquiring and processing satellite imagery was complex and costly, rendering this technology inaccessible to smaller news outlets (Glaser and Brender 1986), the landscape has shifted dramatically with the widespread availability of privately owned EO capabilities.

Nowadays, newspapers and cable news networks can afford the 1980s aspiration of leveraging the power of satellite imagery for news gathering, enriching scrutiny with the ultimate layer of transparency. News media can now use tools such as geographical information systems to gather information otherwise impossible to obtain, especially if facts concern autocratic regimes where freedom of the press is an illusion.

Propaganda as Cynical Actors' Tool

In her work *The Origins of Totalitarianism*, Hannah Arendt describes the ideal subject of totalitarianism as the one for whom it is no longer possible to operate the "distinction between fact and fiction" and the "distinction between true and false" (Arendt 1951: 474). Further to the paradigm of "an era of chaos", contemporary international relations are also characterised by having entered a "post-truth era" where the daily accumulation of facts is fertile ground for propaganda, making it increasingly difficult to arrive at the factual truth regarding any major international relations event.

Throughout history, propaganda has been widely used to gain an advantage in different settings, namely warfare. As far back as ancient China, military strategist Sun Tzu is widely known to have recognised that deception is the basis of all warfare. In the history of the 20th century, we can find examples of the variations that propaganda can undergo depending on the context, whether of peace or war. Traditionally, propaganda has been associated with state actors who use it to serve national security or political objectives, such as protecting sensitive information or promoting the national agenda (Carter and Carter 2023). Typically generated to cause harm, deceive, or mislead stakeholders or constituencies, propaganda may also be part of a strategy to uphold national prestige and influence public opinion domestically and internationally (Carter and Carter 2023).

In the wake of globalisation, propaganda has become a more powerful instrument than ever before due to a few factors. Chief among them is the speed at which propaganda can be produced, thanks to automation (Shu et al. 2020). Additionally, there are tools available that can amplify the propaganda's reach to an unimaginable extent (Shu et al. 2020). All these factors, combined with the emergence of the smartphone, have ushered in a new era for propaganda where the potential to influence public opinion and political outcomes is unprecedented (Carter and Carter 2023).

Briefly, propaganda encompasses different types of information characterised by varying degrees of inaccuracy aimed at manipulating public opinion or swaying decision-making (Carter and Carter 2023). Although depicting the dissemination of information or ideas to influence public opinion or behaviour towards a specific cause, organisation, or individual (Carter and Carter 2023), there is no single concept of propaganda, which moreover coexists with related concepts such as "disinformation" or "fake news" (Hohlfeld et al. 2020). Indeed, propaganda may be subversive, defamatory, or discriminatory, with a key example being an incitement to hatred or propaganda associated with the crime of genocide (see Article 3 of the 1948 Convention on the Prevention and Repression of Genocide). In addition, there is the case of propaganda directed at inciting terrorism, as referred to in Security Council Resolutions 1373 (2001), 1624 (2005), and 2178 (2014). A further reference may be found in the Tallinn Manual, a comprehensive guide on International Law regarding cyber operations, which also acknowledges the use of propaganda and disinformation in section 8 and norm 61 (Schmitt 2013).

Whereas International Law prohibits the practice of annexation and Human Rights abuses, the same cannot be said regarding propaganda. Crucially, the legality of propaganda aimed at safeguarding either national security interests or political objectives is a matter falling essentially within the domestic jurisdiction of any state subject to the limits imposed by the principle of non-interference. Although non-binding, the UN General Assembly Resolution 73/27, passed on 11 December 2018 – one year before the SARS-CoV-2 virus emerged – explicitly recognises the obligation of States to refrain from engaging in defamatory campaigns, vilification, or hostile propaganda aimed at intervening or interfering in the internal affairs of other States.

The COVID-19 pandemic and the conflict in Ukraine have shown how controlling the flow of information can be crucial for cynical actors seeking to shape their preferred narrative and obscure the truth (Sarmina 2020). One of the most common tactics is to generate a massive amount of information, with extensive coverage of important events in the state government, increasing the chances of reaching the target audience, dominating the news cycle and conveying the desired message (Carter and Carter 2023). This practice often sees the term "fake news" associated, which, after almost a decade of academic discussion, remains without a clear definition. The concept of "fake news" can encompass various types of information that are either inaccurate, transformed, inauthentic, false, unreal, unrealistic, imprecise, exaggerated, undocumented, or unsubstantiated (Hohlfeld et al. 2020). Moreover, the term may also refer to deliberately manipulated or adulterated information (disinformation) or unintentionally incorrect information that has been purposefully created to cause harm to individuals, groups, or organisations (Hömberg 2020).

The delegitimisation of renowned Western media outlets with a history of successful scrutiny is also a common tactic autocracies use along with targeted negative coverage to push their agenda (Carter and Carter 2023). For example, autocracies may blame another actor for a problem, creating a negative image in the minds of the masses and diverting attention from the real issues to blame an external entity for the State's problems (Kahl and Wright 2021).

As defined in the Lotus Case by the Permanent Court of International Justice on 7 September 1927, in paragraph 18, non-interference is a fundamental principle in International Law that prohibits any state from interfering in the internal affairs of another state. This principle is further reinforced by the Montevideo Convention on the Rights and Duties of States, which emphasises in Article 5 that the rights of states are inviolable and cannot be infringed upon by other states, and in Article 7, which prohibits states from interfering in the internal or external affairs of another state. In Nicaragua v. United States, the International Court of Justice also affirmed this interpretation of the principle of non-interference (International Court of Justice 1986).

When assessing what may constitute propaganda, a distinction should be made between at least two situations. Firstly, statements by officials and agents of a state that, due to their deceptive nature and malicious intent, qualify as propaganda. Secondly, disinformation produced by public sector media or privately owned but public sector controlled media. As regards the first situation, to the extent that there has been a breach of an obligation under International Law, it may be a situation of international responsibility under Article 4 of the Draft Articles on Responsibility of States for Internationally Wrongful Acts (ASRIWA) relating to the conduct of organs of a state. In addition, in the case of media under state control, the framework of ASRIWA Article 5 (referring to the conduct of persons or entities excising elements of government authority) should be considered to the extent that it is proven that they act under the instructions, direction, or control of the state. Regarding the second situation, in the case of so-called "private" media, it is essential to briefly reference Article 19 of the Universal Declaration of Human Rights, Article 10 of the European Convention on Human Rights, and Article 19 of the International Covenant on Civil and Political Rights, all referring to freedom of expression.

Humankind's First Transparent War: Is a New Type of Warfare Emerging?

In 2014, the Russian Federation invaded and annexed a part of the Ukrainian territory: the Crimean Peninsula. In what became a classic example of hybrid warfare in practice combining kinetic action with cognitive attacks (Pocheptsov 2018), Russia sought to obfuscate the deployment of troops into Ukrainian territory, cynically showing no concern for International Law while invoking International Law (Merezhko 2015; D'Anieri 2023). In 2022, again, Russia cynically invoked International Law to justify further invading Ukraine, claiming it was protecting Russian populations against genocide (D'Anieri 2023).

The 1991 Gulf War ushered in the era of live, 24-hour cable news TV channels. Using satellite technology, a lesser-known cable TV channel called Cable News Network (CNN) was the last media outlet with a crew in Baghdad (Napoli 2020). By setting up a camera pointed at the city's skyline and broadcasting via satellite, CNN became the first TV channel in history to broadcast a war as it happened to the world (Napoli 2020). Thanks to CNN, the 1991 Gulf War became History's first semi-transparent war in that although the novel aspect of live TV coverage of the US-led coalition bombing of Baghdad, all information was mediated by a TV network with CNN going a step further and inaugurating the practice of embedded reporters.

A little-known fact is that a similar revolution in journalism had been proposed a few years before, aiming to leverage satellite technologies for news gathering. Motivated by state oppression at the hands of cynical actors, there were those in the US who looked at satellite imagery as a novel means to level the information playing field (Glaser and Brender 1986). Specifically, the Human Rights situation in South Africa led US authors to question whether gaining independent access to satellite imagery could be a game changer as a news-gathering tool (Glaser and Brender 1986).

Battlefield transparency and the potential consequences were being discussed more than two decades ago, motivated by the consolidation of cable news channels (Kaspar et al. 2001). Likewise, the advent of social media had authors questioning how the conduct of the conflict in the theatre of operations would be digitalised, with another dimension being added as the fighting extended to social media platforms like Twitter (now, “X”) (Patrikarakos 2017). Nonetheless, the missing piece of the transparency puzzle was delivered by commercial space companies in 2022. Thanks to a free press leveraging Earth observation data provided by US-based EO companies, the armed conflict in Ukraine became Humanity’s first Transparent War (de Sousa 2024).

For the first time during an armed conflict, anyone could scrutinise, judge, and support the war via smartphone: war became fully disintermediated. Because the war was also evidently being fought in the information domain, demonstrating legitimacy became a priority for both sides. The war’s disintermediation helps explain political leaders’ high-speed communication dynamics, making achieving gains in the theatre of operations at least as important as achieving gains on the media waves (Sousa 2022).

Drawing from the Ukrainian conflict, a transparent war or transparent armed conflict is understood as a conflict where the power of satellite technologies is fused with smartphones, ubiquitous connectivity and social media. On the one hand, the war is made transparent because technology enables its documentation in extreme detail and frequency by combatants and non-combatants alike. On the other hand, the information gathered is communicated directly through live broadcasts or posts on social media networks. Crucially, as the conflict in Ukraine shows, satellite-based internet hindered attempts to isolate populations from the outside world and led Russia to increase cyber-attacks (Jenkinson 2023).

In sum, transparent war refers to an armed conflict that is fully disintermediated in informational terms regarding events in the theatre of operations. Unlike previous armed conflicts in the 20th and 21st centuries, where information was limited to traditional media sources or official sources and the information flow could be controlled by the belligerents, in a transparent war, information is gathered, processed, and shared independently and readily available to everyone with access to social media.

Ultimately, the transparent war sees a daily accumulation of facts that, in essence, resembles a never-ending information puzzle composed of small pieces and big pieces. Whereas the small pieces of the puzzle consist of disintermediated information in the form of limited social media posts lacking context, the big pieces of information are provided by legacy media outlets fusing traditional reporting that occasionally leverages commercial Earth observation data.

Beyond 2024, the question is whether satellite technology coupled with social media and smartphones will prove a transparency game changer as satellite communication coupled with cable TV proved to be in the past. Resorting to satellite technologies brings the potential for increased transparency, but it remains to be seen how this could positively impact the conduct of hostilities in armed conflicts. More specifically, although EO data has the potential to become a deterrent against bad actors' cynical invocation of International Law and gross violations of Human Rights, autocracies may adapt to a new type of transparent warfare and develop propaganda-type countermeasures.

Humankind's First Transparent Pandemic: Satellites Against the Great Firewall

Paraphrasing Aldous Huxley, the most important lesson history has to teach is that humankind struggles to learn from past experiences, especially the ones with negative consequences. In 2003, the Severe Acute Respiratory Syndrome (SARS) outbreak registered 8,096 total cases and 774 deaths (World Health Organization 2003a). The lesson the world, especially China, supposedly learned from this experience was that timely reporting of health emergencies was crucial to contain the spread of contagious respiratory diseases and prevent them from becoming global pandemics. Two decades later, although the deaths caused by the Severe Acute Respiratory Syndrome Coronavirus 2 (SARS-CoV-2) are on track to stabilise in 2024, the virus is still killing millions, with the total number of deaths up until February 2024 numbering more than 7 million (World Health Organization 2024a).

As it did in 2019, China may be said to have responded to the SARS outbreak as a national security threat. Succinctly, SARS is believed to have been first identified in China in November 2002, but the Chinese authorities tried to conceal its existence, failing to report it to the world immediately as required by the duty of international cooperation (Kahl and Wright 2021). Due to the spread of the disease, a global alert was issued by the World Health Organization (WHO) on 12 March 2003, causing worldwide concern, as China only revealed the extent of the SARS epidemic on 26 March 2003, almost five months after the first outbreak (Kahl and Wright 2021). The Chinese authorities then allowed a WHO team to enter the country and assess the situation on 2 April 2003, at a time when the epidemic had spread to other Asian countries, prompting the WHO to issue a global alert on travel (World Health Organization 2003b) helping to stop the epidemic on 5 July 2003 (World Health Organization 2003c).

In December 2019, when SARS had become a distant memory, the inhabitants of the Chinese city of Wuhan began to exhibit symptoms of an unknown virus, with reports stating that Wuhan residents were already infected with SARS-CoV-2 at least in November 2019 (Kahl and Wright 2021). On 31 December 2019, China informed the WHO about pneumonia cases of unknown cause registered in Wuhan (World Health Organization 2024b). By 3 January 2020, only 44 cases had been reported, 11 of which were said to be severe (World Health Organization 2024b). According to media reports, a market located in Wuhan had been closed on 1 January 2020 for environmental sanitation and disinfection (Kahl and Wright 2021). However, Chinese authorities reported that the cause had not yet been identified, further informing that patients had been isolated and were being treated in Wuhan medical facilities and did not report significant human-to-human transmission (World Health Organization 2024b).

On 1 January 2020, when the transmissibility of SARS-CoV-2 between humans was a known fact in Wuhan – but officially denied by the Chinese government – the official statement emerged that the Wuhan police had taken action against eight citizens accused of spreading rumours about an unknown pneumonia (Kahl and Wright 2021). On 23 January 2021, the Chinese government made the surprising decision to lock down the city of Wuhan and three other cities, initiating a quarantine of more than 100 million people (Kahl and Wright 2021). Between December 2019 and the end of January 2020, there were different messages spread by Chinese government officials, state agents and Chinese state media, which did not correspond to the reality already known officially (Kahl and Wright 2021).

Crucially, Chinese authorities announced the first fatality by SARS-CoV-2 only on 11 January 2020 (World Health Organization 2024a). China reported to WHO on 11–12 January that 41 cases of novel coronavirus infection were detected in Wuhan City between 8 December 2019 and 2 January 2020: "seven cases are severely ill, one death was reported, and six patients have been discharged" (World Health Organization 2024a).

In early 2023, however, media scrutiny leveraging satellite imagery dating back to 2021 revealed an abnormally high influx of cars concentrating outside the parking lots of state-owned crematoriums in China (Oakford et al. 2023), portraying a level of activity inconsistent with the numbers of deaths reported by China for the same period (World Health Organization 2024b). Consequently, what could only be described as state-promoted propaganda was directing international attention from an observable fact: Chinese authorities were concealing a spike in the number of deaths. Even though satellite imagery did not allow a direct connection to be established between abnormal activity in crematorium parking lots and deaths by infections caused by SARS-CoV-2, the pictures enabled an unprecedented level of international scrutiny of Chinese domestic affairs.

The circumstances surrounding the outbreak of SARS-CoV-2 created an environment where state propaganda had considerable opportunity to spread. This included statements made by Chinese state-controlled media, agents and officials that the virus had been intentionally created in a laboratory as a biological weapon and may have been disseminated intentionally (Kahl and Wright 2021). Other claims were that the virus did not originate in China and had been introduced there accidentally or intentionally by the US military, suggesting the US had tampered with the virus in a laboratory in order to create a biological weapon to use against China (Kahl and Wright 2021). In fact, a study conducted by the Atlantic Council, focusing on social media activity originating from the US, China, Russia, and Iran revealed four prevailing narratives surrounding the SARS-CoV-2 pandemic (Atlantic Council 2021). The first narrative posits that the US Army introduced the virus as part of a military competition in Wuhan in 2019 (Atlantic Council 2021). The second narrative suggests that SARS-CoV-2 escaped from a laboratory in Wuhan (Atlantic Council 2021). The third narrative alleges that the US intentionally spread SARS-CoV-2 in China to undermine its economy (Atlantic Council 2021). Lastly, the fourth narrative proposes that the US has genetically engineered a virus to target ethnic groups specifically (Atlantic Council 2021).

By partially contradicting the official narrative while showing Chinese deception, satellite technologies exposed that Chinese officials had been cynically engaging with WHO officials from the onset of infections by SARS-CoV-2 (World Health Organization 2020). With reports questioning whether Chinese authorities had been able to sequence the SARS-CoV-2 virus "about two weeks before the Chinese government publicised the pathogen's sequence" (Field 2024; Strobel 2024), the perception today is that cooperation with the international community had been premised in a degree of bad faith since the onset. In sum, although Chinese authorities resorted to the defence policy toolkit to deploy propaganda to obfuscate facts to safeguard the national interest, commercially sourced EO data enabled an unprecedented level of independent scrutiny.

Conclusion

The international (dis)order is again multipolar and contemporary international relations are "entering an age of chaos" (UN Secretary-General 2024), but the chaos will not occur in complete darkness. Space commercialisation has ushered in an "age of transparency" for human existence. For the past five years, an unprecedented level of transparency regarding the actions of States in domestic affairs and international relations transformed Earth into a glass house where democracies and autocracies co-exist.

Four decades ago, authors in the US questioned whether news organisations should endeavour to access satellites independently as a news-gathering tool. In the past decade, that vision has come true as US-led space commercialisation has enabled easy access to EO data bringing with it the first transparent pandemic. Moving forward, the realisation is that independent media can now scrutinise the domestic and international conduct of states like never before.

Following the armed conflict in Ukraine in 2022, transparent warfare is likely to be the new normal of warfighting in the 21st century. By bringing the ultimate measure of transparency to international relations, US-based EO companies yield the power to expose cynical invocations of International Law and prevent them from becoming the basis for new doctrine. Nevertheless, although the control over the domestic flow of information is weakened in China and Russia, it remains to be seen how significant a deterrent satellite imagery may become for autocracies.

The exceptional level of media and civil society scrutiny of China's response to the pandemic and the Russian invasion of Ukraine clearly demonstrates how space commercialisation is proving a game-changer beyond the economic and scientific domains. Firstly, on the internal or domestic dimension, both the suppression of free speech and state control over information have weakened. Secondly, the shelf life of state propaganda is now shorter, as the conduct of armed conflict is no longer documented exclusively by official sources and societies can now freely access a plethora of disintermediated sources.

In theory, gross Human Rights violations have become more challenging to obfuscate and hide. Looking to the near future, an idealist approach would undoubtedly lead to calls for US-based EO companies to join forces with independent media. Searching for signs of war crimes or genocide could ultimately be promoted as an innovative way to adhere to environmental, social, and governance (ESG) goals. A more balanced approach, however, leads to the conclusion that the most immediate potential of the novel transparency from space is in contributing to law enforcement efforts at the national or domestic level. By focusing on small-scale violations that illicit significant public outrage, such as environmental crime, societal expectations could more easily be met, confounding the expectations raised by increased levels of transparency. A key challenge is that space imagery alone will likely never be enough for law enforcement purposes as *in situ* verification and collection of evidence cannot be dispensed with.

Alas, satellite-based transparency has a fundamental downside. All private companies are driven by profit, shaped by ideology, and limited by the law. US-based EO companies, barring legal obligations such as commercial sanctions, are not beholden to any ethical imperatives other than those they create for themselves. In essence, for US-based EO companies with such concerns, this takes the form of independent ethics boards that advise on whom to sell, what to sell and what to disclose publicly.

Looking to Europe, States increasingly rely on US-based EO companies for their space capabilities, much like their dependence on US Big Tech. Although the European Union has advanced space capabilities, Europe now finds itself virtually deprived of autonomous access to space. Crucially, it lacks EO capabilities that perform at the level of the US privately owned EO satellite constellations. Thus, Europe may hope to influence the developments in EO and related space activities only partially and most likely limited to EU law regulations.

To conclude, and paraphrasing André Malraux, the space-based *age of transparency* will either be American or not at all, leaving international relations to plunge into a dark *age of chaos.* US-based EO companies such as Planet, Maxar or Capela Space would be wise to recall the key lesson Benjamin Franklin Parker imparted to his young nephew Peter, summarised in these six words: *with great power comes great responsibility.*

References

Alamalhodaei A (2023) Apple-backed Globalstar Inks $64M Launch Contract with SpaceX. From: https://techcrunch.com/2023/09/01/apple-backed-globalstar-inks-64m-launch-contract-with-spacex/.

Arendt H (1951) *The Origins of Totalitarianism.* New York: Harcourt.

Atlantic Council (2021) Weaponized: How Rumors About COVID-19's Origins Led to a Narrative Arms Race. From: https://www.atlanticcouncil.org/wp-content/uploads/2021/02/Weaponized-How-rumors-about-COVID-19s-origins-led-to-a-narrative-arms-race.pdf.

Baade B, Burchardt D, Feihle P, et al. (eds) (2020) *Cynical International Law?: Abuse and Circumvention in Public International and European Law.* Heidelberg: Springer Berlin.

BryceTech (2020) 2020 Orbital Launches Year in Review. From: https://brycetech.com/reports.

Carter EB and Carter BL (2023) *Propaganda in Autocracies: Institutions, Information, and the Politics of Belief.* Cambridge: Cambridge University Press.

Corcoran M (2018) Satellite Journalism – The Big Picture (Newsgathering applications of emerging satellite technology). From: https://reutersinstitute.politics.ox.ac.uk/sites/default/files/2018-10/Mark%20Corcoran%2C%20Satellite%20Journalism_1.pdf.

D'Anieri P (2023) *Ukraine and Russia: From Civilized Divorce to Uncivil War.* Cambridge: Cambridge University Press.

Davenport C (2018) *The Space Barons: Elon Musk, Jeff Bezos, and the Quest to Colonize the Cosmos.* New York: PublicAffairs.

de Sousa BR (2024) Cyber Operations Targeting Space Systems. In: de Andrade FACP, Covelo de Abreu SJRSC and Freitas PMF (eds) *Legal Developments on Cybersecurity and Related Fields.* Cham: Springer International Publishing, pp. 39–59.

Deloitte (2024) Deloitte's TMT Predictions 2024 – Three Themes Highlight this year's Report: Generative AI, Sustainability, and Monetization. From: https://www2.deloitte.com/content/dam/insights/articles/ca176625_tmtp24-introduction/DI_TMT-predictions-2024.pdf.

EU (2017) Dependence of the European Economy on Space Infrastructures. From: https://op.europa.eu/en/publication-detail/-/publication/e5450aac-0d4d-11e8-966a-01aa75ed71a1/language-en.

Field M (2024) Did China Keep the COVID Virus Sequence Secret for Weeks? From: https://thebulletin.org/2024/01/did-china-keep-the-covid-virus-sequence-secret-for-weeks/.

Fuller RB (1969) *Operating Manual for Spaceship Earth.* Carbondale: Southern Illinois University Press.

G7 (2021) Carbis Bay G7 Summit Communiqué, Our Shared Agenda for Global Action to Build Back Better, 11–13 June 2021. From: https://www.whitehouse.gov/briefing-room/statements-releases/2021/06/13/carbis-bay-g7-summit-communique/.

Glaser PE and Brender ME (1986) The First Amendement in Space: News Gathering from Satellites. *Issues in Science and Technology* 3(1): 60–67.

Han B-C and Butler E (2015) *The Transparency Society.* Stanford: Stanford Briefs, an imprint of Stanford University Press.

Han B-C and Steuer D (2022) *Infocracy: Digitalization and the Crisis of Democracy.* Medford: Polity Press.

Harrison RG (2013) Unpacking the Three C's: Congested, Competitive, and Contested Space. *Astropolitics: The International Journal of Space Politics & Policy* 11 (3): 123–131.

Harrison T (2023) Understanding the Third Space Age. Metrea Strategic Insights. From: https://metrea.aero/third-space-age/.

Hohlfeld R, Harnischmacher M, Heinke E, Lehner L and Sengel M (eds) (2020) *Fake News und Desinformation: Herausforderungen für die vernetzte Gesellschaft und die empirische Forschung.* Berlin: Nomos Verlag.

Hömberg W (2020) "Fake News", Medienfälschungen, Grubenhunde. Fälschungsfallen im Journalismus und in den Medien. In: Hohlfeld R, Harnischmacher M, Heinke E, Lehner L and Sengel M (eds) *Fake News und Desinformation: Herausforderungen für die vernetzte Gesellschaft und die empirische Forschung.* Berlin: Nomos Verlag.

International Court of Justice (1986) Case concerning military and paramilitary activities in and against Nicaragua (Nicaragua v. United States of America) (Merits).

Jenkinson A (2023) *Digital Blood on Their Hands: The Ukraine Cyberwar Attacks.* CRC Press.

Kahl C and Wright T (2021) *Aftershocks: Pandemic Politics and the End of the Old International Order.* New York: St. Martin's Publishing Group.

Kaspar BM, Strategy AUCf and Technology (2001) *The End of Secrecy: Military Competitiveness in the Age of Transparency.* Center for Strategy and Technology, Air War College, Air University.

Merezhko O (2015) Crimea's Annexation by Russia: Contradictions of the New Russian Doctrine of International Law. *Zeitschrift für ausländisches öffentliches Recht und Völkerrecht* 75: 167–194.

Napoli L (2020) *Up All Night: Ted Turner, CNN, and the Birth of 24-Hour News.* New York: Abrams Press.

Oakford S, Kuo L, Chiang V, et al. (2023) Satellite images Show Crowds at China's Crematoriums as Covid Surges. *The Washington Post*, 09 January 2023. From: https://www.washingtonpost.com/investigations/2023/01/09/china-crematorium-covid-deaths/.

Patrikarakos D (2017) *War in 140 Characters: How Social Media Is Reshaping Conflict in the Twenty-First Century.* New York: Basic Books.

Pelton JN (2017) *The New Gold Rush: The Riches of Space Beckon!* 1st edn. Cham: Springer International Publishing: Imprint: Copernicus.

Planet (2024) Products. From: https://www.planet.com/products/.

Pocheptsov G (2018) Cognitive Attacks in Russian Hybrid Warfare. *Information & Security: An International Journal* 41: 37–43.

Sarmina A (2020) Die Macht der Propaganda im Ukrainekonflikt. In: Hohlfeld R, Harnischmacher M, Heinke E, Lehner L and Sengel M (eds) *Fake News und Desinformation: Herausforderungen für die vernetzte Gesellschaft und die empirische Forschung.* Berlin: Nomos Verlag.

Schmitt MN (2013) *Tallinn Manual on the International Law Applicable to Cyber Warfare.* Cambridge: Cambridge University Press.

Shu K, Wang S, Lee D and Liu H (2020) *Disinformation, Misinformation, and Fake News in Social Media: Emerging Research Challenges and Opportunities.* Cham: Springer International Publishing.

Statista (2024) Number of Smartphone Mobile Network Subscriptions Worldwide from 2016 to 2022, with Forecasts from 2023 to 2028. From: https://www.statista.com/statistics/330695/number-of-smartphone-users-worldwide/.

Strobel WP (2024) Chinese Lab Mapped Deadly Coronavirus Two Weeks Before Beijing Told the World, Documents Show. From: https://www.wsj.com/world/china/chinese-lab-mapped-deadly-coronavirus-two-weeks-before-beijing-told-the-world-documents-show-9bca8865.

UN Secretary-General (2024) Secretary-General's Remarks to the General Assembly on Priorities for 2024. From: https://www.un.org/sg/en/content/sg/speeches/2024-02-07/secretary-generals-remarks-the-general-assembly-priorities-for-2024.

World Health Organization (2003a) Summary of Probable SARS Cases with Onset of Illness from 1 November 2002 to 31 July 2003. From: https://www.who.int/publications/m/item/summary-of-probable-sars-cases-with-onset-of-illness-from-1-november-2002-to-31-july-2003.

World Health Organization (2003b) SARS Outbreak WHO Investigation Team Moves to Guangdong China New Travel Advice Announced. From: https://www.who.int/news/item/02-04-2003-sars-outbreak-who-investigation-team-moves-to-guangdong-china-new-travel-advice-announced.

World Health Organization (2003c) SARS Outbreak Contained Worldwide: Threat Remains and More Research Needed, Says WHO. From: https://www.who.int/news/item/05-07-2003-sars-outbreak-contained-worldwide.

World Health Organization (2020) WHO, China Leaders Discuss Next Steps in Battle Against Coronavirus Outbreak. From: https://www.who.int/news/item/28-01-2020-who-china-leaders-discuss-next-steps-in-battle-against-coronavirus-outbreak.

World Health Organization (2024a) WHO COVID-19 Dashboard: Number of COVID-19 Deaths Reported to WHO (cumulative total). From: https://data.who.int/dashboards/covid19/deaths?n=c.

World Health Organization (2024b) Timeline: WHO's COVID-19 Response. From: https://www.who.int/emergencies/diseases/novel-coronavirus-2019/interactive-timeline#event-0.

6 A Space Brutality

Satellite-Enabled Perpetration of Mass Image-Based Sexual Violence

Valérie Pijlman

Introduction

When thinking of space, one may initially think of Earth, our solar system, the Milky Way galaxy and all things *beyond*. However, in space and on planet Earth, another type of space exists that relies heavily on outer space: cyberspace. By no means exhaustive, but a definition of cyberspace is as follows: 'A notional electronic environment created by computer networks, such as the Internet, in which people communicate and interact' (Colman 2015: 183). In other words, it is the online or 'virtual' world that is created by all links between devices that are connected to or enabled by the Internet. It allows individuals to connect and interact with others around the world, for example, through the use of social media applications.

A popular and sometimes risky activity in cyberspace, particularly amongst youth, is sexting, i.e., 'the sending, receiving, or forwarding of sexually explicit messages, images, or photos to others through electronic means' (Klettke et al. 2014: 45). Although such interactions, when consensual, can be a source of enjoyment and sexual exploration (e.g., Uhls et al. 2017), it can also put individuals at risk of experiencing sexual violence in the cyberspace. 'Online' forms of sexual violence are oftentimes referred to as '*technology-facilitated sexual violence*', or TFSV in short. In key literature, TFSV is considered an umbrella term containing a wide variety of abusive behaviours: '(1) online sexual harassment, (2) gender- and sexuality-based harassment, (3) cyberstalking, (4) image-based sexual exploitation, and (5) the use of a carriage service to perpetrate a sexual assault or coerce an unwanted sexual experience' (Henry & Powell 2018: 196). In recent years, image-based forms of sexual violence, i.e., *image-based sexual harassment and abuse*, have garnered more attention, both from popular media outlets and from scientific research. Image-based sexual harassment includes the sending of one's own unsolicited nude or sexual images to others (e.g., 'dickpics' or 'nudes'; Ringrose et al. 2021), whereas the term image-based sexual abuse refers to three abusive behaviours: 'First, the non-consensual *taking* of nude or sexual images; second, the non-consensual *sharing* of nude or sexual images; and third, *threats to share* nude or sexual images' (Henry et al. 2020: 4). The non-

DOI: 10.4324/9781003437178-7

consensual taking or creating of nude or sexual images refers to public or private situations in which images are taken of a person without one's consent (e.g., upskirting) or digitally fabricated or altered by the perpetrator (e.g., AI-generated deepfakes; Flynn et al. 2022). When speaking of non-consensual sharing or distribution of nude or sexual images, the images could be consensually produced and sent by the victim, either non-consensually taken or created by the perpetrator or stolen from the victims' devices (e.g., by hacking; Henry et al. 2020). Lastly, a perpetrator may threaten victims with the distribution or creation of nude or sexual images to blackmail them into sending money, further images or to engage in other unwanted (sexual) acts (e.g., sextortion; O'Malley 2023). In this chapter, image-based sexual harassment and abuse, or image-based sexual violence (IBSV) for the sake of simplicity, will be the focus for which we will explore to what extent the developments in space have enabled the mass perpetration of IBSV and what difficulties with prosecution and jurisdiction can be anticipated. We will conclude with a reflection on potential prevention measures.

Consequences of Image-Based Sexual Violence

As IBSV comprises a form of sexual violence that is facilitated by technology and perpetrated through cyberspace, experiencing IBSV may have severe consequences for victims' mental health. Recent literature has highlighted that IBSV victimization may lead to the development of, amongst others, post-traumatic stress disorder, depressive and anxiety disorders, and suicidality (e.g., Campbell et al. 2022; Huber 2023a; Srivastava et al. 2023). In fact, in describing the totality of harms experienced by IBSV victims, McGlynn and colleagues (2021) speak of a so-called 'social rupture', making the experience of IBSV in that sense a gross disruption of victims' lives: it violates, amongst others, their privacy, integrity and sense of self. Moreover, it affects victims' daily life as well as their relationships, creating a sense that there is a divide between their life before and after the victimization (e.g., Huber 2023a; McGlynn et al. 2021). For instance, victims' trust in others may have been affected, thereby influencing existing relationships and potentially leading to an inability to engage in new (romantic) relationships (e.g., Bates 2017; Campbell et al. 2022). Victims could also suffer from major life changes, losing or purposefully quitting their jobs or education (e.g., Huber 2023a; Kamal & Newman 2016; Martin 2021), whilst others who are victimized due to (fear of) offline harassment and stalking, are afraid to leave their homes (e.g., Campbell et al. 2022; Huber 2023a). These consequences can lead victims to feel a sense of 'constrained liberty'; victims feel as if they are unable to do the same things that they did prior to their victimization (Huber 2023a; McGlynn et al. 2021). Therefore, problematically, the consequences of IBSV can be far-reaching and all-encompassing for victims' well-being and daily lives, in extreme cases potentially leading to social death (e.g., in ethnic or religious minority groups; Nhass 2019), as happens in other cases of mass sexual violence (e.g., in wartime sexual violence; Rose 2023).

The online nature of IBSV brings about specific problems and difficulties. Although the consequences on victims' well-being are similar to those experienced by victims of contact sexual violence, due to the 'online' nature of the abuse, IBSV and its impact on victims is oftentimes underestimated by the general population and professionals alike (e.g., du Mello Gibbard & Fido 2023; Flynn et al. 2023a; Henry et al. 2018). Next to the minimization, third-parties tend to blame IBSV victims, particularly when images are self-produced by the victim (e.g., Attrill-Smith et al. 2021; Call 2021; Zvi 2022). Another problematic dimension of IBSV, is its potential continuity (Dodge 2019). With the non-consensual sharing of nude or sexual images, a risk exists of continuous (re-)victimization and (re-)traumatization; shared images are difficult to remove from cyberspace (Cecil 2014; Dodge 2019). Victims could experience a sense of uncertainty with regard to the re-emerging of the images, as they feel the images may re-emerge 'anywhere' at 'any' given moment. Therefore, they may consequently experience continuous stress and worry (e.g., Huber 2023a; O'Malley 2023), particularly as it is unknown whether the images can ever be fully removed or where they may still potentially be stored, for instance, on someone's computer, the Dark Web, or Revenge Porn websites (McGlynn et al. 2021). In this light, victims may experience a sense of powerlessness, as they feel to have lost control of their images (e.g., Bates 2017; Martin 2021). Hence, they could feel as if they constantly have to be vigilant and may continually try to find their own images online (Huber 2023a; McGlynn et al. 2021). In relation to this, victims may worry about who has seen the images (Campbell et al. 2022).

Due to the boundlessness of cyberspace, there may also be a boundlessness in audience; 'anyone' could be a spectator from the victims' family to complete strangers. Victims may overtly analyse their social relationships to gauge if anyone has viewed their images (McGlynn et al. 2021). This may strain their relationships, and victims may subsequently withdraw and isolate themselves from these relationships, as well as delete their online presence (Campbell et al. 2022; Huber 2023a; McGlynn et al. 2021). Overall, it becomes apparent that the boundlessness of cyberspace may contribute to victims experiencing so-called 'cycles of abuse' (Regehr et al. 2021), a seemingly never-ending victimization.

Space, Enabler for Mass Perpetration of Image-Based Sexual Violence?

When discussing sex in space, it still appears to be somewhat of an afterthought, which could explain the lack of scientific knowledge on the possibility of sex and human reproduction in space (e.g., Weinersmith & Weinersmith 2023), particularly on difficulties conducting the act itself as well as the long-term effect of being in space on women's reproductive health and the development of their offspring (see D'Souza et al. 2022). However, herein lies the assumption of there being two consenting individuals, whilst non-consensual acts of sexual violence could also, if not already, appear in space. Either way, generally, we do not think of the possibility of sexual violence in space, and specifically not amongst astronauts, as they are considered to be a very small number of 'hand-picked elites of their own countries' and it is assumed 'there will be much less likely instances of serious crime' (Gorove 1991: 46). This also reveals how space crime contends to be considered as crimes in space; not also as crimes through space, back on Earth (see Lampkin & White 2023).

Regarding IBSV however, we ought to consider it such a crime through space, as the mass victimization potential is specifically enabled through space with space technology, in this case satellites. Around two-thirds of the world population are estimated to be Internet users, thereby navigating in the cyberspace (Euroconsult 2023; Inmarsat 2023). Individuals may receive an Internet connection through their modem or router by various ways, depending on where on Earth they are located. In many, often easy-to-reach, developed and highly populated areas, individuals may receive Internet through fibre, cable or mobile broadband connections established on Earth itself, like underwater cables in the Baltic Sea (Inmarsat 2023). However, this does not provide connectivity to the whole world population, as some areas on Earth remain too remote or inaccessible for establishing a communication infrastructure. A satellite-based Internet system has therefore been created and is being used to tackle this problem, with the ability to provide global coverage (Hu & Li 2001; Inmarsat 2023). Nevertheless, although the use of satellite constellations may solve the problem of connectivity, its use does not come without costs, amongst others, due to its impact on the pristine outer space environment (see Barentine et al. 2023; Buchs & Bernauer 2023; Lampkin & White 2023; Lawler 2023; Rawls et al. 2020). To give an example, it has only been recent that Dish Network, an American telecommunications company, has been penalized US $150,000 by the US government for not properly disposing of a satellite and for violating the Federal Communications Commission's (FCC) anti-space debris rule, which regulates satellite waste in space. This marks the first-ever space debris penalty imposed (by the FCC) (Clayton 2023), which is not based on an existing specific space (criminal) law. As such, the example shows that environmental space crime remains a hardly criminologically researched and definitely a non-policed domain. Whilst Internet connectivity is one of the main causes, outer space is more and more saturated with Internet-enabling satellites.

But space debris is not the only harmful crime in space. By providing a global coverage of Internet connectivity, it also enables the perpetration and furthers the victimization through IBSV. By virtue of increasing the connectivity of the global population by satellite Internet connection and thereby increasing access to the cyberspace to 'new' Internet users (e.g., think of India (Tiwari 2023) as well as the entire population that may be reached by satellites (Euroconsult 2023)), who perhaps are unaware of the harms of IBSV, more individuals may become both perpetrators or victims of IBSV. Satellite Internet allows individuals to send or share images in the remotest places, like on an airplane or on a ship on the high seas (Inmarsat 2023). Keeping the development of space tourism in mind (see Reddy et al. 2012), IBSV cannot only be enabled through space but also occur in space once individuals create, send and share images from and to outer space. Put differently, with an increased number of Internet users in the 'boundless' cyberspace on and off Earth, images can be sent by and shared with anyone.

Consequently, by increasing the number of Internet users, satellite Internet impacts the extent of victimization and potentially furthers the continuity of victimization, resulting in the phenomenon of satellite-enabled perpetration of mass image-based sexual violence. As more individuals become Internet users, more individuals could become a victim, a passive bystander or spectator to the abuse (i.e., the boundlessness in audience), or may turn into active bystanders (i.e., perpetrators) by re-sharing or re-posting images to other individuals or websites. In this light, with the increase in perpetration and bystandership resulting from an increase in Internet connectivity, satellite Internet may enable the earlier stated 'cycles of abuse', with more individuals being able to (re-)share the images. Victims may, therefore, never be certain if the abuse will ever be over, potentially experiencing continuous victimization. Moreover, the risk of victimization following data leakage may be increased due to the network security issues of satellite Internet (Cao et al. 2020; Willbold et al. 2023). With more individuals being connected to the Internet, and consequently more data sharing taking place, with a lack of sufficient satellite network security, more individuals may be subjected to IBSV by data leakage or hacking. Not just specific data, but the hacking of a satellite as a whole, is a problematic possibility. In 2022, a decommissioned satellite was taken over by a group of hackers who used it to stream conference talks and movies on hacking (Franceschi-Bicchierai 2022). In this sense, one may think of the problematic applications of the hacking of (decommissioned) satellites for IBSV perpetration. Overall, the specific effects of satellite Internet on IBSV remain speculative and require further research, still the potential effects are concerning. As the consequences of IBSV on victims' well-being are multidimensional, focus ought to be put on redress as well as prevention.

The Aftermath of Mass IBSV: Liability and Prosecution

After experiencing IBSV, victims may primarily desire to have the shared images removed or taken down from the digital platforms on which they are shared (see, e.g., Flynn et al. 2022; Henry et al. 2020; Huber 2023b). Where social media providers should have mechanisms in place for victims to report the abuse with the aim to have images taken down, in practice, this is an often-difficult affair for victims, which may lead to secondary victimization (De Angeli et al. 2023; see Martin (2021) for a victims' account). Furthermore, victims may desire to receive recognition or acknowledgement for their victimization experience, seek redress or exact punishment to the perpetrator (Flynn et al. 2022; Henry et al. 2020, 2023). This may, however, not be an easy feat, due to victims experiencing a wide array of barriers to help-seeking

(Henry et al. 2018; Pijlman et al. 2023). For some, preconceived beliefs about and negative prior experiences with the police and criminal justice system could hinder this process, for example, a fear of victim blaming or not being taken seriously (e.g., Campbell et al. 2022; Flynn et al. 2022; Henry et al. 2020; McGlynn et al. 2021). For others, the endeavour may not be feasible due to a lack of relevant legislation in their respective country, a lack of evidence, particularly with regard to the identity of the perpetrator, or the perpetrator being located in a foreign country (e.g., Henry et al. 2018; Flynn et al. 2022, 2023b). The unknown identity of the perpetrator may be a consequence of anonymity measures used by the perpetrator, such as a VPN, to send or share images (Henry & Witt 2021). To this end, the following question remains: who is liable, specifically for satellite-enabled mass IBSV?

When the actual perpetrator cannot be held liable, one could point to the digital platform on which the IBSV took place, as they provided the opportunity to send or share images which enabled the abuse. However, it becomes a question of 'what came first, the chicken or the egg'. An individual may share the images online on a digital platform (e.g., website or social media application), through this digital platform other individuals may see and share the images. Thus, the IBSV may not have been possible without the digital platform, which would not function or exist without an Internet connection.

As various private companies, including Elon Musk's SpaceX, and state-owned organizations worldwide are actively investing in the development of satellite constellations to establish high-speed Internet connectivity worldwide, they are directly affirming control over the exchange of information globally (Voelsen 2021). Although in light of corporate social responsibility, and in case of state-subsidized or state-owned organizations, these companies and organizations are perhaps naively believed to consider national interest, but this may not seem to be fully applicable in the case of IBSV. This has to do with the fact that the perpetration of IBSV is hardly possible without an Internet connection. Therefore, one could argue that Internet providers are complicit in the perpetration of IBSV by enabling the connection to the Internet. Placing this in a more traditional context, one's thoughts may turn to the well-known example of the 'wet floor' warning sign in a supermarket; the supermarket employee who just mopped the floor, but forgot to put up the 'wet floor' warning sign. If a customer were to slip and fall, it is not the employee as an individual who is liable, but the supermarket company itself, since they have a duty of care. Should satellite Internet providers not take measures to ensure users are safe when making use of their services, much like in the case of creating space debris? And may they be held accountable for having failed to do so, implying a duty of care?

As Internet providers ought to be, and are most likely, aware of the perpetration of cybercrime, particularly where IBSV is concerned, it appears they may enable the perpetration thereof, which due to the availability of satellite Internet, may happen on a grander scale than ever before. It becomes a question whether their duty of care increases, because of the mass-dimension and potential continuity of IBSV through satellite Internet. Given how international law regulates to an extent the activities in space, in particular, through the United Nations Treaty on Principles Governing the Activities of States in the Exploration and Use of Outer Space, including the Moon and Other Celestial Bodies (1967), or the Outer Space Treaty in short, a duty of care in relation to satellite-enabled mass IBSV could be implied for states where satellite Internet providers are registered. The Treaty states in Article VI that:

> States Parties to the Treaty shall bear international responsibility for national activities in outer space, including the Moon and other celestial bodies, whether such activities are carried on by governmental agencies or by non-governmental entities, and for assuring that national activities are carried out in conformity with the provisions set forth in the present Treaty. The activities of non-governmental entities in outer space, including the Moon and other celestial bodies, shall require authorization and continuing supervision by the appropriate State Party to the Treaty.
> (United Nations Office for Outer Space Affairs 2008: 5)

This indicates strongly that the responsibility and liability for the national activities taken by governmental as well as non-governmental agencies lie with the state itself (Morozova & Vasyanin 2019; Voelsen 2021). Although it is unsure whether Internet providers directly have a duty of care, making it difficult to point out direct responsibility in the case of satellite-enabled mass IBSV, the relevant government may monitor such indirect responsibility (and thus legal accountability) as to nationally enforce compliance of satellite Internet providers to the Outer Space Treaty.

If that would be possible, mass IBSV victims may be able to hold the state responsible for the IBSV experienced through satellite Internet connection, especially when IBSV is criminalized in the state's legislation. If IBSV is not criminalized in the state's legislation, victims may be able to appeal to the violation of their fundamental or human rights. Perhaps this can be based on a recent case; the US youth who won a lawsuit against the state of Montana for the violation of 'their constitutional right to "a clean and healthful environment," as well as their rights to dignity, health and safety, and equal protection of the law' (United Nations Human Rights Office of the High Commissioner 2023).

Following the widespread consequences for victims' lives, experiencing IBSV may be considered a violation of Article 8 of the 1950 European Convention for the Protection of Human Rights and Fundamental Freedoms, the right to respect for private and family life (O'Connell 2020), and similarly, a violation to Article 17 of the 1966 United Nations International Covenant on Civil and Political Rights, which states: 'No one shall be subjected to arbitrary or unlawful interference with his privacy, family, home or correspondence, nor to unlawful attacks on his honour and reputation' (United Nations

Human Rights Office of the High Commissioner 2023). With regard to European Union law, one could turn to the 2000 Charter of Fundamental Rights of the European Union. A victim of satellite-enabled mass IBSV may file for a violation of their right to mental and physical integrity, as stated in Article 3, as well as, their right to respect for private and family life (Article 7), and the protection of personal data (Article 8). In relation to this, one could also think of the 2016 General Data Protection Regulation (EU) 2016/679 (GDPR), which describes in Article 1.2 its aim to protect the fundamental rights and freedoms of human beings, particularly the right to the protection of their personal data. So, the body of space law could relate to cybercrime, in case of mass IBSV, and the body of cyberlaw could relate to IBSV as crime through space; making mass IBSV a clear example of how satellite crimes in general are at the (legal) outer space/cyberspace nexus (Potter 1995).

With the risk of continuous victimization, a relevant right for victims of IBSV in the GDPR appears to be the right to erasure, or in other words, to be forgotten (Article 17). However, with the transnational nature of IBSV and lack of cooperation of digital platforms, Article 17 may not be sufficient in providing victims with justice or redress (Cater 2021; Nguyen 2021). Therefore, an appeal to their fundamental and human rights may potentially be more fruitful, which appears particularly important as currently only a small number of victims take their case to the police, and an even smaller number bring their case to court (see, e.g., Flynn et al. 2023b; Klettke et al. 2019; Powell & Henry 2019). Perhaps the abovementioned legal opportunities, appealing to fundamental and human rights, may encourage victims to take their case to the police and may enable victims to receive justice and redress.

A Stitch in Space, Saves the Case: Concluding Thoughts on Mass IBSV Prevention

Finally, the focus moves to the prevention of mass IBSV. Nowadays, various social media companies (e.g., Facebook) have developed means to detect, filter and remove the content which violates their community guidelines, for example, through human content moderators or AI tools (Henry & Witt 2021). However, it is not only social media companies, but also satellite Internet providers and governments, who have a duty of care as argued in the previous section, and thus also the ethical and moral responsibility to focus on the prevention of mass IBSV. In light of the prevention of victimization, Henry & Witt (2021) highlight that the moderation of user-generated content is normally done 'ex post', meaning the content is checked only after the posting of the content; i.e., after the victimization has already occurred. Therefore, content moderation can hardly be seen as a preventative measure. Perhaps AI tools can be applied to filter out data 'ante post', for example, in the satellite Internet connection itself, before the data reaches its intended destination. However, it may be difficult for AI tools to recognize when images are non-consensually shared or being sent (Henry & Witt 2021).

Therefore, a more general route may be taken focusing on nude or sexual images in general. AI tools may be used to recognize someone attempting to share or send content that contains nudity or is of sexual nature, and send a warning message to discourage the individual from sending or posting it. In a randomized-controlled experiment, Prichard and colleagues (2022) showed warning messages to men who visited a fake website which stated to provide free pornography to those who uploaded a sexual image of a woman. Individuals who were shown a warning message indicating that the sharing of sexual images of minors is illegal, were less likely to try and continue to access the fake website than those who were not. Thus, a warning message may be effective in dissuading individuals from sending and sharing nude or sexual images in general, and especially when they are aware of their Internet connection being satellite-enabled.

Overall, matters surrounding the governance of cyberspace and space remain unclear (Eriksson & Giacomello 2022). Although on Earth, various actions are taken in light of accountability and prevention, they are not yet effective, still Henry and Witt state that:

> Governments around the world are introducing new regulatory regimes to attempt to better hold technology companies accountable for hosting harmful content on their platforms. At the same time, technology companies are themselves taking more proactive steps to tackle this problem.
> (Henry & Witt 2021: 750)

Therefore, perhaps we ought to consider the involvement of supranational organizations, such as the United Nations' International Telecommunications Union (ITU). The ITU has a global membership which includes 193 member states, as well as over 900 companies, universities and organizations (ITU n. d.). Thus, they may have the appropriate far-reaching platform to regulate cyberspace (in space) and encourage the collaboration between private and public entities in developing prevention measures for IBSV. Still the ability (and perhaps responsibility) for governance of cyberspace (i.e., digital platforms and Internet providers) in space, and the abovementioned potential liability, may in light of the Outer Space Treaty, first and foremost lie with the nation-states (see Henry & Witt (2021) for a recommendation on state-based regulation of digital platforms).

Thus, IBSV can be regarded as a space crime, as the global coverage of satellite Internet may lead to mass victimization and perpetration. Considering the serious and potentially long-term consequences of IBSV victimization on victims' well-being, as well as the risk of continuous victimization, steps need to be taken with regard to minimizing the harm, allowing victims to access proper means of redress and justice, as well as to prevent the abuse. With the reflections described in this chapter, we hope to ignite further empirical and legal research into this overlooked yet important subject. Although the Internet, and cyberspace in general, have become an integral part of many individuals' daily lives, increased connectivity through outer space in combination with the absence of, or insufficient, governance and regulations, should not come at the expense of individuals' well-being.

References

Attrill-Smith A, Wesson CJ, Chater ML & Weekes L (2021) Gender differences in videoed accounts of victim blaming for revenge porn for self-taken and stealth-taken sexually explicit images and videos. *Cyberpsychology: Journal of Psychosocial Research on Cyberspace* 15(4): Article 3. https://doi.org/10.5817/CP2021-4-3.

Barentine JC, Venkatesan A, Heim J, Lowenthal J, Kocifaj M & Bará S (2023) Aggregate effects of proliferating low-Earth-orbit objects and implications for astronomical data lost in the noise. *Nature Astronomy* 7(3): 252–258. https://doi.org/10.1038/s41550-023-01904-2.

Bates S (2017) Revenge porn and mental health: A qualitative analysis of the mental health effects of revenge porn on female survivors. *Feminist Criminology* 12(1): 22–42. https://doi.org/10.1177/1557085116654565.

Buchs R & Bernauer T (2023) Market-based instruments to incentivize more sustainable practices in outer space. *Current Opinion in Environmental Sustainability* 60: 101247. https://doi.org/10.1016/j.cosust.2022.101247.

Call C (2021) Perceptions of image-based sexual abuse among the American public. *Criminology, Criminal Justice, Law & Society* 22(3): 30–43.

Campbell JK, Poage SM, Godley S & Rothman EF (2022) Social anxiety as a consequence of non-consensually disseminated sexually explicit media victimization. *Journal of Interpersonal Violence* 37(9–10): NP7268–NP7288. https://doi.org/10.1177/0886260520967150.

Cao H, Wu L, Chen Y, Su Y, Lei Z & Zhao C (2020). Analysis on the security of satellite internet. In Lu W, Wen Q, Zhang Y, Lang B, Wen W, Yan H, Li C, Ding L, Li R & Zhou Y (eds.) *Cyber Security: 17th China Annual Conference, CNCERT 2020*, Beijing, China, August 12, 2020, Revised Selected Papers 17. Singapore: Springer, pp. 193–205. https://doi.org/10.1007/978-981-33-4922-3_14.

Cater L (2021) How Europe's privacy laws are failing victims of sexual abuse. *Politico*. From: https://www.politico.eu/article/how-europe-privacy-laws-are-failing-victims-of-sexual-abuse/.

Cecil AL (2014) Taking back the internet: Imposing civil liability on interactive computer services in an attempt to provide an adequate remedy to victims of non-consensual pornography. *Washington & Lee Law Review* 71(4): 2513–2556.

Charter of Fundamental Rights of the European Union, December 18, 2000. From: https://www.europarl.europa.eu/charter/pdf/text_en.pdf..

Clayton A (2023) US government issues first-ever space debris penalty to Dish Network. *The Guardian*. From: https://www.theguardian.com/science/2023/oct/02/fcc-space-debris-fine-dish-network-satellite.

Colman AM (2015) *A Dictionary of Psychology*, Fourth Edition. Oxford: Oxford University Press.

D'Souza S, Haghgoo N, Mankame K, Mummigatti S & Saadi A (2022) Safe spaceflight for women: Examining the data gap and improving design considerations. *Journal of Space Safety Engineering* 9(2): 154–159. https://doi.org/10.1016/j.jsse.2022.02.010.

De Angeli A, Falduti M, Menendez-Blanco M & Tessaris S (2023) Reporting non-consensual pornography: clarity, efficiency and distress. *Multimedia Tools and Applications* 82(9): 12829–12858. https://doi.org/10.1007/s11042-022-14291-z.

Dodge A (2019) Nudes are forever: Judicial interpretations of digital technology's impact on "Revenge Porn". *Canadian Journal of Law and Society* 34(1): 121–143. https://doi.org/10.1017/cls.2019.4.

du Mello Gibbard G & Fido D (2023) A comparison of judgements of image-based and physical sexual abuse: A pilot study. *Journal of Concurrent Disorders* 5(1): 92.

Eriksson J & Giacomello G (2022). Cyberspace in space: fragmentation, vulnerability, and uncertainty. In Cavelty MD & Wenger A (eds) *Cyber Security Politics*. London: Routledge, pp. 95–108. https://doi.org/10.4324/9781003110224-8.

Euroconsult (2023) 2.6 billion remained unconnected to broadband at the end of 2022 – a $74 billion untapped opportunity. From: https://www.euroconsult-ec.com/press-release/2-6-billion-remained-unconnected-to-broadband-at-the-end-of-2022-a-74-billion-untapped-opportunity/.

European Convention for the Protection of Human Rights and Fundamental Freedoms, November 4, 1950. From: https://www.echr.coe.int/documents/d/echr/convention_eng.

Flynn A, Powell A, Scott AJ & Cama E (2022) Deepfakes and digitally altered imagery abuse: A cross-country exploration of an emerging form of image-based sexual abuse. *The British Journal of Criminology* 62(6): 1341–1358. https://doi.org/10.1093/bjc/azab111.

Flynn A, Cama E, Powell A & Scott AJ (2023a) Victim-blaming and image-based sexual abuse. *Journal of Criminology* 56(1): 7–25. https://doi.org/10.1177/26338076221135327.

Flynn A, Powell A & Hindes S (2023b) Policing technology-facilitated abuse. *Policing and Society* 33(5): 575–592. https://doi.org/10.1080/10439463.2022.2159400.

Francheschi-Bicchierai L (2022) Hackers took over a commercial satellite to broadcast hacker movies. From: https://www.vice.com/en/article/y3pwqx/hackers-took-over-a-commercial-satellite-to-broadcast-hacker-movies.

Gorove S (1991) Space stations: Problems of jurisdiction and conflicts of law. *Journal of Space Law* 19(1): 45–52.

Henry N & Powell A (2018) Technology-facilitated sexual violence: A literature review of empirical research. *Trauma, Violence, & Abuse* 19(2): 195–208. https://doi.org/10.1177/1524838016650189.

Henry N & Witt A (2021) Governing image-based sexual abuse: Digital platform policies, tools, and practices. In Bailey J, Flynn A & Henry N (eds) *The Emerald International Handbook of Technology-Facilitated Violence and Abuse*. Emerald Publishing Limited, pp. 749–768. https://doi.org/10.1108/978-1-83982-848-520211054.

Henry N, Flynn A & Powell A (2018) Policing image-based sexual abuse: Stakeholder perspectives. *Police Practice and Research* 19(6): 565–581. https://doi.org/10.1080/15614263.2018.1507892.

Henry N, Gavey N, McGlynn C & Rackley E (2023) 'Devastating, like it broke me': Responding to image-based sexual abuse in Aotearoa New Zealand. *Criminology & Criminal Justice* 23(5): 861–879. https://doi.org/10.1177/17488958221097276.

Henry N, McGlynn C, Flynn A, Johnson K, Powell A & Scott AJ (2020) *Image-Based Sexual Abuse: A Study on the Causes and Consequences of Non-Consensual Nude or Sexual Imagery*. London: Routledge. https://doi.org/10.4324/9781351135153.

Hu Y & Li VO (2001) Satellite-based internet: A tutorial. *IEEE Communications Magazine* 39(3): 154–162. https://doi.org/10.1109/35.910603.

Huber A (2023a) 'A shadow of me old self': The impact of image-based sexual abuse in a digital society. *International Review of Victimology* 29(2): 199–216. https://doi.org/10.1177/02697580211063659.

Huber AR (2023b) Image-based sexual abuse: Legislative and policing responses. *Criminology & Criminal Justice* 0(0). https://doi.org/10.1177/17488958221146141.

Inmarsat (2023) Satellite internet: connected from space. From: https://www.inmarsat.com/en/insights/corporate/2023/satellite-internet-connected-from-space.html.

ITU (n.d.) About International Telecommunication Union (ITU). From: https://www.itu.int/en/about/Pages/default.aspx.

Kamal M & Newman WJ (2016) Revenge pornography: Mental health implications and related legislation. *Journal of the American Academy of Psychiatry and the Law Online* 44(3): 359–367.

Klettke B, Hallford DJ & Mellor DJ (2014) Sexting prevalence and correlates: A systematic literature review. *Clinical Psychology Review* 34(1): 44–53. https://doi.org/10.1016/j.cpr.2013.10.007.

Klettke B, Hallford DJ, Clancy E, Mellor DJ & Toumbourou JW (2019) Sexting and psychological distress: The role of unwanted and coerced sexts. *Cyberpsychology, Behavior, and Social Networking* 22(4): 237–242. https://doi.org/10.1089/cyber.2018.0291.

Lampkin J & White R (2023) *Space Criminology: Analysing Human Relationships with Outer Space*. Palgrave Macmillan Cham. https://doi.org/10.1007/978-3-031-39912-1.

Lawler S (2023) Bright satellites are disrupting astronomy research worldwide. From: https://www.nature.com/articles/d41586-023-03610-5.

Martin N (2021) Image-based sexual abuse and deepfakes: A survivor turned activist's perspective. In Powell A, Flynn A & Sugiura L (eds) *The Palgrave Handbook of Gendered Violence and Technology*. Palgrave Macmillan Cham, pp. 55–72. https://doi.org/10.1007/978-3-030-83734-1.

McGlynn C, Johnson K, Rackley E, Henry N, Gavey N, Flynn A & Powell A (2021) 'It's torture for the soul': The harms of image-based sexual abuse. *Social & Legal Studies* 30(4): 541–562. https://doi.org/10.1177/0964663920947791.

Morozova E & Vasyanin Y (2019) International Space Law and satellite telecommunications. *Oxford Research Encyclopedia of Planetary Science*. https://doi.org/10.1093/acrefore/9780190647926.013.75.

Nguyen TNA (2021) European 'right to be forgotten'as a remedy for image-based sexual abuse: A critical review. *KnowEx Social Sciences* 2(01): 59–72. https://doi.org/10.17501/27059901.2021.2105.

Nhass H (2019) Niet jouw schuld! *Kennisplatform Integratie & Samenleving*. https://www.kis.nl/sites/default/files/niet-jouw-schuld-verkenning-shame-sexting.pdf.

O'Connell A (2020) Image rights and image wrongs: image-based sexual abuse and online takedown. *Journal of Intellectual Property Law & Practice* 15(1): 55–65. https://doi.org/10.1093/jiplp/jpz150.

O'Malley RL (2023) Short-term and long-term impacts of financial sextortion on victim's mental well-being. *Journal of Interpersonal Violence* 38(13–14): 8563–8592. https://doi.org/10.1177/08862605231156416.

Pijlman V, Eichelsheim V, Pemberton A & de Waardt M (2023) "I did not want to make a bigger deal out of it than it was": A mixed-method study on the help-seeking behavior of victims of image-based sexual harassment and abuse. [Manuscript submitted for publication].

Potter M (1995) The outer space cyberspace nexus: Satellite crimes. *Journal of Space Law* 23(1): 55–56.

Powell A & Henry N (2019) Technology-facilitated sexual violence victimization: Results from an online survey of Australian adults. *Journal of Interpersonal Violence* 34(17): 3637–3665. https://doi.org/10.1177/0886260516672055.

Prichard J, Scanlan J, Krone T, Spiranovic C, Watters P & Wortley R (2022) Warning messages to prevent illegal sharing of sexual images: Results of a randomised controlled experiment. *Trends and Issues in Crime and Criminal Justice* 647: 1–15. https://search.informit.org/doi/10.3316/informit.565933568461626.

Rawls ML, Thiemann HB, Chemin V, Walkowicz L, Peel MW & Grange YG (2020) Satellite constellation internet affordability and need. *Research Notes of the AAS* 4(10): 189. https://doi.org/10.3847/2515-5172/abc48e.

Reddy MV, Nica M & Wilkes K (2012) Space tourism: Research recommendations for the future of the industry and perspectives of potential participants. *Tourism Management* 33(5): 1093–1102.

Regehr K, Birze A & Regehr C (2021) Technology facilitated re-victimization: How video evidence of sexual violence contributes to mediated cycles of abuse. *Crime, Media, Culture* 18(4): 597–615. https://doi.org/10.1177/17416590211050333.

Regulation (EU)2016/679, on the protection of natural persons with regard to the processing of personal data and on the free movement of such data, and repealing Directive 95/46/EC (General Data Protection Regulation). European Parliament and Council. From: https://eur-lex.europa.eu/legal-content/EN/TXT/HTML/?uri=CELEX:02016R0679-20160504.

Ringrose J, Regehr K & Whitehead S (2021) Teen girls' experiences negotiating the ubiquitous dick pic: Sexual double standards and the normalization of image based sexual harassment. *Sex Roles* 85(9): 558–576. https://doi.org/10.1007/s11199-021-01236-3.

Rose S (2023) Stigmatization and social death of survivors of wartime sexual violence. *Global Studies Quarterly* 3(2): ksad021. https://doi.org/10.1093/isagsq/ksad021.

Srivastava A, Rusow J, Schrager SM, Stephenson R & Goldbach JT (2023) Digital Sexual violence and suicide risk in a national sample of sexual minority adolescents. *Journal of Interpersonal Violence* 38(3–4): 4443–4458. https://doi.org/10.1177/08862605221116317.

Tiwari S (2023) How satellite internet can transform digital connectivity in India. From: https://www.ey.com/en_in/technology/how-satellite-internet-can-transform-digital-connectivity-in-india.

United Nations Human Rights Office of the High Commissioner (2023) "This is about our human rights:" U.S. youths win landmark climate case. From: https://www.ohchr.org/en/stories/2023/08/about-our-human-rights-us-youths-win-landmark-climate-case.

United Nations International Covenant on Civil and Political Rights (no date) International Covenant on Civil and Political Rights, December 16, 1966. From: https://www.ohchr.org/en/instruments-mechanisms/instruments/international-covenant-civil-and-political-rights.

United Nations Office for Outer Space Affairs (2008) United Nations Treaty on Principles Governing the Activities of States in the Exploration and Use of Outer Space, including the Moon and Other Celestial Bodies, October, 1967. From: https://www.unoosa.org/res/oosadoc/data/documents/2008/stspace/stspace11rev_2_0_html/st_space_11rev2E.pdf.

Uhls YT, Ellison NB & Subrahmanyam K (2017) Benefits and costs of social media in adolescence. *Pediatrics* 140(Supplement 2): S67–S70. https://doi.org/10.1542/peds.2016-1758E.

Voelsen D (2021) Internet from space: How new satellite connections could affect global internet governance (SWP Research Paper, 3/2021). Berlin: Stiftung Wissenschaft und Politik, Deutsches Institut für Internationale Politik und Sicherheit. https://doi.org/10.18449/2021RP03.

Weinersmith K & Weinersmith Z (2023) Space billionaires should spend more time thinking about sex. From: https://www.nytimes.com/2023/11/05/opinion/space-billionaires-sex.html.

Willbold J, Schloegel M, Vögele M, Gerhardt M, Holz T & Abbasi A (2023) Space Odyssey: An experimental software security analysis of satellites. *IEEE Symposium on Security and Privacy.* From: https://publications.cispa.saarland/3934/1/SatSec-Oakland22.pdf.

Zvi L (2022) The double standard toward female and male victims of non-consensual dissemination of intimate images. *Journal of Interpersonal Violence* 37(21–22): NP20146–NP20167. https://doi.org/10.1177/08862605211050109

7 Space Victimology

Out of the Dark and into the Light

Jack Lampkin

Introduction: A History of Criminology and Outer Space

This book is primarily concerned with the topics of crime, criminal justice and ethics as they pertain to anthropogenic interactions with outer space. Historically, criminology has had very little to say about outer space issues (Eski, 2023; Rothe and Collins, 2023). There are, however, a few noteworthy exceptions. In 1968, Professor Luis Kutner proposed outer space as a location for housing terrestrial law-breakers, freeing Earth-based communities of dangerous individuals. More recently, in 2006, Dr Julian Hermida wrote in detail about a series of interpersonal crimes that took place at an astronaut training facility in Russia, whereby a female American astronaut was physically and sexually assaulted by Russian comrades during a prolonged isolation trial. These interesting publications, however, have remained on the criminological periphery for many years, with Kutner's (1968) work in particular mostly viewed as radical at best and nonsensical at worst.

Despite the historical lack of criminological attention concerning outer space, a few criminologists have recently started to grapple with such issues. This is likely due to the recent expansion of space related industries, mainly driven by the private sector, which has sparked public interest and renewed enthusiasm in space exploration activities.

To be more specific, Takemura (2019) and Lampkin (2021) have theorised about the impact of anthropogenic activities on outer space environments from the perspective of astro-green criminology. They argue that human actions in outer space are harmful to the pristine natural landscapes and atmospheres that exist there. The main contributors to space-based environmental harm (from the author's perspective) are orbital debris accumulation, mining in outer space, and the emissions pollutions associated with space travel, space tourism, rocket launching, and space industry supply chains (Lampkin, 2021).

While activities like rocket launching and orbital debris directly affect planet Earth, some influential work has considered environmental harms further afield. Gorman (2019), for instance, has framed space junk from the perspective of cultural anthropology, arguing that materials in space are

DOI: 10.4324/9781003437178-8

either scientifically valuable (if still operational), or culturally significant and representative of immense human scientific achievement and capability (if now defunct). Lampkin and Wyatt (2022) took a similar view, but additionally advocated for the protection of culturally significant areas in outer space through the creation of extraterrestrial heritage sites. These could, for instance, be implemented for the purposes of protecting the momentous Apollo 11 lunar landing sites which represented the first time humans stepped foot on another world.

In a similar vein, Rothe and Collins (2023) recently published a seminal article in the prestigious *British Journal of Criminology* regarding environmental harm and space weaponisation. Also situating their work within the perspective of green criminology, the authors called for more research and engagement with outer space issues from green-critical criminologists. In particular, their focus was on the weaponisation of outer space, military weapon development, and the impact that such behaviours have on the natural world.

Following this train of destructive thought, Eski (2023) recently utilised a green-critical criminological perspective to imagine how runaway climate change may lead to human extinction in his quintessential book *A Criminology of the Human Species*. In Chapter Six specifically, Eski (2023) discusses the self-annihilating nature of humans, and the extent to which this could be extended into outer space. From Eski's (2023) perspective, the tendency of humans to engage in mass exploitation activity means that such behaviour could be extended into space – the final frontier. Here, humans could disturb landscapes irreparably, threaten (basic) life forms that exist upon other celestial bodies, and even prevent future extraterrestrial life from emerging as a result. This may be thought of as a form of *ecocide* or *astrocide*, whereby environs are either destroyed, or prevented from materialising, as a direct consequence of anthropogenic activity.

Perhaps the most significant text to date, however, relating to criminology and outer space is Lampkin and White's (2023) book simply entitled *Space Criminology*. This manuscript brought together the current astro-green criminological notions of crime and harm in outer space as it relates to the natural world (orbital debris, space mining, emissions pollutions), as well as future issues of an orthodox criminological nature. This included chapters deliberating the policing of space crimes and issues around living and working in outer space. These are the first writings of what we might call an *applied space criminology* – thoughts surrounding how we can practically address unwanted behaviours in outer space environs for the benefit of humans and the protection of extraterrestrial landscapes.

It is clear, then, that academic interest in criminology, the environment, and outer space, is gaining some momentum. This is perhaps a reflection of the general heightened interest in outer space in the collective social consciousness; driven by the private sector and state-corporate space exploration contracts. However, even in these academic works, the focus has largely been on either the natural world (i.e. orbital debris and space mining), or how to conceptualise and prepare for crimes occurring in outer space in the near future. Largely omitted (thus far) from such debates is consideration of the victim of space crimes, broadly conceived. This chapter will attempt to provide the first discussion of victimological issues specifically pertaining to anthropogenic interactions with outer space along with the first definition of what constitutes a 'space victim' and a 'space crime'. Doing so will help protect future space victims from experiencing the negative impacts of space crimes. Or, even better, to prevent victimisation occurring at all.

The main purpose of advocating for an explicit *space victimology* is avoiding victims of space crimes from suffering the same fate as their terrestrial counterparts. Whereas sustained criminological thought and research can be traced back to the enlightenment era of the 18th Century, the focus of early criminology was mostly restricted to consideration of the criminal and attempting to explain why certain people commit heinous acts. Even when early victimological studies began to surface in the mid 20th Century, these are now largely regarded as obsolete, owing to the antagonistic nature of early theories of victimhood. The writings of von Hentig (1948), Amir (1967), Mendelsohn (1974) and Wolfgang (1957), for instance, mostly blamed the victim for their own victimisation, rather than other contributory factors. Western feminist and radical victimological scholars successfully reverted criminological attention away from victim-blaming and back towards offender-focused theories by the end of the 20th Century (Fisher et al., 2016). However, unfortunately, into the 21st Century, criminologists and victimologists are yet to extend their gaze beyond Earth which risks similar victim alienation. This chapter intends to plug this victimological research gap by campaigning for special consideration of the 'space victim' within academic and organisational research on outer space. Doing so will ensure the unique needs, requirements, and perhaps *rights* of space victims are acknowledged, considered and, ideally, prioritised.

As such, the remainder of this chapter will attempt to scope out a 'space victimology' including consideration of potential issues, areas of study and avenues for future research. This will provide future academic criminologists and victimologists with a platform for thinking, theorising, researching and, ultimately, preventing and protecting future victims of space harms and crimes.

In order to do this, the chapter will be split into three sections. The first will consider victims of traditional criminological issues, focussing largely on 'street-based', interpersonal crimes that mirror orthodox criminological focus in terrestrial environments. The second section will consider what might be called *astro-green victimology*. Those harms and crimes committed in outer space that directly impact the environments and landscapes of extraterrestrial locales and any living organisms that may exist there. These two sections mirror contemporary thought regarding crime, harm and outer space which is generally split into *environmental justice* issues (astro-green criminology) and *human justice* issues (space criminology). Although there are overlaps between the two (i.e. actions that harm the environment can constitute violations of existing criminal laws that are consequently of an orthodox criminological appeal), considering victims of environmental harm and traditional harm independently may help academics to theorise and research in a more clear and meaningful way at this nascent stage. In that regard, the third section will move on to discuss the application of existing criminological and victimological theories to outer space in an attempt to identify fruitful avenues for future research.

Orthodox Space Criminology and Victimisation

Before delving into the nuances of space criminology and victimology, it is important to state what is meant by a space crime and, thus, a space victim. While the terms 'crime' and 'victim' have been long debated in criminology, the conceptualisation of crimes and victims in the outer space context is potentially slightly less contentious because all such arguments must have in them some element of the extraterrestrial. Therefore, the definition of a *space crime* proposed in this chapter is:

> Any unwanted action or behaviour that results in harm to people, property, non-human species or landscapes/ecologies. These actions or behaviours could take place in outer space (for instance on space stations, spacecraft, or ground-based settlements) or on planet Earth, as long as they relate in some way to outer space activities (for instance, crimes that take place during astronaut training). Examples of space crimes include those that are terrestrial-based (i.e. state-corporate corruption, tax evasion, or fraud relating to outer space missions and endeavours), or they could be based off-Earth (i.e. assault or battery occurring within a human-constructed Martian or lunar colony).

Conversely, a *space victim* is defined here as:

> Someone or something *harmed* in some way by an activity or behaviour committed in outer space (i.e. the victim of a sexual assault aboard an inhabited space station), or on planet Earth (i.e. an employee within a space company who is treated unfairly or in contravention of domestic laws or regulations).

Space victimology should be concerned with both the prevention of space crime (and thus, victimisation), as well as how to address crime and harm that has already occurred. After all, while prevention is a productive endeavour in terms of impacting the overall levels of crime and victimisation in a particular environment or society, some crime will always occur. The thought that crime is a ubiquitous element of all human societies is widely accepted as a criminological truth, owing to the existence of criminal or deviant behaviour across human history. As William Barnes Jr (1998: 631) put it so eloquently and succinctly: ‘crime will always exist despite all our best efforts’.

This means that space victimology has four key elements. Preventing space crimes on Earth and in outer space, as well as addressing those space crimes that have already materialised on Earth or in outer space. This is perhaps more easily understood through the consideration of a visual aid, provided in Figure 7.1.

Presenting these issues as a matrix is appropriate because some of these issues overlap. For instance, if a country were to intentionally launch a missile from Earth and destroy the military satellite of another nation that was otherwise in-orbit many miles above Earth’s surface, this would be classified as a crime that occurred both on and off-Earth. Such differentiations are largely irrelevant to the study of space victimology because it is the victim, and the victim’s rights and needs, that should be the focus of study, rather than the geography of where the crime occurred. However, due to the unique ability of humans to exist and interact with environments beyond Earth, it is envisioned that a matrix will help to focus criminologists on what are essentially a very wide net of potential space crimes. Consider, for instance, the victimisation of a human cosmonaut verbally abused by another in outer space. This is a very different criminological issue than, for instance, exploring the financial accounts of a company accused of evading taxes. With this in mind, the below list expands upon the above space crime focus matrix by offering a comprehensive – and, strangely, largely hypothetical list – of potential space crimes that could occur on and off-Earth. This will help to break down a seemingly vast subject area into more manageable, singular issues. Furthermore, like before, the intention of doing this is to invite criminologists and victimologists in different areas of study to consider how their work might apply to outer space. Doing so will ensure that researchers are engaging with the full breadth of deviance across the criminological spectrum, but also that all potential victims are accounted for and considered equally. Of course, this list is not,

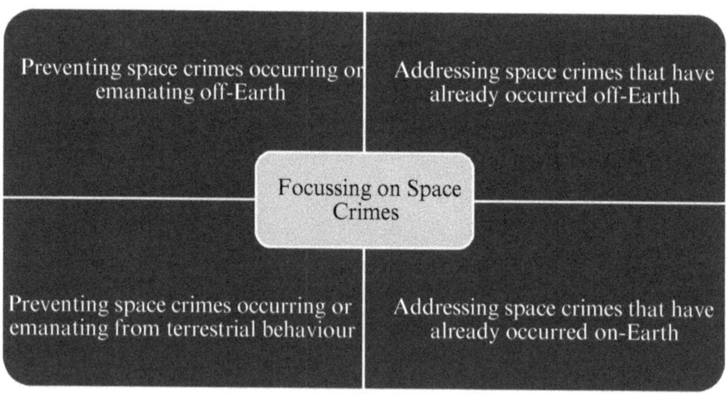

Figure 7.1 The Space Crime Focus Matrix

and cannot be, exhaustive. On the contrary, future space criminologists are likely to identify additional issues that could be included, owing to the evolving nature of space exploration and anthropogenic interactions with outer space.

To break such issues down even further, space crime issues can be separated into micro, macro and meso-level issues. Again, doing so enables researchers to identify and situate their specific research area into the broader field of study. Micro-issues here relate to 'the study of human behaviour in contexts of everyday direct interaction' (Giddens, 1997: 883). Conversely, macro-level refers to social issues 'at the national and global level' (Sallum Jr, 2005: 23), with meso-level issues situated somewhere in between (Serpa et al., 2019). While such delineations are useful in the context of considering a whole area of study with its individual nuances, it is appropriate to recognise that micro- and meso-level issues are often affected by macro issues, and, likewise, macro-issues may be impacted by high-profile singular issues that occur at a micro-level. Despite such contestations, the list demonstrates both the interpersonal nature of some space crimes, and the higher-level, state-corporate manner of others.

Micro-Level

- Interpersonal crimes in space: assault; physical, verbal or sexual abuse; rape; threats of violence; theft; burglary; robbery; mugging; murder; poisoning.
- Domestic crimes while living in space: inter-partner abuse; child abuse; crimes induced by drugs or alcohol.
- Public order crimes in space: riot; affray; causing harassment, alarm or distress; criminal damage and destruction of property; vandalism; graffiti; arson; littering.
- Racially or religiously aggravated crimes.

Macro-Level

- Acts of terror.
- Military warfare and space weaponisation; anti-satellite testing.
- Restrictions on access to space: space racism; space sexism; space ableism.
- Major environmental damage (such as from construction or mineral excavation).
- Mass murder or genocide.
- Professional and organised crime.
- State, corporate and white-collar crime.

Meso-Level

- Cybercrimes committed on-Earth.

- Cybercrimes committed in space: onboard computer systems set-up in spacecraft, aboard space stations, or while habiting a celestial body.
- Minor environmental damage; health and safety violations.
- Place-based crime; handling stolen goods.
- Money laundering; corruption; tax evasion; fraud; identity theft.
- Torture; war-like tactics; interrogation; treason.

Most Western criminologists would likely agree that the behaviours identified in the above list of conduct are criminal at best and deviant or unwanted at the very least. However, it must be stated that what constitutes a crime (and therefore a victim), varies over time and place.

It is also worth considering that there are many criminal behaviours omitted from the list, behaviours that are frequently regarded as criminal on Earth, but have little relevance when applied to outer space. For instance, crimes of human trafficking, slavery, and those typically (but not exclusively) associated with the lower-classes (the underclass, conditions of poverty, begging, shoplifting, gang crime, prostitution, pornography) are unlikely to be readily applicable to outer space issues due to early restrictions on the type and number of persons able to gain access to space. Historically, access to outer space has been dominated by white, well-educated, American, middle-aged men. In fact, this is the only type of person to have stepped foot on the Moon so far (at the time of writing, November 2023).

This is not to say that victimologists should ignore all prior criminological research and theory relating to the lower classes when considering the application of criminological and victimological theory to outer space. In fact, some theories usually applied to such explanations of crime may actually be applicable to outer space given some careful thought. For instance, anomie and strain theories as initially proposed by the likes of Durkheim and Merton, were originally devised as a way to explain behaviours of people unable to attain goods and services through legitimate means (Zhang, 2019). While humans living and visiting space now and, in the future, may have access to everything they need in terms of shelter and sustenance, they may experience strain through occupying confined spaces for long periods of time. Such strains could include lack of human and familial contact, an inability to access luxuries, or a constant requirement to work or be on high alert. Such isolation and pressure may lead to stress, anxiety, depression and other such mental health disorders, increasing the risk of harm to self and others, possibly increasing the propensity for criminal offending behaviours.

It is also worth stating at this stage that these behaviours often do not represent crimes in the outer space context, particularly when conducted in off-Earth locales. This is because no person, company or nation-state owns outer space, or any of the celestial bodies contained therein. At least in theory.

To understand this situation requires a discussion of pertinent international law. In the 1960s a so-called *Cold War* emerged between the United States (U.S.) and former Soviet Union. As both nations engaged in a race to do certain things in space (i.e. visit and walk on the Moon; image the Moon's far side), the international community was left in a vulnerable position. Only the U.S. and Soviet Union had the money, resources and technical capability to push forward in space exploration. This led to anxieties about the misuse of space for weaponisation, celestial appropriation, and – ultimately – world domination. This apprehension resulted in the drafting (and wide acceptance) of the *1967 Outer Space Treaty*, designed primarily to prevent the national appropriation of celestial bodies. Although no person or nation-state has laid ultimate claim to a celestial body, there have been several interesting legal developments regarding the appropriation of outer space matter and regions of space.

For instance, through domestic law, the U.S., in 2011, passed the *Lunar Legacy Landing Act* which attempted to classify the Apollo 11 lunar landing site as a U.S. National Park (Lampkin and Wyatt, 2022). While this legislation was successfully enacted, it is only binding on people and companies based in the United States where laws are enforced domestically upon U.S. citizens only. While there are strong arguments to protect such regions which do possess important cultural significance, there is nothing legally to prevent another nation driving across the Apollo 11 landing site in a lunar rover, destroying the first footprints and experiments that took place off-Earth.

Another interesting legal consideration comes from the possession and ownership of material taken from extraterrestrial locations. Lampkin and White (2023) make reference to the case of the *United States vs One Lucite Ball Containing Lunar Material*. They recall that:

> Moon rock had been given by the U.S. to Honduras as a gift. The Moon rock was subsequently stolen and eventually sold back to an American resident in Florida. As observed earlier in this book, the 1967 Outer Space Treaty prohibits the appropriation of celestial bodies, but establishing property rights for the returned Moon rock was essential in enabling the court to deem the rock to be 'property' of someone and, therefore, able to be stolen and the case pursued in the criminal courts (DiMaria, 2016). In this case, the Moon rock was determined to be property in the form of a gift to Honduras, and the Florida citizen had to forfeit the item, later returned to Honduras.

Such cases clearly highlight the criminological (and thus victimological) relevance of human–space interaction. They also attest that many space-crime issues have some relationship to the environment. This is perhaps one reason why environmental issues, space environmentalism and astro-green criminological research have preceded other, more conventional, forms of criminological enquiry. It is to these ecological concerns that this chapter will now turn.

Astro-Green Victimology

While traditional notions of victimhood largely centre on human victimisation (Walklate, 2012), radical and critical criminologists have extended the criminological gaze to encompass non-human victims such as animals (Hall, 2014; White, 2018) and natural environments (Williams, 1996). As stated previously, these conceptions have been recently applied to outer space through the development of an *astro-green criminology* – the consideration of the impact of anthropogenic practices on terrestrial and celestial environments. Such research includes matters relating to the pollution generated through launching rockets (Kokkinakis and Drikakis, 2022), the impact of space tourism activities on the Earth environment and atmosphere (Toivonen, 2022), as well as proposals to conduct space mining and extraction activities in off-Earth locales (Anderson et al., 2019).

However, mostly absent from these discussions is consideration of the victims to such harmful activity. This section will consider victimology from the perspective of outer space environs, what might be called an *astro-green victimology.*

In order to delineate different types of astro-green victimology, this section will draw upon the seminal work of Professor Rob White (2008) who discussed different classifications of justice in his quintessential text *Crimes Against Nature: Environmental Criminology and Ecological Justice.* Here, White (2008) suggests the use of three specific terms to differentiate dissimilar types of anthropogenic environmental impact: environmental justice, species justice and ecological justice. Firstly, *environmental justice* relates to human victims, future human generations, and human rights. In short, it encompasses areas of study that concern human victimisation broadly conceived, including social, physical and emotional harm or injury. Applied to outer space, an astro-environmental justice angle would consider the human impact of environmentally destructive state-corporate or otherwise anthropogenic space practices that impact humans directly (physically), or their property, surroundings and environment in which they reside or rely. Secondly, *species justice* concerns the rights of animals and, subsequently, the impact of human activity on non-human species (such as on their ability to survive and thrive, access to shelter, food and water, etc.). Applied to outer space, an astro-species justice approach would consider non-human victims which could include basic microbial life forms existing off-Earth and any animals that we find, or take with us, into space. Finally, *ecological justice* refers to the general health of the overall environment including humans, non-humans, plants and ecosystems. An astro-ecological justice approach would therefore recognise the impact of space-related human actions on the overall health of planet Earth and other celestial bodies including their landscapes, natural terrain and atmospheres.

Considering space victimology from the perspectives of environmental, species and ecological justice enables a broad understanding and remit capable of covering most (if not all) forms of environmental harm pertaining to outer space. Ultimately, however, there is relatively little ecological diversity on the surfaces of nearby celestial bodies. As a result, much space victimisation is likely to be of a human nature. By considering existing, well-established, anthropogenic victimological theories, we may be able to prepare more fully for future physical space exploration activity in the years to come.

Applying 'Old' Theories to 'New' Landscapes

Due to word count restrictions and the vast array of literature that has accumulated in criminology over the past 300 years, it is impossible (and unhelpful) to apply all criminological and victimology theories, perspectives and insights to human interactions with outer space. Indeed, some theories are not relevant and are consequently wholly inapplicable and impractical. For instance, it is difficult to envisage how Lombroso's classic biological theory of offending behaviour and notion of the *born criminal* is in anyway transferable to outer space settings. Similarly, while discussions about the reasons for recidivism following incarceration dominate much criminological discourse, the absence of prisons in outer space means such discussions cannot be applied – not least until Kutner's (1968) radical vision materialises.

There are, however, some traditional criminological and victimological theories that have more relevant outer space applications. For instance, some aspects of strain theory may be used to understand how the pressures of outer space isolation may impact the human psyche and, consequently, offending behaviour. Furthermore, Marxism, feminism and explanations of white-collar crime may help us to understand why only some people (usually white, middle-aged, American men) dominate the statistics of those who have travelled into outer space.

In a similar vein, there are a number of victimological theories and perspectives that could be applied to human activities in outer space. Considering the focus of this chapter on space victimology, it is these theories that the remainder of the chapter will focus on. Table 10.1 gives an outline of how some criminological/victimological theories may be applicable to outer space, with the ones highlighted in bold being discussed in further detail later on in this section.

The overview in Table 10.1 is by no means exhaustive. It does, however, constitute a useful list of orthodox criminological and victimological theories that may be the most easily applicable to understanding victimisation relating to outer space. There are undoubtedly other applicable theories relevant now, and also theories yet to materialise, that could be directly applicable to understanding space victimisation. Due to word count restrictions, the remaining focus will be on just two of the theories identified in the overview (state-corporate crime theory and eco-global criminology), in order to flesh

Table 10.1 The application of criminological and victimological theories in outer space

Criminological/Victimological Theory	*Outer Space Application*
Biological, Biosocial and Psychological Theories	• The impact of a person's biology and surroundings (upbringing) on their offending behaviour. • Psychological theories, in particular, may be applicable to outer space due to exposure to isolation and stress over long periods.
Eco-Global Criminology	• Encourages greater global collaboration and consensus on environmental matters. Because space venturing is limited to just a few countries (mainly the U.S., Russia and China), the rest of the world has little say in how outer space is used or protected.
Life-Style Exposure Theory	• 'Different lifestyles expose people to varying levels of risk for personal victimisation' (Fisher et al., 2016: 37). Space-goers are therefore vulnerable to space-specific crimes and victimisation.
Opportunity Theory	• If there is an opportunity to commit crime, this makes it more likely. If there is an opportunity to commit a crime in outer space, a person may seize that opportunity.
Routine Activity Theory	• In space, there may be a motivated offender, a suitable target, and the absence of a capable guardian.
Social Learning Theory	• Astronauts may learn from each other establishing what is acceptable behaviour in outer space. This could lead to the learning of criminal or deviant behaviours in space societies.
State-Corporate Crime Theory	• As states and corporations are the major investors in outer space, they are the most likely to 'offend', abuse their power, or contribute to the victimisation of people and places.

out some preliminary thoughts and understandings of space crime. However, it must be noted that this chapter is merely a starting point for other scholars to grapple with the theorisation of deviant human actions in, and related to, outer space in the future. Further research is desperately needed if we are to better understand the nuances of space crime and, as a result, prepare for dealing with such behaviour in a way that minimises victimisation, while also protecting and safeguarding victims that bear the brunt of space harms.

State-Corporate Crime Theory

White-collar crimes, also known as crimes of the powerful, or state-corporate crimes were first investigated by the eminent American criminologist Edwin Sutherland. He defined white-collar crime as 'crime committed by a person of respectability and high social status in the course of his occupation' (Sutherland, 1949: 9). While this definition is clearly targeted at business professionals and corporate executives, the same principles of power and occupational offending can be applied in a space context. Specifically, those who currently have the power, money and resources to invest in space exploration activities are some of the World's richest people. For instance, Elon Musk of Tesla and Twitter fame is also CEO of SpaceX, a company primarily concerned with the development of satellite mega-constellations, and also the human colonisation of Mars. Likewise, Amazon founder Jeff Bezos is the CEO of Blue Origin, a space exploration company currently developing private space-stations. Finally, billionaire Richard Branson, CEO of Virgin, has developed a business venture – *Virgin Galactic* – providing space tourism initiatives.

While the Cold War of the 20th Century was largely a political battle between powerful nation-states (the U.S. and former Soviet Union), contemporary space ventures are either state driven, business driven, or a state-corporate collaboration. This means there is more power, money, creativity and determination than ever before. This ambition, however, can ignite greed, corner-cutting, malpractice and, ultimately, criminal and deviant behaviour. For instance, comparisons have already been made between the aerospace and aviation sectors regarding corruption in the designation of high-value contracts. For instance, Lampkin and White (2023) write about the financial penalties (US$3.5 billion) issued to Airbus as a result of institutionalised bribery. The authors quote Boakye et al. (2022: 14) who suggest that 'Airbus was indicted for failure to prevent payment of bribes to secure high value contracts for their Commercial and Defence and Space Divisions in five jurisdictions including Ghana, Sri Lanka, Malaysia, Taiwan and Indonesia.' Unfortunately, Airbus is not alone in engaging in this type of behaviour, which only serves to highlight the possibility for future corruption within outer space development and, consequently, the importance of future research in this area.

Linked to these considerations of white-collar crime is the impact states and corporations have within the aerospace sector and the monopolisation of money and power in the hands of just a few actors. Eco-global criminology is an emerging area of green-critical criminology that advocates for more inclusive conversation and collaboration across national borders to create a greater degree of knowledge-sharing and, ultimately, input and say on ecologically destructive practices. Applied to outer space, eco-global criminology would advocate for a greater distribution of power and wider involvement in aerospace decision-making processes in order to create a more inclusive space sector that addresses the wants and needs of the many, not the few.

Eco-Global Criminology

As White (2022: 118) eloquently puts it, *eco-global criminology* emphasises the merit inherent in 'an enhanced appreciation of the need for collaboration involving individuals, groups and countries from different geographical regions. Such collaborative efforts are vital in regards to research, investigation and exposure of environmental harm and victimisation.' Regarding outer space endeavours, Lampkin and White (2023) explore annual space launch statistics in order to expose the monopolisation of space ventures, which are dominated by just a few countries. They highlight that 562 of 648 (86.73%) total orbital launches between the years of 2018 and 2022 were conducted by just three nations: the U.S., China and Russia (primary participants). Most of the remaining launches were conducted by the European Space Agency (38, or 5.86%), India (22, or 3.4%), Japan (16, or 2.47%) or Iran (7, or 1.08%) (secondary participants). Israel, North Korea and South Korea each had one orbital launch in the study period, collaboratively representing the remaining three launches (0.46%) (tertiary participants) (Lampkin and White, 2023).

It is clear, then, that there is a contemporary dominance of outer space interaction which falls into the hands of just a few, mainly 'northern' nation-states, who have ultimate control over what happens in outer space. Eco-global criminology affirms the inherent advantages in a broader access to space, and wider inclusion in space development discussions, in the best interests of both humanity and extraterrestrial landscapes.

On a positive note, there are signs that activities in space are diversifying. For instance, by the time this book is published, NASA plans to have landed the first woman and first person of colour on the Moon through their Artemis lunar infrastructure development programme (Creech et al., 2022). Furthermore, in 2022, NASA astronaut Nicole Aunapu Mann became the first native American woman to work aboard the International Space Station. Additionally, the International Space Station is, ultimately, an advertisement for the great science that can occur in outer space given collaborative transnational investment and cooperation.

Despite these recent advances, outer space remains dominated by powerful people and nation-states. Without an eco-global effort in considering the outer space environment, extraterrestrial landscapes, orbits, atmospheres and resources will remain vulnerable to exploitation and development by powerful white-collar actors.

Conclusion: Shall We Shine a Light on Space Victimisation?

As Nils Christie (1977: 1) once said, 'maybe we should not have any criminology'. Perhaps the same can be said of space criminology and astro-victimology. After all, crime is a social construct with no ontological reality. It has developed as a way for society to describe, manage and respond to unwanted behaviours of its citizens. The infancy of space law and the absence of jurisdiction and ownership of outer space makes theorising deviancy in space endeavours very difficult.

This chapter, though, has highlighted the possibilities of victimisation in (and related to) outer space. Following the radical and critical criminological traditions, perhaps we should consider the terminology of 'space criminology' more meaningfully. Reflecting on both the absence of law and crime in outer space, perhaps 'space deviance' or 'space zemiology' would be more fitting lexical choices for this area of study, with a focus on *harm* more suitable than *crime*. Such discussions about the place of outer space within the criminological and victimological arenas need to take place.

Finally, it is imperative that criminological discussions about outer space do not remain 'trapped in a realm of (theoretical) ideas' (Schillmoller and Pelizzon, 2013: 14 in Lampkin, 2020: 61). Akin to the precautionary principle of environment law, debates about space crime and victimisation should happen now, at the perceived beginning of mass outer space exploration and exploitation, rather than waiting and responding to harms that have already occurred. Consequently, an applied approach to space criminology will help to bring issues pertaining to astro-victimisation out of the dark and into the light.

References

Amir M (1967) 'Victim Precipitated Forcible Rape', *The Journal of Criminal Law, Criminology, and Police Science*, 58 (4): 493–502.

Anderson SW, Christensen K and LaManna J (2019) 'The Development of Natural Resources in Outer Space', *Journal of Energy & Natural Resources Law*, 37 (2): 227–258.

Barnes Jr WL (1998) 'Revenge on Utilitarianism: Renouncing a Comprehensive Economic Theory of Crime and Punishment', *Indiana Law Journal*, 74: 627–651.

Boakye D, Siaw D and Sarpong D (2022) 'The Airbus Bribery Scandal: A Collective Myopia Perspective', *European Management Review* (2022): 1–17.

Christie N (1977) 'Conflicts as Property', *British Journal of Criminology*, 17 (1): 1–15.

Creech S, Guidi J and Elburn D (2022) *Artemis: An Overview of NASA's Activities to Return Humans to the Moon*. [Online]. Available at: https://ntrs.nasa.gov/api/citations/20210026673/downloads/2022%20IEEE%20Aero%20Creech%20Artemis%20Overview%20V3.pdf (Accessed: 13th July, 2023).

DiMaria S (2016) 'Starships and Enterprise: Private Spaceflight Companies' Property Rights and the US Commercial Space Launch Competitiveness Act', *St. John's Law Review*, 90 (2): 415–440.

Eski Y (2023) *A Criminology of the Human Species*. London: Palgrave.

Fisher BS, Reyns BW and Sloan III JJ (2016) *Introduction to Victimology*. Oxford: Oxford University Press.

Giddens A (1997) *Sociology*. Lisbon: Fundagio.

Gorman A (2019) *Dr Space Junk vs the Universe: Archaeology and the Future*. Cambridge: The MIT Press.

Hall M (2014) 'Environmental Harm and Environmental Victims: Scoping Out a "Green Victimology"', *International Review of Victimology*, 20 (1): 129–143.

Hermida J (2006) 'Crimes in Space: A Legal and Criminological Approach to Criminal Acts in Outer Space', *Annals of Air and Space Law* (XXXI): 1–19.
Kokkinakis IW and Drikakis D (2022) 'Atmospheric Pollution from Rockets', *Physics of Fluids*, 34 (5): 1–13.
Kutner L (1968) 'A World Outer Space Prison', *Denver Law Review*, 45 (5): 702–718.
Lampkin JA (2020) *Uniting Green Criminology and Earth Jurisprudence*. Oxon: Routledge.
Lampkin JA (2021) 'Mapping the Terrain of an Astro-Green Criminology: A Case for Extending the Green Criminological Lens Outside of Planet Earth', *Astropolitics: The International Journal of Space Politics and Policy*, 18 (3): 238–259.
Lampkin JA and White R (2023) *Space Criminology: Analysing Human Relationships*. London: Palgrave.
Lampkin JA and Wyatt T (2022) 'Widening the Scope of "Earth" Jurisprudence and 'Green' Criminology? Towards Preserving Extra-Terrestrial Heritage Sites on Celestial Bodies'. In: Gacek J and Jochelson R (eds) *Green Criminology and the Law*. London: Palgrave Macmillan.
Mendelsohn B (1974) 'The Origin of the Doctrine of Victimology.' In: Drapkin I and Viano E (eds) *Victimology*. Lexington: D.C. Heath and Company, pp. 3–11.
Rothe DL and Collins VE (2023) 'Planetary Geopolitics, Space Weaponization and Environmental Harms', *The British Journal of Criminology*, p.azad003.
Sallum Jr B (2005) 'The Future of Social Sciences: Sociology in Question', *Sociologia, Problemas e Praticas*, 48: 19–26.
Schillmoller A and Pelizzon A (2013) 'Mapping the Terrain of Earth Jurisprudence: Landscape, Thresholds and Horizons', *Environment and Earth Law Journal*, 3: 1–32.
Serpa S and Ferreira CM (2019) 'Micro, Meso and Macro Levels of Social Analysis', *International Journal of Social Science Studies*, 7 (3): 120–124.
Sutherland EH (1949) *White Collar Crime*. Hinsdale, IL: Dryden Press.
Takemura N (2019) 'Astro-Green Criminology: A New Perspective Against Space Capitalism', *Toin University of Yokohama Research Bulletin*, 40: 7–16.
Toivonen A (2022) 'Sustainability Dimensions in Space Tourism: The Case of Finland', *Journal of Sustainable Tourism*, 30 (9): 2223–2239.
von Hentig H (1948) *The Criminal and His Victim*. United States: Yale University Press.
Walklate S (ed.) (2012) *Handbook of Victims and Victimology*. Oxon: Routledge.
White R (2008) *Crimes Against Nature: Environmental Criminology and Ecological Justice*. Oxon: Routledge.
White R (2018) 'Green Victimology and Non-Human Victims', *International Review of Victimology*, 24 (2): 239–255.
White R (2022) *Theorising Green Criminology: Selected Essays*. Oxon: Routledge.
Williams C (1996) 'An Environmental Victimology', *Social Justice*, 23 (4): 16–40.
Wolfgang ME (1957) 'Victim-Precipitated Criminal Homicide', *Journal of Criminal Law, Criminology, and Police Studies*, 48: 1–11.
Zhang J (2019) 'The Strain Theory of Suicide,' *Journal of Pacific Rim Psychology*, 13: 1–15.

8 Connecting the Analogue Dots

Insights into the Future of Space Crime, Criminal Justice, and Ethics Now

Rebecca Kaiser

Guardians of the Ethical Galaxy: Space Analogues' Role

The exploration of outer space poses unique challenges, not only in the realms of science and technology but also in the understanding of how ethics, criminality and justice will function in such an environment. Drawing from a growing body of research, this chapter delves into the altered social interactions and behavioural variances experienced by individuals in the isolated, confined, and extreme (ICE) environments that mimic conditions in space, such as Antarctic research stations. This chapter highlights the implications of discoveries from analogues for the future of spacefaring and explores how to answer the pressing questions of what types of crimes may occur in space, how they may be reacted to and dealt with, and the judicial and ethical considerations that space analogues can identify for future consideration.

Space analogues have provided insights into a multitude of areas such as psychology, sociology, and mental health that have assisted in providing a safer and more beneficial environment for those currently residing in space. Alongside the human factors research, areas such as environmental science, resource management and medical research have all benefited from research conducted in space analogue environments (Azua-Bustos et al. 2022; van Ombergen et al. 2021; Yan & England 2001).

By examining the research conducted in these analogue environments, this chapter underscores the relevance of such studies in shaping the future of space exploration, our understanding of crime, criminal justice, and ethics in outer space, highlighting the wealth of questions that still need to be asked before humanity can venture to the stars safely. It will also further emphasise the crucial nature of acknowledging the parallels between space analogues, the unique behaviours within them, and the challenges that may be encountered in genuine space missions.

As we step into a new era of space travel and look towards its future, the insights gained from the space analogues that exist in the present will be instrumental in shaping the policies, protocols and procedures, and training strategies that will provide a safe, positive, and fruitful experience for

DOI: 10.4324/9781003437178-9

spacefaring teams. This chapter will serve as a guide to the exciting world of space analogues and the invaluable resource they represent for the future of humanity's journey to the stars.

Grounded in Reality: Space Analogues as Crucial Proving Grounds

Space analogues are terrestrial settings that are either specifically designed to imitate the conditions experienced in space, such as specifically built module habitats or those that are more naturally occurring such as Antarctica. These space-like environs have provided invaluable testing grounds for studying the physical, psychological, and social aspects of space habitation and exploration. These analogues enable us to bridge the gap between theoretical possibility and practical reality when it comes to life beyond Earth. To understand the applicability of analogues to space habitation, I will explore the similarities between the two that allow researchers to have confidence in the fidelity of their results.

An 'isolated' environment (in the circumstances considered here) is one that is generally far removed from an inhabitant's normal home and regular support system (Kaiser & White 2022). Such an environment is also generally restricted to a remote/isolated geographical location, restricting its inhabitants from easily entering or exiting the location, even in the event of an emergency (Kaiser & White 2022). The 'confined' aspect of an ICE is often due to the need to limit oneself to an artificial structure as a result of the nature of the surrounds and/or circumstances. The 'extreme' part of an isolated, confined, and extreme environment is often defined as such by the lengths a human must go to, to be able to survive within it for any length of time (Suedfeld 2012). It is often the extreme environment that prompts the need for physical confinement. Examples of ICE environments that are easily recognisable are Antarctic research stations and spacefaring environments like the International Space Station (ISS). Both environments would prove fatal to anyone attempting to inhabit them without extensive training and technological support such as sealed habitation modules and artificial climate control.

These shared unique circumstances lead those who work within them to display similarly shared responses to their environment. These range from involuntary physiological responses to adaptive social and behavioural reactions (Leon et al. 2011). Newer research, such as that of Kaiser and White (2022) suggests that these differences extend beyond the current understanding, leading to behavioural adaptation's that have only been marginally studied. It is these adaptations that indicate that the study of crime, criminal justice, and ethics in these environments is vital to devising strategies that will aid in maintaining a safe and achievably harmonious situation aboard a vessel that is engaged in long-term space travel.

Of ICE environments, Antarctica has seen the lion's share of ICE environment behavioural research prior to the last decade (Heinicke & Arnhof 2021). Polar conditions are seen as similar enough to the conditions in space to be viewed as a pre-ready, viable and high-fidelity analogue. Other ICE environments and situations that have hosted space-applicable research include saturations diving, military settings, caves habitation, desert habitats, and remote island studies (Heinicke & Arnhof 2021; Wang et al. 2020)

Arctic and Antarctic stations – The polar regions are a natural analogue for outer space conditions. Researchers in this area have uncovered how ICE conditions influence many areas of human behaviour such as group dynamics, decision-making and the consequences of the unique behavioural variations that have been found to occur (Marcinkowski, Bell, & Roma 2021; Corneliussen 2017; Kaiser & White 2022; Leon et al. 2011; Marcinkowski et al. 2021; Palinkas & Suedfeld 2021). Due to the varied and extensive research carried out in the Arctic and Antarctic zones, viable structures to support human life already exist. This is a more cost-effective and feasible avenue for many researchers who want to conduct research that is applicable to spacefaring over attempting to secure the extensive resources needed for the process of building and establishing a functional artificial environment dedicated to a single research goal.

Desert analogues – Desert regions on Earth, such as the Atacama Desert in Chile, serve as valuable analogues for the arid and extreme environments of Mars. These areas share similarities in terms of low precipitation, extreme temperatures, and challenging terrain. Researchers conduct field tests and experiments in these deserts to understand the potential conditions of future Martian missions (Azua-Bustos et al. 2022).

Underwater habitats – Underwater habitats like the Aquarius Reef Base in Florida and the NEEMO (NASA Extreme Environment Mission Operations) project provide a unique environment for simulating weightlessness, isolation, and confinement. Aquanauts live and work in these underwater habitats for extended periods, conducting experiments and testing equipment in preparation for space missions.

Isolated islands – Remote islands, like Devon Island in the Canadian Arctic and Hawaii's Mauna Loa, are used to simulate isolated and extreme environments. Researchers conduct Mars analogue missions on Devon Island, while the HI-SEAS (Hawaii Space Exploration Analog and Simulation) project on Mauna Loa focuses on long-duration isolation and confinement studies (Kizzia 2015).

Closed ecosystems – Closed ecosystems, such as the Biosphere 2 facility in Arizona, mimic the life support systems of spacecraft. They are designed to study self-sustaining environments and test regenerative technologies for long-duration space missions.

Spacecraft simulations – Space agencies like NASA and the ESA operate spacecraft simulations on Earth. These facilities replicate the interior of spacecraft, allowing astronauts to train for specific missions and practice working in the confined quarters of a spacecraft.

Mars Analogue Dedicated Missions – Artificial environments like the Mars Desert Research Station (MDRS) and the Mars Society's Mars Arctic 365 mission, simulate life on Mars. Crew members live in Mars-like habitats, wear space suits, and conduct experiments to prepare for future Martian missions.

Space analogues provide a unique bridge between the theoretical understanding of space exploration and the practical reality of life beyond Earth. Findings from research (Kaiser & White 2022; Kizzia 2015) in these analogue environments are instrumental in shaping our approach to space exploration, criminal justice, and ethics and highlight the importance of applying lessons from analogue habitats to future lunar and Martian habitats. The insights garnered from analogue research can help us anticipate challenges, develop effective policies, and devise innovative solutions for maintaining order and harmony during space missions.

Icy Discoveries: How ICE Environments Shape Us

The study of human behaviour in ICE environments has a long and well-established history. From as early as the 1950s scientists have looked to Antarctica as the ideal place to conduct research for the betterment of our spacefaring future (Jessup & Taubenfeld 1959; van Ommbergen et al. 2020).

More recent investigation has broadened its scope to encompass psychosocial research from the Arctic, the Antarctic and space stations (Yan & England 2001). Such research has shown that the interpersonal tension caused by being in an isolated and confined environment to increase as stress increases, significantly and continuously (Palinkas 2001). It is not uncommon to see studies on the effects of polar habitats and space stations and their necessary isolated and confined environments making mention of social monotony and increased levels of anger, irritability, depression, and insomnia as a result (Palinkas 2001; Yan & England 2001). It is difficult to find new or novel social opportunities on an Antarctic station during a winter-over, it is exceptionally more difficult still to find them while on a space station.

In ICE environments, it is highly challenging to avoid trespassing on another's personal space. Past research has identified that crowding and violations of territory or personal space are significant factors in environmentally triggered aggression (Palinkas 2001). There is a well-established but complex link between increased social strain and stressors, criminal behaviour, and harm that is the underlying theme of many criminological and sociological theories (White & Perrone 2015).

It is when these transgressions are triggered or facilitated by stressors stemming from the environment of the perpetrator that we can see there is the opportunity to apply criminological theory to extreme and unusual environments. In an isolated and confined environment such as a space station, these strains are impossible to avoid (Palinkas 2001).

Understanding how community bonds form and influence behaviours is also crucial for the success of space missions, where self-reliance, teamwork, and resilience are vital. As referenced in studies by Bell et al. (2019), this dynamic can inform mission planning, crew selection, and leadership strategies in outer space. Research conducted in ICE environments also notes the rapid and deep establishment of community bonds (Kaiser & White 2022; Stuster et al. 2000). However, this sense of tight-knit group identity is a double-edged sword.

This is because a sense of team identity can also lead to the dismissal of any authority outside of the immediate group, such as mission control back on Earth. Historic instances, like the 'rebellion' on an Australian Antarctic station in the 1990s (Spielmann 1996), provide critical insights into the dynamics that lead to such conflicts and the importance of balancing community autonomy with external oversight. What is seen as an infamous example of the rejection of external authority is the 'mutiny' that occurred aboard the Skylab 4 (Palinkas & Suedfeld 2021). Contrary to popular media portrayal, the mutinous actions of a difficult team aboard the Skylab 4 was, in fact, an exhausted crew stating that they needed a day of rest after 40 days of continuous work, with three of those days intended to have been rest days that the crew had been instructed to work through by mission control. Acknowledging this conflicting perception of circumstances and its results later enabled what came to be known as the 'first sensitivity session' in space that resulted in implementing programs and work changes that led to the so-called mutinous crew becoming one of the most productive Skylab crews to venture forth into space.

Space analogues have highlighted other communication challenges between communities inhabiting the ICE environment and those supervising from a distant base as was seen in the Skylab 4 incident. Discord and conflict between these two groups are not uncommon, and lessons learned from analogues underscore the importance of effective and efficient communication protocols to address disputes and maintain overall mission success.

Future communication during long distance space missions, such as to Mars, will likely experience communications delays of 20 minutes or more per direction. This makes research on the impact this delay could have to be of considerable importance (Hagemann et al. 2023). A common finding of research in this area is that the more significant the time delay between crew members and their support teams, the more conflict and frustration developed between the two groups (Basner et al. 2014). This research also found that increased crew autonomy and improvements in communication training helped mitigate some of the adverse effects of communication delays.

Supporting these findings from ICE analogues, research conducted on three astronauts aboard the ISS specifically measuring the effect of communication delays on task performance found that crew well-being was significantly reduced in communication delayed tasks compared to control tasks (Kintz et al. 2016). This study provides a high-fidelity result that demonstrates the connection between previous analogue findings and actual situation results regarding both the association between communication delay, stress, and frustration, and the benefit of heightened crew autonomy.

Crimes Beyond Earth: Space Analogues – The First Insight

A truly intriguing and complex aspect of future space exploration concerns the potential for criminal activity (Lampkin & White 2023). Considering this, questions arise as to the exact nature of these potential crimes in space. At present, we have only a single, documented, case of an alleged crime being committed in space. Dubbed the 'first crime in space' (Baker 2019), Astronaut Anne McClain was aboard the ISS when she accessed the personal bank account of her estranged spouse, Summer Worden. This resulted in a complaint from Worden to both the Federal Trade Commission and NASA's Office of Inspector General, accusing McClain of a number of crimes linked to the accessing of her account. While the subsequent investigation cleared McClain of wrongdoing, this was the first instance of alleged criminal activity to have taken place on the International Space Station (Williams 2021).

Further media articles have also provided insight into some of the behaviours that take place in the ICE environments frequently used as space analogues. The analogue space program HI-SEAS, conducted in the Hawaiian lava fields of Mauna Loa volcano, has seen the occurrences of instances of small rule-breaking such as scheduling an extravehicular activity EVA just to get outside, or arguments over dessert (Kizzia 2015). These are the sort of instances frequently overlooked during scientific study as being inconsequential to the research taking place. Such instances, however, indicate an area of behaviour whose investigation should be paramount when looking towards a future where a group of people are restricted to a similar environment, facing similar and further still extreme situations.

Reports on prior analogues have also conveyed instances of more serious crimes and conflicts. An analogue study simulating the conditions of a space station conducted in Moscow in 1999 resulted in a bloody fistfight between two male participants, with a female participant being forcefully kissed by the crew commander (Kizzia 2015).

If the inhabitants of artificial ICE environments, whose danger is only simulated, are driven to conflict and rule-breaking due to their environmental circumstances, we must ask how those on a long-term space mission will likewise act out and, as a result, what the consequences of such behaviour might be. To begin to find answers to questions such as these, we can look to environments such as Antarctica.

Antarctica has been host to a great deal of human inhabitation and research for over 50 years and so presents many of the instances that we can draw upon to demonstrate the significance of the need to recognise and invest in this area of research.

An incident that took place in 2018 involving a stabbing between Russian Antarctic expeditioners Oleg Beloguzov and Sergey Savaitsky, starkly brings to light the potential for serious harm between ICE inhabitants. Reportedly, Beloguzov stabbed Savaitsky over the latter's continued 'spoiling' of the endings of the books that Beloguzov had taken from the station library

(Rousseau 2016). The two men had reportedly been in constant conflict in the months leading up to the incident, but the nature and instances of this previous conflict were not reported. This incident is not the first time that Antarctic expeditioners have resorted to violence to settle their grudges. The story of the 'chess murder' is notorious amongst Antarctic expeditioners. The murder is said to have taken place on the Vostok station in 1959 (Bennett 2016). Allegedly the loser of a chess game was upset enough over the games result to fatally attack the winner with an ice axe, leading to the subsequent ban on chess in Russian Antarctic stations. While never explicitly confirmed by Russian authorities, this alleged incident can nevertheless illustrate the plausible consequences of conflict in such ICE environments.

Alongside bodily harm and murder, the past has shown that arson is also not outside the realm of plausible consequences. After being ordered to spend a second winter in Antarctica in 1984, the leader and medical doctor of an Argentinian Station known as Brown station, found he could not cope with the prospect and was reported to have waited until the last ship was due to depart for Argentina before setting the entire station alight, forcing the whole crew to return on the departing ship (Roe 2022).

These three cases sit alongside other instances of physical altercation and harm. Such behaviours include assault by a station cook with a hammer, indecent exposure, the rebellion of 15 station staff that required a mediator to be flown to an Australian station during the dead of winter (Rousseau 2016), animal abuse, and environmental damage (Joyner & Chopra 1988). Consequently, it is clear that many varied harms can permeate from ICE environments. As they have occurred in Earth-based settlements and research stations, it is likely that they will also occur in isolated and confined environments off-Earth.

The only unexplained death in Antarctica's history, the passing of Australian astrophysicist Rodney Marks from acute methanol poisoning in May 2000, falls into the void of potential crime. This incident could well be a case of icy murder, but the uncooperative nature of the different agencies at play in the investigation ensured that a true conclusion may never be reached. This case shows us that it is not just individual people who are capable of transgressive behaviour in ICE environments, and so are not the only aspect of life in space that will need to be explored thoroughly before mission commencement. The case of Rodney Marks is a cautionary tale for the handling of any future instances of this nature that may occur and a stark reality to consider when planning policy and procedure for future long-term space inhabitation (Debczak 2019).

In a book written about his time on the southern ice, Nicholas Johnson (2005) retells the varied instances of crime, conflict, deviance, and transgression he witnessed at both expeditioner and corporate level, during his multiple years in Antarctica. This autobiography demonstrates that those in control of the daily aspects of station life in Antarctica can be severely damaging to those within it. The book points out that those who have a high degree of impact on the lives of expeditioners, and so can harm them greatly, are not necessarily just the expeditioners themselves.

Private companies such as those Johnson mentions in his book, have the potential to make an expeditioner's life truly miserable. Many of these criminal instances mentioned occurred when an expeditioner reached the end of their ability to cope and found themselves unable to escape the machinations of the pervading bureaucracy that ruled their time on the ice (Johnson 2005).

Another distinct aspect of life in isolated environments, as demonstrated by Johnson's book and research based in Antarctic stations, is the alteration of sexual norms. Behaviours that might be considered deviant or transgressive in broader society are often seen as a normal part of life in an ICE environment (Kaiser & White 2022; Nash & Neilsen 2020). The topic of sex and intimacy in such situations is complex, as are the dynamics of evolving relationships in isolated environments. The research conducted in analogues (Kaiser & White 2022), including studies on altered sexual norms and relationships, prompts questions that were previously considered contentious or irrelevant. Addressing sensitive subjects related to sexual and intimate relationships and establishing ethical norms and policies can help prevent crew conflicts and relationship breakdowns during extended space missions. Understanding these behavioural variations is essential to fostering harmonious and productive spacefaring communities where the potential for conflict and relationship breakdown can be minimised. These insights will challenge conventional beliefs about sexual and intimate relationship norms, and policies will need to be carefully considered for the context of space habitation and exploration.

The presence of alcohol in isolated environments has been a subject of significant interest. Since the early explorers set foot on Antarctica, alcohol has been a part of the culture in these remote and isolated locations. However, research (Kaiser & White 2022) has revealed that the presence of alcohol is linked to problematic behaviours in analogue environments thus presenting a dual challenge. On one side of the equation lie the potential risks associated with alcohol consumption in space environments, where the consequences of misconduct can be dire. This consideration is balanced against the varied cultural importance of alcohol as a social lubricant and a means of relaxation in isolated environments. Evaluating these considerations will be crucial in shaping the cultural norms and policies for future space missions.

Criminal Justice: Bridging the Ice and the Void

Past experiences in space analogues raise questions about how justice and order will be maintained in outer space. The development of legal frameworks for space travel are a pressing concern, and the insights gained from analogues may help shape these frameworks. Space missions, whether in space analogues or actual space travel, will require a robust framework for dealing with crimes and criminal behaviour.

Due to the small amount of research in this area, we have limited examples of the application of criminal justice in ICE environments. Research conducted with Antarctic expeditioners has shown that self-policing behaviours can evolve in response to the surroundings. Expeditioners studied by Kaiser and White (2022) were found to apply unique punitive actions that were distinctively suited to the environment. Isolating the offender or 'othering' them were found to be common responses to breaches of established behavioural norms.

Othering is the process where those who are seen to be the law-abiding, respectable, majority of a group or community, differentiate some of its members as being less, different, or somehow incompatible (White, Perrone, & Howes 2019). The result of this in an Antarctic base leaves the expeditioner treated as an outsider. This research by White, Perrone, & Howes (2019) found this process happened not just as a punishment for transgressive behaviour, but also for more benign disagreements such as contention over the best sporting team.

The second disciplinary technique applied to those found to act beyond the established behavioural code was to isolate such individuals from the 'main group'. Physical isolation as punishment was found to be enforced at varying levels of severity. At times, the offender was isolated from other people or areas, or restricted to a certain area. This physical distancing and isolation behaviour was found to be implemented from a group standpoint (Kaiser & White 2022). Where needed, the offender was instead isolated from the group by the actions of the group themselves.

This system of punishment and deterrence is uniquely effective in this situation. Being isolated, both physically and socially, deprives the offender of the sense of safety that they would otherwise experience as a member of the community.

Such self-policing behaviour is a clear example of community-level enforcement that may be replicated in space. Isolation from Earth's judicial influences may necessitate innovative approaches to maintaining order and resolving conflicts in outer space settings, and exploring justice or mediation possibilities prior to their requirement is vital. Understanding how these mechanisms operate in space analogues provides valuable insights into the potential dynamics of future space communities. This also links to the value of considering ethics in the space context, to which this chapter will now turn.

Pioneering Ethics: How Space Analogues Can Plot the Path to a Moral Cosmos

An underestimated but highly significant offering of space analogues is the role they could play in building ethical guidelines for future space missions. As we consider the implications of altered social norms, relationships, and the potential for crime in space, it becomes evident that the development of a clear ethical framework is essential. Analogues offer a controlled environment for studying these issues, helping us create ethical guidelines that will govern behaviour and relationships in space.

The HI-SEAS missions in Hawaii have provided situations that show examples of the potential space analogue environments present for the development of ethical frameworks while raising questions for future consideration. During an early iteration of the HI-SEAS missions, one crew member required evacuation due to a medical issue (Koren 2018). This situation ran the risk of breaking the simulation. After much contemplation, the team decided to pretend the crew member had died and so would be buried in the Martian sand where they would not decay, in the hopes of bringing the body back to Earth later for proper burial. The crew acted out the entire scenario, allowing the crew member to leave for the medical care they required without breaking the rules of Martian survival. Another crew encountered internal conflict when it came to dealing with the appropriate response to an injured crew member, with clashing priorities causing dissention and a protracted time elapsing before a solution was agreed upon. These unique situations have been handled in unorthodox ways that provide insight as to how such future problems might be considered in situations such as long-term space travel, like the loss of a crew member or the serious injury of one (Koren 2018).

The unique adaptations that have been seen to occur in ICE environments indicate that the scientific community needs to engage in a reconsideration of approach when thinking about how the current theories and structures of crime, criminal justice, and ethics will fit into the future conditions of humanity's space travels.

Conducting research applicable to outer space often involves international collaboration, reflecting the global nature of space exploration. Experiences acquired in these analogues can inform discussions on international agreements, protocols, and norms for outer space activities. Ensuring that ethical, legal, and behavioural standards are consistent across nations is crucial for maintaining peace and cooperation in space. The knowledge shared and generated in these analogues contributes to discussions on international space policies, including ethical and legal considerations. The establishment of universally accepted norms and principles is essential for the peaceful and responsible use of outer space.

Space Analogues Today Shape Our Interstellar Tomorrow: Charting the Path Forward

As I have established thus far, space analogues can provide valuable insights into the ethical and legal challenges of space exploration. As we contemplate the unique dynamics of isolated and confined space missions, analogues provide case studies for drafting space-specific codes of conduct and regulations. This is especially important as we expand our presence beyond Earth's orbit and encounter diverse cultural perspectives on ethics and justice. The lessons from terrestrial and space analogues can guide the development of behavioural guidelines for space travellers. As we prepare for extended missions to the Moon and Mars, guidelines become instrumental in maintaining crew cohesion, promoting mental well-being, and preventing conflict. Analogues assist in creating protocols for addressing interpersonal issues and criminal or unethical behaviour in space. Space policies and regulations must address the risks associated with space exploration.

Analogues provide valuable data on potential risks related to human behaviour in confined and isolated settings (Kaiser & White 2022). These insights enable the development of risk assessment and mitigation plans to ensure the safety and success of space missions. The problem-solving skills developed in space analogues are crucial for addressing unexpected challenges during space missions. Analogues present participants with scenarios that require creative thinking, adaptability and resourcefulness, preparing them for the dynamic nature of space exploration.

In our quest to explore space, humanity will be met with a myriad of challenges, some of which will be unexpected and unforeseen. By utilising the opportunities for research that space analogues present us, issues regarding crime, criminal justice, and ethics need not be prominent amongst those unexpected trials. Space analogues are critical in shaping the future of space exploration, the knowledge gained from these analogues influences the development of ethical guidelines, technological advancements, and the preparation for potential challenges in outer space.

Space analogues, by simulating the isolation, confinement, and extreme conditions of spacefaring, provide us with invaluable insights into human behaviour and adaptation that could only be otherwise theorised on. Conducting research in space analogue environments will provide us with a deeper understanding of what crimes may be committed in space, the prospective complexities of criminal justice in such an environment, and the ethical consideration that will be needed to guide our extraterrestrial goals.

Conclusion

This chapter traversed the Earth, from pole to pole, to outline the fascinating realm of space analogues and the unparalleled opportunities that they offer for research and insight in the areas of crime, criminal justice, and ethics for outer space. It has brought into the light the crucial importance of adapting our understanding and approach to these areas of research in view of the unique nature of the space environment.

It is through these analogues that we can conduct research into crime, criminal justice, and ethics in outer space not simply as a theoretical exercise but instead in a reliable testing ground that will provide tangible findings which are relevant and applicable to our future in outer space.

As the citizens of Earth embark on their next great step into the cosmos, the lessons learned from analogue research will prove instrumental to their safety and success. Insights gained from space analogues will inform policies and protocols that are characterised by fairness, integrity, and shaped to the realities of space conditions. They will provide the groundwork for future judicial structures and ensure that when the people of Earth reach the stars, they have taken with them an ethical ethos that humanity can be proud of.

References

Azua-Bustos A, González-Silva C & Fairén AG (2022) The Atacama Desert in Northern Chile as an analog model of Mars. *Frontiers in Astronomy and Space Sciences* 8.

Baker M (2019) NASA astronaut Anne McClain accused by spouse of crime in space. From https://www.nytimes.com/2019/08/23/us/astronaut-space-investigation.html.

Basner M et al. (2014) Psychological and behavioral changes during confinement in a 520-day simulated interplanetary mission to Mars, *PLoS ONE* 9(3): 89.

Bell S, Brown S & Mitchell T (2019) What we know about team dynamics for long-distance space missions: A systematic review of analog research. *Frontiers in Psychology* 10: 811.

Bennett J (2016) How Antarctic isolation affects the mind. From https://canadiangeographic.ca/articles/how-antarctic-isolation-affects-the-mind/.

Corneliussen JG et al. (2017) Individual traits, personal values, and conflict resolution in an isolated, confined, extreme environment. *Aerospace Medicine and Human Performance* 88(6): 535–543.

Debczak M (2019) Death at the South Pole: The mystery of Antarctica's unsolved poisoning case. From https://www.mentalfloss.com/article/579732/mysterious-death-rodney-marks-scientist-who-was-poisoned-antarctica.

Hagemann V et al. (2023) Communication quality affects performance of astronauts and support teams through increased workload: Insights from the AMADEE-20 analog Mars mission. *Acta Astronautica*210: 162–175.

Heinicke C & Arnhof M (2021) A review of existing analog habitats and lessons for future lunar and Martian habitats. *REACH*: 21–22.

Jessup PC & Taubenfeld HJ (1959) *Controls for Outer Space and the Antarctic Analogy*, Columbia University Studies in International Organization: no. 1, Columbia University Press, New York.

Johnson N (2005) *Big Dead Place: Inside the Strange and Menacing World of Antarctica*. Feral House.

Joyner CC & Chopra SK (1988) *The Antarctic Legal Regime*. Martinus Nijhoff Publishers.

Kaiser R & White R (2022) What happens on the ice, stays on the ice: Transgression and layered silence in Antarctica. *Current Issues in Criminal Justice* 34(3): 239–254.

Kintz NM et al (2016) Impact of communication delays to and from the International Space Station on self-reported individual and team behavior and performance: A mixed-methods study. *Acta Astronautica* 129: 193–200.

Kizzia T (2015) Moving to Mars. From https://www.newyorker.com/magazine/2015/04/20/moving-to-mars.

Koren M (2018) When a Mars simulation goes wrong. From https://www.theatlantic.com/science/archive/2018/06/mars-simulation-hi-seas-nasa-hawaii/553532/.

Lampkin JA & White R (2023) *Space Criminology: Analysing Human Relationships with Outer Space*. London: Palgrave.

Leon GR, Sandal GM & Larsen E (2011) Human performance in polar environments. *Journal of Environmental Psychology* 31(4): 353–360.

Marcinkowski MA, Bell ST & Roma PG (2021) The nature of conflict for teams in isolated, confined, and extreme environments. *Acta Astronautica* 181: 81–91.

Nash M & Nielsen H (2020) Gendered power relations and sexual harassment in Antarctic science in the age of #MeToo. *Australian Feminist Studies* 35(105): 261–276.

Palinkas AL (2001) Psychosocial issues in long-term space flight: Overview. *Gravitational and Space Biology Bulletin* 14(2): 25–33.

Palinkas LA & Suedfeld P (2021) Psychosocial issues in isolated and confined extreme environments. *Neuroscience & Biobehavioral Reviews* 126: 413–429.

Roe J (2022) True crime in Antarctica. The history of illegal activities on the ice continent. From https://mpora.com/outsiders/history-of-crime-in-antarctica/.

Rousseau B (2016) Cold cases: Crime and punishment in Antarctica. From https://www.nytimes.com/2016/09/29/world/what-in-the-world/antarctica-crime.html.

Spielmann P (1996) FBI agents to visit Antarctica in rare investigation of assault. From https://www.spokesman.com/stories/1996/oct/14/fbi-agents-to-visit-antarctica-in-rare/.

Stuster J, Bachelard C & Suedfeld P (2000) The relative importance of behavioral issues during long-duration ICE missions. *Aviation, Space, and Environmental Medicine* 71(9): 17–25.

Suedfeld P (2012) Extreme and unusual environments: Challenges and responses. In SD Clayton (ed.), *The Oxford Handbook of Environmental and Conservation Psychology*, Oxford University Press.

van Ombergen A, Rossiter A & Ngo-Anh TJ (2021) "White Mars" – nearly two decades of biomedical research at the Antarctic Concordia station. *Experimental Physiology* 106(1): 6.

Wang Z et al. (2020) Sleep quality and its impacts on quality of life among military personnel in remote frontier areas and extreme cold environments. *Health and Quality of Life Outcomes* 18(1): 1–10.

Williams K (2021) Space crime continuum: Discussing implications of the first crime in space. *Boston University International Law Journal* 39: 79.

White R & Perrone S (2015) *Crime, Criminality and Criminal Justice*, 2nd edn, Melbourne:, Oxford University Press.

White R, Perrone S & Howes L (2019) *Crime, Criminality and Criminal Justice*, 3rd edn, Docklands: Oxford University Press.

Yan XW & England ME (2001) Design evaluation of an Arctic Research Station: From a user perspective. *Environment and Behaviour* 33: 449–470.

9 "If There Was an Observer on Mars, They Would Probably Be Amazed That We Have Survived This Long"[1]

Environmental Decline, Elite Escapes, and Space Colonies

Anita Lam, Nigel South and Avi Brisman

Introduction

Although defining (environmental) harm in outer space is a matter of increasingly pressing concern, there is a risk that adopting a narrowly myopic focus, bent on adding "new" sites of criminal activity and definitions of harm, will obscure the connections between the so-called "space race 2.0" and the activities of the richest 1% on Earth. This chapter takes a green cultural criminology approach to argue that "space criminology" should not be considered an entirely "new" field of exploration and builds on our past work on elite escapes, bunkerization, and ocean cruising, as well as popular fiction and other academic works that have identified space as the next location for human activities related to resource extraction and elite consumption. We begin with an overview of "space criminology" and its attendant "astro-green criminology", situating them in relationship to green criminology, in particular, and green cultural criminology, more specifically. From here, we turn to a discussion of climate breakdown, the "polluter elite", and bunkerization, before contemplating outer space-as-frontier. We conclude by emphasising the need to focus on terrestrial anthropogenic environmental issues as we extend our criminological gaze to the heavens.

Space Criminology and Green Cultural Criminology

While acknowledging that space criminology has no "history of the discipline" and no antecedent criminological studies from which to draw, Lampkin and White (2023), nevertheless, provide an outline of space criminology's initial configurations. These contours map insights from existing criminological theories and methods onto various forms of space crime, raising questions about how to define "harm" and determine legal jurisdiction(s) in outer space (see Sachdeva 2023; see also Eski 2023), as well as assign responsibilities related to policing, regulation, and governance (e.g., Mills 2011; see also Fountain 2003). Although Lampkin and White (2023: 168) concede that space criminology is

DOI: 10.4324/9781003437178-10

not a "new" criminology per se, they still consider outer space as a "new topic area worthy of concerted and systematic attention by criminologists and those interested in matters of harm, morality and justice". In particular, green criminologists have been identified as particularly well-suited to studying environmental harms in outer space. For example, Rothe and Collins (2023), well-known for their scholarship on crimes of the powerful and aiming to contribute to research in green criminology, offer a critical examination of some anticipated environmental harms of space weaponization in the context of planetary geopolitics and the space arms race. Lampkin and Wyatt (2023) argue that green-critical criminologists should pay attention to anthropogenic orbital space debris as a form of harmful space junk—what Groombridge (2013: 396) called "widespread deliberate organized transnational pollution". Lampkin and McClanahan (2023) take a green criminological approach to examine the emergence of extraterrestrial mining, with the hope that research and debate on this phenomenon will help prevent future social and environmental harms in the "invisible environment" of outer space.

In the wake of these studies of anthropogenic environmental harms in outer space, there have also been calls to develop "astro-green criminology"—a term first proposed by Takemura (2019)—as a critical perspective to anchor space criminology. Lampkin (2020: 242) defines "astro-green criminology" as "the theoretical and practical study of space-related environmental harms and crimes that are facilitated by human actions. These harms can be Earth-based, atmospheric, or extraterrestrial and may create human victims, non-human victims, and ecological victims both on Earth and in outer space." Lampkin (2020: 252) argues that while an "astro-green criminology" could take a precautionary approach to explore "the potential of future environmental harms, such as those apparent with space tourism or extraterrestrial colonization," undertaking such studies "may run the risk of also being considered beyond the purview of criminology and green criminology alike". If criminology is indeed a "rendezvous discipline" or "rendezvous subject" (see Brisman & South 2020; South et al. 2013), with elastic boundaries that are drawn and redrawn over time to ensure its ongoing relevance and liveliness, then only a failure of imagination and interdisciplinary risk-taking will maintain such rigidly bound purviews.

When it comes to imagining outer space, there has been no shortage of popular fantasies of human exploration and exploitation of extraterrestrial territories (e.g., Deudney 2020; Geppert 2012). To analyse popular imaginings of present and future human activities, green cultural criminology offers a flexible and evolving perspective for the consideration of how speculative constructions of anthropogenic environmental harms intersect with harmful patterns of consumption, both on- and off-planet. Because green cultural criminologists have examined both the dynamics of consumption and media representation of environmental harms and disasters on Earth (Brisman & South 2014; McClanahan et al. 2017), they have the capacity to interrogate the imaginations and underlying mindsets that currently fuel efforts to "conquer" outer space. When criminologists treat outer space as some cutting-edge, "new" area for exploration, they are implicitly treating it as a *frontier*—that is, as a site for experimenting with applications of old theories and methods, or as a scene for new types of crimes (Eski 2023: 73). Although new categories of crime can emerge from human activities in outer space, in much the same way as new crimes emerged from processes of globalization and engagement with the Internet (e.g., Rothe & Friedrichs 2015; Rothe & Kauzlarich 2022), treating outer space as a frontier can come with additional socio-political and cultural implications that should not be ignored by criminologists. In adopting a frontier mentality, criminologists begin to mimic "space advocates that view space as a vast new frontier where human expansion knows no ultimate limits, where painful choices can be solved or left behind" (Deudney 2020: 12). Such a frontier mentality, as we argue below, is dangerous: it disconnects the contemporary space race from the activities of the richest 1% of humans on Earth, and diverts crucial attention away from ongoing environmental harms that continue to destroy our planet.

Climate Breakdown, the Polluter Elite, and Bunkerization

Construed as "the final frontier" in popular science fiction of the 1960s, space became a site for "boldly go[ing] where no [hu]man has gone before"; starships would "explore strange new worlds—to seek out new life and new civilizations" (title sequence of the original *Star Trek* television series from the late 1960s). In the twenty-first century, however, exploratory missions of space voyages in speculative fiction have been re-conceived as missions central to saving humanity *from* a ravaged Earth (often from humanity's own doing). In the animated Pixar film *WALL-E* (2008), starliners cruise through space as temporary human refuges, while millions of trash compactor robots on Earth attempt to clean up the mountains of garbage that have laid waste to the planet. Designed like massive luxury cruise ships, with "all the trappings of a typical cruise liner, such as shops, restaurants and swimming pools" (Pixar Fandom Trivia n.d.), the spaceships continue to entertain Earth's last humans.

Although both popular fiction and academic scholarship have long pinpointed space as the next location for human activities and, as importantly, next-level human achievements, it would be dangerous to minimize the connections between contemporary dreams of space travel and humans' increasingly troubled relationship with the Earth. In contrast to the first space race in the mid-twentieth century, itself an extension of the Cold War between the United States and the Soviet Union, the contemporary space race, dubbed "space race 2.0" by some (e.g., Clark 2023), is intimately and urgently connected to today's climate catastrophe.

Where the Earth was "[o]nce victorious", triumphing over humans who foolishly attempted to conquer it, it "is now a victim" (Serres 1995: 11), made fragile by destructive human activities that have contributed to the current climate breakdown. In 2023, the Earth experienced its hottest year on record:

every day was 1°C warmer than pre-industrial levels, such that "[t]emperatures during 2023 likely exceed[ed] those of any period in at least the last 100,000 years" (Samantha Burgess, Copernicus Climate Change Service deputy director, quoted in Carrington 2024). With almost half the days in 2023 being 1.5°C warmer (Copernicus Climate Change Service 2023), Earth was 1.48°C hotter when compared to pre-industrial temperatures—a figure that comes close to exceeding the 1.5°C temperature target threshold set by the Paris Agreement under the United Nations Framework Convention on Climate Change (adopted in 2015). Scientists predict that the 1.5°C will likely be passed for the first time in 2024 (Carrington 2024; Milman 2024), which will further devastate Earth's already damaged ecosystems that support biodiverse life. According to Richardson and colleagues (2023), "anthropogenic perturbations of the global environment" have already become so significant that the "Earth is now well outside of the safe operating space for humanity". There are now deep and widespread scientific concerns that planetary disequilibrium and anthropogenic global heating will cause more frequent extreme weather events and destroy more lives and livelihoods across the world. Note that not all humans have had the kind of economic power or social lifestyles to create such extreme climate hazards. In fact, only one-quarter of the world's population can be held responsible for nearly three-quarters of global carbon emissions (Donald 2024).

Deemed the "polluter elite", the richest 1% of humans on Earth contribute more carbon emissions than the poorest of humanity (Watts 2023). Since the 1990s, the wealthiest 1% have burned through more than twice as much carbon in their daily lives as the bottom half of humanity; in 2019, they were responsible for 16% of global carbon emissions, the same amount of emissions associated with the poorest 66% in the world (Oxfam 2023). While the rich can afford to fly in their private jets or sail abroad in their yachts, people with the lowest incomes can rarely afford to drive (Oswald et al. 2020). The widely divergent outcomes of such income inequality are captured in acclaimed science-fiction writer Kim Stanley Robinson's climate fiction novel, *The Ministry for the Future* (2020). Although Robinson made his name writing about humanity's future in space (e.g., the colonization of Mars in "the Mars Trilogy"), his most recent work imagines how a utopian future might unfold when the world (finally) responds to Earth's climate crisis. Climate scientists, CEOs of green technology start-ups, politicians, and tech entrepreneurs have since turned to his speculative fiction as a possible road map for averting biosphere collapse (Alter 2022). While Robinson acknowledges that he is "partially responsible" for the popular fantasy of space exploration (quoted in Alter 2022), now adopted by ambitious tech billionaires, he also holds the richest humans on Earth responsible for their contributions to the ongoing climate emergency in *The Ministry for the Future*. When the "three richest people in the world possess more financial assets than all the people in the forty-eight poorest countries added together", global wealth disparity has reached a point to "now suggest [that] this is the most wealth-inequal moment in human history" (Robinson 2021: 74). This wealth inequality, in turn, breeds different kinds of climate criminals, some with outsized influence on governments. In particular, in the novel, the elite class is singled out as roughly "a hundred people walking this Earth, who if you judge from the angle of the future like you're supposed to do, they are mass murderers" (Robinson 2021: 99). If they woke up one day without their assets, "[t]heir ability to murder the future would be much reduced" (Robinson 2021: 100). Even as members of the "polluter elite" are responsible for immense suffering in vulnerable communities endangered by climate emergency, now and in the future, they continue to have an outsized influence on responses to anthropogenic climate change. And, as Solnit (2023) writes, "billionaires are a menace to the rest of us: their sheer political size warps our public life. Disproportionately older, white and male, they function as unelected powers, a sort of freelance global aristocracy who are too often trying to reign over the rest of us."

While the richest five men in the world have doubled their fortunes since 2020, almost five billion people globally have become poorer during the same period, making hunger and hardship a daily reality for the vast majority of people worldwide (Oxfam 2024). With climate breakdown, "driven by the super-rich" (Oxfam 2024: 9), this extreme widening of inequality has coincided with a sharp narrowing of economic imagination and a disconcerting lack of coordinated, collective action in and between nation-states. As Davis (2010: 37–38) had predicted, coordinated global action, especially on behalf of the poor and the unborn, would seem like a realistic outcome only if it could be "shown that privileged groups possess[ed] no preferential 'exit' option" and that the mitigation of greenhouse gas emissions could be "achieved without major sacrifices in northern hemispheric standards of living—none of which seem likely". As accelerated investments have been made to facilitate and support the "selective adaptation [of] Earth's first-class passengers" (Davis 2010: 38), forms of private consumption and "bunker-fever" (Mahdawi 2024) have appeared in anticipation of surviving a world devastated by nuclear war, disease, and climate catastrophe.

Previously, we have described "bunkerization" as a specific elite response to such converging global crises (Lam et al. 2023). Concerned with the buying, selling, and building of artificial environments, which are designed to provide defensive and protective responses to the political, military, and ecological threats of the Anthropocene (South & South 2021), bunkerization emerged during COVID-19 as a form of "inverted quarantine" (Lam et al. 2023). Drawing on Szasz's (2007) concept of "inverted quarantine", which describes individualistic, consumeristic acts of self-protection against toxic threats, we underscored how the idea of bunkerization highlights privileged responses or reactions to contemporary crises (e.g., climate emergencies, wars, and pandemics), "whereby wealthy elites, when faced with the prospects of a damaged, threatened and dying world, have tended to search for sanctuary, escape, and alternative living environments" (Lam, South, & Brisman 2024).

"Being a billionaire tends to isolate you from the rest of humanity" (Solnit 2023), precisely because extreme wealth can be used to build fortified bunkers to enable a fortunate few to survive and thrive after an apocalyptic catastrophe. While the idea of gated communities and fortress cities was a response to both middle- and upper-class demand for enhanced social and spatial segregation, away from the possibility of social mingling and mixing with different classes, crowds, and ethnicities (Davis 2006; see also Brisman 2004), the contemporary survival bunker represents a much more individualized, modular form of self-protection, where "winners" in this exclusive society not only "take all" (Young 1998: 81; see also Young 1999), but they also *survive it all* by leaving the rest of us behind. For Rushkoff (2022a), these "super-rich" elites have "succumbed to a mindset where 'winning' means earning enough money to insulate themselves from the damage they are creating by earning money in that way. It's as if they want to build a car that goes fast enough to escape from its own exhaust." Similarly, O'Connell (2020) laments that "[t]he bunker, purchased and tricked out by the individual consumer, is a nightmare inversion of the American dream. It's a subterranean abundance of luxury goods and creature comforts, a little kingdom of reinforced concrete and steel, safeguarding the survival of the individual and his family amid the disintegration of the world." Thus, "bunkers have become the ultimate status symbol among the 1%" (Mahdawi 2024), as billionaires outfit their survival bunkers with flammable moats and water cannons to ensure that they remain protected from those they deem undesirable (Schmidt 2024). Notably, the demand for doomsday bunkers increased significantly during the COVID-19 pandemic (Schiffer 2020), and following Russia's invasion of Ukraine (Garrett 2022), as anxieties skyrocketed in the face of multiple existential threats. Interestingly, Garrett (2022) reports that "Ukraine has at least 5,000 publicly accessible bomb shelters, many of which have been upgraded since Russia annexed Crimea in 2014. In stark contrast, the US has few public shelters. This disparity can be traced back to the cold war, when the Soviet Union invested heavily in public defence infrastructure," while the United States—and this should come as no surprise—"placed the burden of nuclear protection on private citizens". No longer limited to fringe prepper communities (for a discussion, see, e.g., Mills 2019, 2021; Mills & Fleetwood 2019; see generally Brisman 2020), doomsday bunkers are now being bought by affluent consumers as part of their preparations for surviving large-scale disasters in the near future (e.g., Fleming 2020). It follows, then, that news about Mark Zuckerberg's expensive, top-secret compound in Hawaii, with its massive underground bunker (see Mahdawi 2024; Scrimgeour 2023), "caused a buying frenzy" among wealthy elites, making it seem like World War 3 is nigh (Ron Hubbard, CEO of Atlas Survival Shelters, quoted in Schmidt 2024).

As the super-rich make plans to build elaborate bunkers in preparation for global societal, political and/or environmental collapse, often in isolated, remote locations with frontier histories (for example, Peter Thiel, the billionaire venture capitalist and founder of PayPal and Palantir Technologies, had originally planned to build his "doomsday" bunker in New Zealand (see

Rushkoff 2022a, 2023; Schmidt 2024; Scrimgeour 2023)), their bunkerization rests upon a specific "Mindset" (Rushkoff 2023)—an elite approach to life, technology and business that depends upon "a belief that with enough genius and technology, they can rise above the plane of mere mortals and exist on an entirely different level, or planet, altogether". This commitment to escape is justified by an "insulation equation" (Rushkoff 2022b: 20) that takes social divisions to the extreme, meaning that while "low-level billionaires" will isolate and shelter on Earth in their massive survival bunkers, "[o]nly trillionaires will actually make it into space to terraform planets" (Rushkoff 2022b: 20). Because near-trillionaires, Jeff Bezos and Elon Musk, are seriously considering space colonies (or what we would consider as "space bunkers") as a solution to escaping Earth's problems of poverty, pollution, and nuclear war (see, e.g., Rushkoff 2022a; Slezak & Solon 2017; Solon 2018; Woolf 2016), we examine in the next section how these visions for space colonization treat outer space as a frontier for new conquests and resource extraction, apparently ignoring all lessons we might have learned about the kinds of intergenerational human and environmental harms that have resulted from colonization (and settler colonialism) on Earth.

Outer Space-as-Frontier

Worried that all reasonable sources of energy will be depleted on Earth within a couple of centuries, Jeff Bezos declared in 2019 that he will find and build a new home for the human species in space (Powell 2019). Sustained by continuous sunshine and resources from across the solar system, his imagined space colonies will be "very large structures, miles on end, and they [will] hold a million people or more each" (Bezos quoted in Powell 2019). Housing trillions of people on the Moon, these millions of space colonies will allow inhabitants to harvest space resources for themselves and those remaining on Earth. Bezos' ambitious plan draws from the frontier vision of Gerard O'Neill, an American physicist known in the 1970s for his detailed conceptualizations of space colonization. Citing O'Neill's book *The High Frontier* (1977) as a major influence, Bezos believes that "[w]e want to go to space to save the Earth", not as part of a "'Plan B' idea that we want to go into space so we have a backup planet" (quoted in Powell 2019). In contrast, Elon Musk's "Martian Dream" revolves around saving the human species by colonizing the red planet in anticipation of "an inevitable extinction event on Earth" (Musk quoted in Woolf 2016). According to Musk, it will be "important to get a self-sustaining base on Mars because it's far enough away from earth that it's more likely to survive than a moon base"; his plan to build a city of one million people on Mars by 2050 will be necessary, because if "there's a third world war we want to make sure there's enough of a seed of human civilisation somewhere else to bring it back and shorten the length of the dark ages" (Musk quoted in Solon 2018; see also McFall-Johnsen and Mosher 2020). For early Mars settlers, Musk predicts that it "will be like

Shackleton's ad for Antarctic explorers: 'Difficult, dangerous, a good chance you'll die, excitement for those who survive'" (quoted in Solon 2018). Musk's reference to the so-called "heroic age" of Antarctic exploration—a venture that was exclusively male and white in the twentieth century—not only reflects "a conjured phantasm of idealised masculinity—the man who provides, the man who protects" (O'Connell 2020), but also brings into focus the mentality of settler colonialism underlying plans for future space colonization. Once conceived as the "last place on Earth" to be discovered and "conquered" by major European nations (Brazzelli 2017; see also Howkins 2010), Antarctica has now been tagged as "White Mars" by the European Space Agency (David 2024). When likened to Antarctica, Mars represents an "ideal settler colony" for tech billionaires (as well as powerful nation-states and multinational corporations), where their expansionist desires can be expressed without the unsettling legacies of imperial displacement, exploitation, and death of Indigenous peoples.

By suggesting that plans for space colonization—whether on the Moon or on Mars—can be situated within a longer history of settler colonial projects, we seek to draw attention to the troubling frontier mentality that underpins these future plans in outer space. With the National Aeronautics and Space Administration (NASA) aspiring to provide American leadership in the enterprise of exploration and development of outer space through partnerships with tech billionaires,[2] it is important to document the agency's rationales for settling the solar system, and how they, too, treat outer space as a new frontier to conquer. In its vision for the next fifty years in space, NASA (2023: 3) writes,

> Five centuries after Columbus opened access to 'The New World' we can initiate the settlement of worlds beyond our planet of birth. The promise of virgin lands and the opportunity to live in freedom brought our ancestors to the shores of North America. Now space technology has freed humankind to move outward from Earth as a species destined to expand to other worlds. … The settlement of North America and other continents was a prelude to humanity's greater challenge: the space frontier.

"As formerly on the western frontier, now similarly on the space frontier", government support—according to NASA (2023: 4)—will be crucial for "open[ing] broad access to new lands". In connecting the western frontier of the past to the space frontier of the future, NASA seems to be extending the specifically American myth of the frontier to outer space.[3] Originally formulated by Frederic Jackson Turner (1893/2008), the frontier thesis—that is, the proposition that the success and fulfilment of American character are the result of "frontier experiences"—endures in NASA's plan for space exploration and colonization. Yet the agency's romanticization of the frontier is troubling. Not only does NASA seem to forget that Native Americans were already living in the areas designated as "frontiers" in the American West, but NASA's report also glosses over the harmful consequences of colonization.

When Columbus "opened access to 'The New World'", disease, environmental destruction, war and genocide ensued. The arrival of Europeans in the Americas was not just an exciting, world-changing discovery for imperial powers; it was an Earth-changing event. According to Lewis and Maslin (2015), the beginning of the Anthropocene can be dated back to 1610, a year when a decades-long decrease in atmospheric carbon dioxide found its nadir. Such a decrease was the result of the long-term and large-scale "onward effects of the arrival of Europeans in the Americas"—more specifically, the deaths of 50 million Indigenous peoples in the Americas due to "exposure to diseases carried by Europeans, plus war, enslavement and famine" (Lewis & Maslin 2015: 177, 175; see also Stannard 1992).

When NASA and the tech billionaires frame outer space as a frontier, they presume that "spacescapes" are not only playgrounds for space cowboys, but also passive, empty sites that have no other purpose than human sustenance and resource provision (Messeri 2017). Gorman (2019: 111) reminds us of an old but futuristic vision that foretells the continuation of human insatiability and destructiveness, so that even as we try to reach out the stars to tap "green" energy we hollow out and extract resources from planets to the point of their implosion:

> Taking up an idea first raised in 1937 by the science-fiction writer Olaf Stapledon, physicist Freeman Dyson imagined a swarm of satellites designed to harvest the maximum amount of energy from the Sun, to power a future industrial culture with energy needs much greater than our own. The satellites would orbit the Sun in spherical formation, perhaps eventually expanding to enclose the entire solar system. However, to obtain enough materials to build the satellites, whole planets would have to be dismantled.

In the current space race 2.0, enormous quantities of money and natural resources are being blown into space. Some valuable outcomes, such as advanced communications, have enlivened civilization and brought benefits to many. But at the same time—in our new outer space frontier, as Gorman (2019: 109) suggests—"thousands of satellites, rocket bodies and pieces of junk" are floating free in an "organically evolved cultural landscape, or spacescape, structured by function, gravity, the space environment, political and scientific agendas, and technological traditions through time and across terrestrial cultures". And the old satellites, beyond the pull of gravity and enticement of re-entry, become the "dead" of space, circulating "among the living, haunting the Earth. Their destructive capacity, through collisions with the still-speaking satellites" turning them into "zombiesats that crash through orbits without purpose" (Gorman 2019: 109). Sadly, none of this is helping to make the use of outer space sustainable (Palmroth 2022) let alone support the sustainability of planet Earth.

Conclusion

New directions in criminology are exciting and stimulating. The "discipline" or "field" or "subject" needs them in order to remain relevant, reflect change, and provide a basis for continued debate and engagement. A "space criminology" is, in some ways, a quintessential example of this. Rather than engage in space expansionism, which is based on a deep-seated belief that humanity has a "cosmic mandate to inseminate the universe with the human" (Lavery 1992: 3), we are concerned—like Eski (2023: 79)—that humans will instead exploit space and "pollute [it] and other celestial bodies." Indeed, as Eski (2023: 76) cautions, "[t]he human species should avoid making the same environmental mistakes in space as it has on Earth, making sure to protect space wilderness, instead of considering it another 'frontier' to exploit". Although others may outline ways to protect space "wilderness," we, ourselves, remain unwaveringly *Earthbound*, precisely because "never before have our society's most powerful players assumed that the primary impact of their own conquests would be to render the world itself unlivable for everyone else" (Rushkoff 2022b: 11). Instead of escaping Earth to establish new frontiers, our own space manifesto calls upon humanity to devote attention to *this planet* while we still have one. For criminologists, there is still much work to do in addressing, first and foremost, anthropogenic environmental issues on planet Earth, and then, perhaps, how these might spill over into extraplanetary colonization and resource extraction. How we—as criminologists, indeed, as *humans*—ultimately imagine, treat, and take responsibility for *this* planet will determine whether we can avert the apocalyptic endgame presumed by the richest 1%.

Notes

1 Noam Chomsky, quoted in McDonald (2010).
2 For example, NASA paid Jeff Bezos' spaceflight company, Blue Origin, US$130 million to develop a new commercial space station in 2021 (Heilweil 2021), and another US$3.4 billion to build a lunar lander for astronauts in 2023 (Roulette 2023). Meanwhile, Elon Musk's SpaceX has earned up to US$15.3 billion in awarded contracts since 2003 (Fernholz 2023), and another US$2.9 billion from NASA in 2021 to build a commercial Moon lander (Satter et al. 2021).
3 Indeed, in his essay on "end-time real estate" and "doomsday luxury accommodation," the Irish author and journalist, Mark O'Connell (2020) refers to "a frontier past and an apocalyptic future" as "the US's twin obsessions".

References

Alter A (2022) A Sci-Fi Writer Returns to Earth: "The Real Story is the One Facing Us". *The New York Times*, 11 May. From: https://www.nytimes.com/2022/05/11/books/kim-stanley-robinson-sci-fi.html.

Brazzelli N (2017) Heroic and Post-Colonial Antarctic Narratives. In Dodds K, Hemmings AD & Roberts P (eds) *Handbook on the Politics of Antarctica*. Cheltenham: Edward Elgar Publishing, pp. 69–83.

Brisman A (2004) Double Whammy: Collateral Consequences of Conviction and Imprisonment for Sustainable Communities and the Environment. *William & Mary Environmental Law & Policy Review* 28(2): 423–475.

Brisman A (2020) Cultural Criminology and Narrative Criminology's Shared Interests. More than just Criminological Verstehen. *Tijdschrift over Cultuur & Criminaliteit* 10 (3): 14–34. doi:10.5553/TCC/221195072020010003002.

Brisman A & South N (2014) *Green Cultural Criminology: Constructions of Environmental Harm, Consumerism, and Resistance to Ecocide.* London: Routledge.

Brisman A & South N (2020) The Growth of a Field: A Short History of a "Green" Criminology. In Brisman A & South N (eds) *Routledge International Handbook of Green Criminology*, 2nd edn. Abingdon, Oxon, UK, and New York: Routledge, pp. 39–51.

Carrington D (2024) 2023 Smashes Record for World's Hottest Year by Huge Margin. *The Guardian*, 9 January. From: https://www.theguardian.com/environment/2024/jan/09/2023-record-world-hottest-climate-fossil-fuel.

Clark S (2023) Space Race 2.0: Why Europe is Joining the New Dash to the Moon. *The Guardian*, 26 November. From: https://www.theguardian.com/science/2023/nov/26/europe-moon-space-race-esa-india-russia-china-japan.

Copernicus Climate Change Service (2023) The 2023 Annual Climate Summary: Global Climate Highlights 2023. From: https://climate.copernicus.eu/global-climate-highlights-2023.

David L (2024) Can Antarctica Serve as a Model for International Cooperation on the Moon? Space.com, 3 March. From: https://www.space.com/moon-base-antarctica-international-cooperation.

Davis M (2006) *City of Quartz*. London: Verso.

Davis M (2010) Who Will Build the Ark? *New Left Review* 61: 29–46.

Deudney D (2020) *Dark Skies: Space Expansionism, Planetary Geopolitics, and the Ends of Humanity.* New York: Oxford University Press.

Donald R (2024) Human "Behavioural Crisis" at Root of Climate Breakdown, Say Scientists. *The Guardian*, 13 January. From: https://www.theguardian.com/environment/2024/jan/13/human-behavioural-crisis-at-root-of-climate-breakdown-say-scientists#:~:text=%E2%80%9CWe%E2%80%98ve%20socially%20engineered%20ourselves,deliberate%20exploitation%20of%20human%20behaviour.

Eski Y (2023) *A Criminology of the Human Species: Setting an Unsettling Tone.* Cham: Palgrave Macmillan.

Fernholz T (2023) Elon Musk's SpaceX and Tesla Get Far More Government Money than NPR. *Quartz*, 13 April. From: https://qz.com/elon-musks-spacex-and-tesla-get-far-more-government-mon-1850332884.

Fleming J (2020) Bunker with a Bowling Alley: How the Rich are Running from Coronavirus. *Los Angeles Times*, 23 March. From: https://www.latimes.com/business/real-estate/story/2020-03-23/rich-are-running-from-coronavirus.

Fountain L (2003) Creating Momentum in Space: Ending the Paralysis Produced by the "Common Heritage of Mankind" Doctrine. *Connecticut Law Review* 35: 1753–1787.

Garrett BL (2022) "Our Top Search Term is Nuclear": US Bunker Sales Soar as Anxiety over Russia Rises. *The Guardian*, 29 March. From: https://www.theguardian.com/us-news/2022/mar/29/bunker-sales-soar-anxiety-over-russia-rises-ukraine.

Geppert AC (ed.) (2012) *Imagining Outer Space: European Astroculture in the Twentieth Century.* Basingstoke: Palgrave Macmillan.

Gorman A (2019) Ghosts in the Machine: Space Junk and the Future of Earth Orbit. *Architectural Design* 89(1): 106–111.

Groombridge N (2013) Matter All Over the Place: Litter, Criminology and Criminal Justice. In South N & Brisman A (eds) *Routledge International Handbook of Green Criminology.* Abingdon, Oxon, UK, and New York: Routledge, pp. 394–408.

Heilweil R (2021) NASA Gave Jeff Bezos Money to Build his Office Park in Space. *Vox*, 3 December. From: https://www.vox.com/recode/2021/10/27/22747509/blue-origin-orbital-reef-office-park-bezos.

Howkins A (2010) *Appropriating Space: Antarctic Imperialism and the Mentality of Settler Colonialism.* In Mar TB & Edmonds P (eds) Making Settler Colonial Space. London: Palgrave Macmillan, pp. 29–52.

Lam A, South N & Brisman A (2023) Convergence of Crises: COVID-19, Climate Change and Bunkerization. *Crime Media Culture* 19(3): 327–344. https://doi.org/10.1177/17416590221120581.

Lam A, South N & Brisman A (2024) Green Cultural Criminology: Foundations, Variations and New Frames. In *The Oxford Research Encyclopedia of Criminology and Criminal Justice.* https://doi.org/10.1093/acrefore/9780190264079.013.749.

Lampkin JA (2020) Mapping the Terrain of an Astro-Green Criminology: A Case for Extending the Green Criminological Lens Outside of Planet Earth. *Astropolitics* 18 (3): 238–259.

Lampkin JA & McClanahan BW (2023) Astronomical Withdrawals: A Green Criminological Examination of Extreme Energy Mining on Extraterrestrial Objects. *Crime, Law and Social Change* 81: 365–384. https://doi.org/10.1007/s10611-023-10123-9.

Lampkin J & White R (2023) *Space Criminology: Analysing Human Relationships with Outer Space.* Cham: Palgrave Macmillan.

Lampkin JA & Wyatt T (2023) An Astro-Green Criminological Examination of Orbital Space Debris. *Criminology & Criminal Justice.* https://doi.org/10.1177/17488958231169124.

Lavery D (1992) *Late for the Sky: The Mentality of the Space Age.* Carbondale: Southern Illinois University Press.

Lewis SL & Maslin MA (2015) Defining the Anthropocene. *Nature* 519: 171–180.

Mahdawi A (2024) The Latest Billionaire Trend? Doomsday Bunkers with a Flammable Moat. *The Guardian*, 25 February. From: https://www.theguardian.com/commentisfree/2024/feb/25/billionaire-doomsday-bunkers-end-times.

McClanahan B, Brisman A & South N (2017) Green Criminology, Culture and Cinema. In Brown M & Carrabine E (eds) *Oxford Encyclopedia of Crime, Media, and Popular Culture.* New York: Oxford University Press.

McDonald A (2010) The NS Interview: Noam Chomsky. *The New Statesman*, 13 September. From: https://www.newstatesman.com/long-reads/2010/09/war-crimes-interview-obama.

McFall-Johnsen M & Mosher D (2020) Elon Musk Says He Plans to Send 1 Million People to Mars by 2050 by Launching 3 Starship Rockets Every Day and Creating "a Lot of Jobs" on the Red Planet. *Business Insider*, 17 January. From: https://www.businessinsider.com/elon-musk-plans-1-million-people-to-mars-by-2050-2020-1?r=US&IR=T.

Messeri L (2017) We Need to Stop Talking about Space as a "Frontier." *Slate*, 15 March. From: https://slate.com/technology/2017/03/why-we-need-to-stop-talking-about-space-as-a-frontier.html.

Miller RW (2001) Astroenvironmentalism: The Case for Space Exploration as an Environmental Issue. *Electronic Green Journal* 1(15): 1–7.

Milman O (2024) Global Heating Will Pass 1.5C Threshold This Year, Top Ex-Nasa Scientist Says. *The Guardian*, 8 January. From: https://www.theguardian.com/environment/2024/jan/08/global-temperature-over-1-5-c-climate-change.

Mills K (2011) Who Will Own Outer Space? Governance over Space Resources in the Age of Human Space Exploration. In Landfester U, Remuss NL, Schrogl KU, Worms JC (eds), *Humans in Outer Space: Interdisciplinary Perspectives*. Studies in Space Policy, vol 5. Springer: Vienna. https://link.springer.com/chapter/10.1007/978-3-7091-0280-0_2.

Mills MF (2019) Preparing for the Unknown…Unknowns: "Doomsday" Prepping and Disaster Risk Anxiety in the United States. *Journal of Risk Research* 22(10): 1267–1279.

Mills MF (2021) Obamageddon: Fear, the Far Right, and the Rise of "Doomsday" Prepping in Obama's America. *Journal of American Studies* 55(2): 336–365.

Mills MF & Fleetwood J (2019) Prepping and Verstehen: A Narrative Criminological Perspective. *Tijdschrift over Cultuur & Criminaliteit* 9(3): 30–47.

NASA (2023) *Pioneering the Space Frontier: The Report of the National Commission on Space – An Exciting Vision of Our Next Fifty Years in Space*. From: https://www.nasa.gov/wp-content/uploads/2023/03/383341main_6020-2020090814.5.the20report20of20the20national20commission20on20space.pdf.

O'Connell M (2020) Real Estate for the Apocalypse: My Journey into a Survival Bunker. *The Guardian*, 17 March. From: https://www.theguardian.com/news/2020/mar/17/real-estate-for-the-apocalypse-my-journey-into-a-survival-bunker.

O'Neill GK (1977) *The High Frontier: Human Colonies in Space*. New York: Morrow.

Oswald Y, Owen A & Steinberger JK (2020) Large Inequality in International and Intranational Energy Footprints Between Income Groups and Across Consumption Categories. *Nature Energy* 5: 231–239.

Oxfam (2023) Climate Equality: A Planet for the 99%. From: https://policy-practice.oxfam.org/resources/climate-equality-a-planet-for-the-99-621551/.

Oxfam (2024) Inequality Inc.: How Corporate Power Divides Our World and the Need for a New Era of Public Action. From: https://policy-practice.oxfam.org/resources/inequality-inc-how-corporate-power-divides-our-world-and-the-need-for-a-new-era-621583/.

Palmroth M (2022) Space Junk is a Danger to Human Life and Future Exploration. *The Engineer*, 27 July. From: https://www.theengineer.co.uk/content/opinion/comment-space-junk-is-a-danger-to-human-life-and-future-exploration/.

Pixar Fandom Trivia (n.d.) *Axiom*. From: https://pixar.fandom.com/wiki/Axiom.

Powell CS (2019) Jeff Bezos Foresees a Trillion People Living in Millions of Space Colonies. Here's What He's Doing to Get the Ball Rolling. *NBC News*, 15 May. From: https://www.nbcnews.com/mach/science/jeff-bezos-foresees-trillion-people-living-millions-space-colonies-here-ncna1006036.

Richardson K et al (2023) Earth Beyond Six of Nine Planetary Boundaries. *Science Advances* 9(37): https://doi.org/10.1126/sciadv.adh2458.

Robinson KS (2021) *The Ministry for the Future*. New York: Orbit.

Rothe DL & Collins VE (2023) Planetary Geopolitics, Space Weaponization and Environmental Harms. *The British Journal of Criminology* 63(6): 1523–1538.

Rothe DL & Friedrichs DO (2015) *Crimes of Globalization*. Abingdon, Oxon, UK, and New York: Routledge.

Rothe DL & Kauzlarich D (2022) *Crimes of the Powerful: White-Collar Crime and Beyond*. Abingdon, Oxon, UK, and New York: Routledge.

Roulette J (2023) Bezos' Blue Origin Wins NASA Contract to Build Astronaut Lunar Lander. Reuters, 19 May. From: https://www.reuters.com/technology/space/nasa-name-second-company-build-astronaut-lunar-lander-2023-05-19/.

Rushkoff D (2022a) The Super-Rich "Preppers" Planning to Save Themselves From the Apocalypse. *The Guardian*, 4 September. From: https://www.theguardian.com/news/2022/sep/04/super-rich-prepper-bunkers-apocalypse-survival-richest-rushkoff.

Rushkoff D (2022b) *Survival of the Richest: Escape Fantasies of the Tech Billionaires*. New York: W.W. Norton & Company.

Rushkoff D (2023) "We Will Coup Whoever We Want!": The Unbearable Hubris of Musk and the Billionaire Tech Bros. *The Guardian*, 25 November. From: https://www.theguardian.com/books/2023/nov/25/we-will-coup-whoever-we-want-the-unbearable-hubris-of-musk-and-the-billionaire-tech-bros.

Sachdeva GS (2023) *Crimes in Outer Space: Perspectives from Law and Justice*. Singapore: Springer.

Satter R, Jin H & Vengattil M (2021) "NASA Rules," Musk Says as SpaceX Wins $2.9 Billion Moon Lander Contract. Reuters, 16 April. From: https://www.reuters.com/technology/spacex-wins-us-contract-spacecraft-send-astronauts-moon-washington-post-2021-04-16/.

Schiffer Z (2020) Home, Sweet, Home? *The Verge*, 13 May. Available at: https://www.theverge. com/2020/5/13/21255777/doomsday-shelters-bunkers-coronavirus-pandemic-covid-19-conspiracy-theory (accessed 5 July 2024).

Schmidt I (2024) Billionaires' Survivalist Bunkers Go Absolutely Bonkers with Fiery Moats and Water Cannons. *The Hollywood Reporter*, 12 February. From: https://www.hollywoodreporter.com/lifestyle/lifestyle-news/bunkers-billionaires-survive-apocalypse-cost-features-1235822762/.

Scrimgeour G (2023) Inside Mark Zuckerberg's Top-Secret Hawaii Compound. *WIRED*, 14 December. From: https://www.wired.com/story/mark-zuckerberg-inside-hawaii-compound/.

Serres M (1995) *The Natural Contract*. Ann Arbor: University of Michigan Press.

Slezak M & Solon O (2017) Elon Musk: SpaceX Can Colonise Mars and Build Moon Base. *The Guardian*, September 29. From: https://www.theguardian.com/technology/2017/sep/29/elon-musk-spacex-can-colonise-mars-and-build-base-on-oon.

Solnit R (2023) Billionaires Are Out of Touch and Much Too Powerful. The Planet Is in Trouble. *The Guardian*, 20 November. From: https://www.theguardian.com/commentisfree/2023/nov/20/billionaires-great-carbon-divide-planet-climate-crisis.

Solon O (2018) Elon Musk: We Must Colonise Mars to Preserve Our Species in a Third World War. *The Guardian*, 11 March. From: https://www.theguardian.com/technology/2018/mar/11/elon-musk-colonise-mars-third-world-war.

South D & South N (2021) Bunkerization: Elite Preparedness and Retreat in the Anthropocene. In Carillo J & Garner C (eds) *City Preparedness for the Climate Crisis*. Cheltenham: Edward Elgar, pp. 234–243.

South N, Brisman A & Beirne P (2013) A Guide to a Green Criminology. In South N & Brisman A (eds) *Routledge International Handbook of Green Criminology*. Abingdon, Oxon, UK, and New York: Routledge, pp. 27–42.

Stannard DE (1992) *American Holocaust: Columbus and the Conquest of the New World*. New York: Oxford University Press.

Szasz A (2007) *Shopping Our Way to Safety: How We Changed from Protecting the Environment to Protecting Ourselves*. Minneapolis: University of Minnesota Press.

Takemura N (2019) Astro-Green Criminology: A New Perspective Against Space Capitalism. *Toin University of Yokohama Research Bulletin* 40: 7–17.

Turner FJ (1893/2008) *The Significance of the Frontier in American History.* UK: Penguin.

Watts J (2023) Richest 1% Account for More Carbon Emissions than Poorest 66%, Report Says. *The Guardian*, 20 November. From: https://www.theguardian.com/environment/2023/nov/20/richest-1-account-for-more-carbon-emissions-than-poorest-66-report-says.

Woolf N (2016) SpaceX Founder Elon Musk Plans to Get Humans to Mars in Six Years. *The Guardian*, 28 September. From: https://www.theguardian.com/technology/2016/sep/27/elon-musk-spacex-mars-colony.

Young J (1998) Inclusive and Exclusive Society. In Ruggiero V, South N & Taylor IR (eds) *The New European Criminology: Crime and Social Order in Europe.* London: Routledge, pp. 64–91.

Young J (1999) *The Exclusive Society.* London: Sage.

10 Harnessing Science Fiction to Reimagine Criminal Justice in Space

Opportunities and Risks

Erika Nesvold

Introduction

The potential for criminal behaviour occurring in space will approach inevitability if humans continue to expand our civilization into space. How should we address this behaviour? Transporting residents accused of a crime back to Earth for trial and/or punishment may be an option in settlements located close enough that such a trip is practical and not excessively expensive, such as the Moon. This process would resemble historical responses to violent crime in Antarctica, a continent often identified as a useful parallel and predictor for how humans will live together in space, given its extreme isolation, physically dangerous environment, and richness in scientific and mineral resources. In 2018, for example, a Russian scientist working at the Bellinghausen research station assaulted a colleague with a knife, then surrendered to the station manager. The attacker was then transported back to St Petersburg, Russia, for arrest and trial, avoiding the need for a local criminal justice solution in Antarctica (Schaub 2018).

However, as human outposts in space expand outward from our home planet, shifting the burden of criminal justice back to Earth will become impractical, and eventually impossible. Many futurists imagining the problems of crime in space therefore propose the simple solution of bringing our modern-day terrestrial justice systems with us into space: in other words, constructing prisons in space.

Astrobiologist Charles Cockell (2016: 27), for example, has proposed a design for a panopticon-like extraterrestrial containment facility (which he terms "Exoconfac") with the purpose of "maintain[ing] criminals under conditions where they can be made safe or carry out useful functions for the settlement". He argues that criminality in space is inevitable, as it would be in any human society, but then concludes that "the need for an extraterrestrial prison follows" directly from the presence of criminality, without considering other potential solutions for addressing criminal behaviour in a community. Instead, Cockell focuses on the benefits of building a prison in the space environment rather than on Earth (namely, the reduced risk of escape or rioting due to the deadly conditions just outside the walls) and the importance of designing a "humane" environment and providing "productive work".

DOI: 10.4324/9781003437178-11

Prisons also appear in space in a variety of forms in science fiction stories. Narratives set aboard spaceships or stations usually rely on the model of a naval-style "brig": a small, featureless cell, often outfitted with a futuristic, transparent force field rather than metal bars, that can be used for a number of purposes: to apply temporary punishment to crew members (e.g., Kolbe 1998); to house passengers being transported to a more stationary prison facility or a planet-side courtroom for trial; or to incarcerate prisoners of war (the Death Star Detention Level in Lucas' 1977 *Star Wars*; see also Rymer 2005). These cells are designed as temporary holding areas, and plotlines occasionally present the difficult decision of whether to confine a character in the brig indefinitely or to find an alternate method of addressing their crime and protecting the rest of the crew (or perhaps, the rest of civilization) from the danger they present (e.g., Bole 1996; Ferrara 1994).

Other science fiction worlds imagine conditions aboard prison ships—like the spaceship *Purgatory* in the *Mass Effect 2* video game or the *Astral Queen* in the *Battlestar Galactica* television series (Kroeker 2004)—or on the surface of entire planets or moons that have been converted into penal colonies. The recent television series *Andor*, set in the *Star Wars* universe, was praised for its portrayal of the monotonous, dehumanizing experience of life in an imperial prison labour facility on an ocean-covered planet (Kain 2022). After escaping the facility itself, the characters must contend first with the surrounding sea, and then with the challenge of finding a ship to leave the planet (Haynes 2022). Escapees from the underground space prison in the 2004 film *The Chronicles of Riddick*, on the other hand, are threatened by the deadly heat on the day side of the fictional planet Crematoria (Twohy 2004).

While these imagined space prisons are designed to address law-breaking or harmful behaviour among people living away from Earth, space has also been suggested as a useful overflow site to house the Earth's expanding prison population. The year before the first human set foot on the Moon, law professor Luis Kutner (1968: 702) proposed a "World Outer Space Prison": an orbital space station created and maintained by the United Nations to "hold the entirety of the planet's population of felons". Kutner (1968: 704) also suggests using convicted prisoners as the first human colonists on the Moon, to give them job training and experience while providing them with "a special importance to themselves and to the Earth". But like Cockell, he notes that the primary benefit of a space prison over a terrestrial facility would be the significantly reduced risk of escape.

Science fiction has also extensively explored the idea of penal colonies in space to incarcerate convicted residents of Earth. One of the most well-known examples is Robert Heinlein's 1966 novel *The Moon is a Harsh Mistress*, in which the Moon houses over three million "Loonies": transported prisoners from Earth and their free-born descendants. It is not only the lack of transportation that keeps the lunar inhabitants from escaping to Earth: Heinlein imagines that the low lunar gravity inflicts "irreversible physiological changes" that prevent Loonies from being able to withstand Earth gravity. George

Zebrowski's 1999 novel *Brute Orbits* imagines prisons built inside hollowed-out asteroids and launched onto decades-long orbits around the sun, while the more recent English-language French film *Lockout* is primarily set on a maximum-security prison in orbit around the Earth (Leger & Mather 2012).

Why do visions of our future in space frequently resort to the model of incarceration as the primary response to criminal behaviour? Science fiction as a genre frequently uses its unusual setting to provide commentary on today's society, so naturally space prisons are useful for exploring the complexities of our own carceral system, as well as themes like isolation, anarchy, and justice. Science fiction prisons can extrapolate the worst aspects of our contemporary prisons to their extreme, to illustrate these problems for the reader—a tool that will be discussed further later in this chapter.

The space environment also shares several characteristics with prisons on Earth, as well as with terrestrial landscapes where humans have historically constructed prisons and transported prisoners to for incarceration. A space prison would be isolated and surrounded by a deadly environment, and would therefore be difficult to escape from, like island prisons on Earth. The surfaces of moons or planets are also vast, empty landscapes, as Europeans considered North America and Australia to be during their policies of transportation (despite the presence of Indigenous people living in these landscapes). Of course, this also conveniently puts the imprisoned population "out of sight, out of mind": As Angela Davis (2003: 16) describes it, a prison "functions ideologically as an abstract site into which undesirables are deposited, relieving us of the responsibility of thinking about the real issues afflicting those communities from which prisoners are drawn in such disproportionate numbers". Space is not only physically distant from us, but its environment is so alien to our everyday experience that it can achieve an *abstractness* far beyond terrestrial prisons.

This common trope also reflects the difficulty of imagining non-carceral alternatives for criminal justice—even in a fictional world where spaceships and force fields are commonplace—for a futurist raised in today's societies, given how entrenched prisons are in the dominant cultural and legal approaches to criminal justice. As English Literature professor Ariel Wetzel (2014: 2) noted in their dissertation on prisons in science fiction, "penitentiaries, since their advent some three centuries ago, have become timeless and taken-for-granted in the popular imaginary; it is difficult to imagine a society without them, even in science fiction".

The Practical and Ethical Challenges of Space Prisons

For humans living in permanent or long-term space settlements, the physical characteristics of the space environment itself will present both practical and ethical challenges to adopting a terrestrial criminal justice system based on incarceration. The practical challenges of a space prison will be especially dire, as space settlements will likely be extremely fragile and experience a scarcity of both resources and labour for many generations. The first challenge will be finding the physical space to construct or repurpose an area that can be used for incarceration: while habitats built on the surface of a planet or moon or as free-floating space stations will have large quantities of surrounding "empty space" to expand into, extending an existing habitat will require the construction or excavation (in the case of subsurface habitats) of areas that must also be sealed, pressurized, and shielded from radiation. Once constructed, the habitability of the prison will need to be maintained with the same kinds of life support already required for the rest of the settlement: a steady supply of power for heating, cooling, and lighting; air and water recycling; waste processing; food; exercise equipment to maintain physical health in microgravity; medical treatment; etc. Aside from these physical resources, the settlement will also need to contribute additional labour to the prison, first for the construction of the prison itself and then for any guarding and life support maintenance needs. The labour supply of the settlement will also be diminished due to the loss of incarcerated settlers from the labour pool, unless forced labour is incorporated into the criminal justice system.

There will also be ethical problems with incarceration as a criminal justice policy in space, as there are on Earth. For example, incarceration exposes incarcerated people to violence and sexual abuse (Novisky & Peralta 2020), imprisoned populations are often exploited for low-cost or dangerous labour (American Civil Liberties Union 2022); and marginalized groups in a society tend to be disproportionately imprisoned by the society's criminal justice system (Pettit & Gutierrez 2018). As with many ethical and social justice issues in society, these kinds of problems will only be exacerbated in and by the space environment. For example, incarceration can have detrimental effects on the physical and mental health of incarcerated people, especially in the presence of added factors like distance from their friends and family (Edgemon & Clay-Warner 2018), solitary confinement (Grassian 2006), a lack of contact with the natural world (Morris & Izenberg 2023), and other physical discomforts and deprivation (e.g., Soble et al. 2020). All of these factors will already be present in a space settlement, potentially affecting the mental health of all residents: isolation from Earth; disconnection with plant and wildlife, fresh air, a view of the sky, and terrestrial day/night cycles; potentially limited food, water, and air; physical discomfort from non-terrestrial gravity, temperature extremes, and radiation exposure. Human space exploration planners already worry about the effects these factors will have on the physical, psychological, and social health of space travellers; a prison built in a community that already experiences these factors will have them built into the prison itself.

Incarcerated people are also, by definition, cut off from society as a whole. This can serve a variety of purposes: protecting society from harmful future behaviour, providing an unpleasant deterrent for people considering committing crimes, and isolating convicted people from criminal influences while they are rehabilitated. However, this social isolation also damages mental health and makes it difficult for people who have completed their prison terms to re-enter and reconnect with society, which increases the probability of recidivism. Residents of a space settlement will already experience social isolation, due to their extreme distance from the rest of the human species back on Earth, exacerbated by the communication delay due to the speed of light, which will prevent real-time communication for any space settlements farther away than the Moon. Adding on additional isolation by separating convicted people into prisons away from the rest of the settlement's population would compound these effects. As on Earth, these ethical concerns are also practical ones: if the mental health damage and social isolation inflicted upon incarcerated people also increases their chances of recidivism (compared with non-carceral justice approaches), this decreases the effectiveness of prisons, regardless of the ethical concerns.

Science Fiction Perspectives on Criminal Justice: An Opportunity

Despite the frequency of carceral tropes in mainstream science fiction involving space criminal justice systems, this genre of storytelling also provides a unique opportunity not only for imagining alternative approaches to criminal justice in a future space community, but also for developing non-carceral justice solutions here on Earth today. The exercise of designing a space criminal justice system is, in a sense, an act of creating science fiction, even if it takes the form of serious planning rather than entertainment: This process requires that we first establish circumstances and technologies that do not exist today (but may in the future) and then extrapolate possible social consequences. These sociological thought experiments can guide our approach to criminal justice in the near future on Earth as well as humanity's more distant future in space.

There are no prisons in space today. Thus, the challenge for prospective designers of space communities differs from that of criminologists and criminal justice activists on Earth: space planners have the opportunity to implement new criminal justice systems from scratch, rather than needing to advocate for the reform or abolishment of an existing (carceral) system. This can provide futurists with a kind of psychological flexibility to imagine solutions that would otherwise appear too radical in the context of reform in our modern terrestrial justice systems.

Science fiction already takes advantage of this psychological flexibility to not only criticize our current criminal justice systems, but to propose non-carceral alternatives. Criticisms often take the form of using inhumane justice systems to illustrate the villainy of fictional alien societies. This tool is frequently used in the *Star Trek* universe, in plots in which the protagonists survive and escape gruelling prison conditions while commenting on the injustice of the systems (e.g., Beaumont 1997; Landau 1996; Meyer 1991). Other science fiction narratives imagine humans themselves creating these dystopian punishments, often extrapolating current trends to the extreme: the novel *Brute Orbits* imagines prisoner transportation not to a distant prison colony, but to the empty depths of space (Zebrowski 1999); *Alien 3* and the *Andor* television series portray prison labour camps on remote moons (Fincher 1992; Haynes 2022); the films *Fortress* and *Lockout* are set in for-profit prisons in the near future (Gordon 1992; Leger & Mather 2012).

But as Wetzel's (2014: 4) analysis of space prisons in science fiction observes, while science fiction "normalizes the existence of the prison…it also critiques the prison and begins to think of ways out of all the problems a prison embodies and represents". We can see this latter work performed in abolitionist speculative fiction, which explores alternatives to prisons using science fiction settings. On the inhabited fictional moon of Anarres in Ursula K. Le Guin's classic 1974 novel *The Dispossessed*, children learn about prison in history class but find the concept too alien to grasp; public reprimand from the community is used to discourage unwanted behaviour, and "murderers and chronic work-quitters" may ask to move to an island Asylum for therapy and protection from revenge. Works of feminist utopian fiction like Marge Piercy's *Woman on the Edge of Time* (1976) or Starhawk's *The Fifth Sacred Thing* (1993) envision future communities on Earth, rebuilt after climate disasters and war, in which the response to violent behaviour is mental health treatment, negotiated restitution, shunning, or ultimately exile, rather than imprisonment. Even in Heinlein's *The Moon is a Harsh Mistress* (1966, 163), which portrays a libertarian utopia forming among the freed prisoners on the Moon rather than a collectivist one, imagines public opinion as a more effective response to violence than imprisonment: "If you eliminate a man other than through self-defense, you pay his debts and support his kids, or people won't speak to you, buy from you, sell to you".

Social justice activists have also used science fiction anthologies to gather diverse voices to criticize today's pressing social challenges, including our carceral criminal justice system, and imagine better solutions. For example, the 2015 anthology *Octavia's Brood: Science Fiction Stories from Social Justice Movements* includes stories that explore possibilities for justice that range from banishment to community-moderated relationship-building between a killer and her victim's family (brown & Imarisha 2015). Inspired by *Octavia's Brood*, criminologist Phil Crockett Thomas edited and published a 2022 anthology titled *Abolition Science Fiction* with stories authored by UK-based prison abolition and transformative justice activists and researchers. One of the co-editors of *Octavia's Brood*, adrienne maree brown, observed in a concluding chapter of the volume that "Science fiction is the perfect 'exploring ground,' as it gives us the opportunity to play with different outcomes and strategies before we have to deal with the real-world costs" (brown & Imarisha 2015: 279).

Prison abolitionist Walidah Imarisha (2015: 3), the other co-editor of *Octavia's Brood*, has argued that "all organizing is science fiction. When we talk about a world without prisons; a world without police violence…we are talking about a world that doesn't currently exist. But collectively dreaming up one that does means we can begin building it into existence". Designing space settlements, of course, is also an act of creating science fiction, of imagining a world that does not yet exist to prepare us to build it into existence. Space settlement advocates can therefore take advantage of this science fiction setting to perform the same kind of work in imagining better criminal justice systems than the ones we have today on Earth.

This work should also include the question of how to design a society that reduces or even eradicates the incidence of survival crimes by ensuring that the material needs of all residents of a space settlement are met. This goal may appear fantastical and naively utopian here on Earth, with our massive, globally distributed population and high levels of inequality accumulated over generations. But no one is underfed or unhoused in space today, providing us with a potentially brief window in which we can create an off-Earth population free of the cycle of generational poverty. The earliest space settlements will have no natural sources of air or water, let alone food or habitable shelter, so designers will need to decide from the start what kind of values to build into the architecture and life support of the habitats: in an environment where being too poor to afford air would mean death in a matter of minutes, will poverty be tolerated at all? Or will the basic survival needs of every resident be guaranteed? Today's advocates for transformative justice methods as an alternative to imprisonment argue that we must transform society itself, not just our carceral systems, to bring justice and safety to communities. Creating new communities in space will be transformative by definition; the question that remains is whether we will transform our social systems in just and equitable ways during this transition.

The Risks of Reimagining Justice in Space

There are, however, risks involved in using the context of space settlement to consider alternatives to our modern-day carceral criminal justice system. Hypothetical space settlements are situated at a comfortable distance from today's society, spatially but also in terms of time and reality. As Wetzel (2014: 18) notes, it is "this temporal displacement that allows SF [science fiction] to be more critical than realist, contemporary prison narratives". But this distance can also manifest as an emotional disconnection, making it easier for futurists to dismiss the ethical challenges of their vision of a space criminal justice system in favour of a purely utilitarian approach.

In fact, science fiction frequently explores the idea that the very nature of the space environment—its extreme uninhabitability and scarcity of resources—will force its inhabitants to make difficult and even unjust choices in order to ensure the survival of a human community teetering on the knife's edge of extinction. A classic example of this concept is Tom Godwin's 1954 short story "The Cold Equations", in which a spaceship pilot discovers a young stowaway and ultimately decides to jettison her via the airlock in order to preserve enough fuel to deliver much-needed medicine to a desperate human outpost. "It was a law not of men's choosing but made imperative by the circumstances of the space frontier", the narrative explains, absolving himself of moral culpability due to the circumstances of the ship's mission and the "cold equations" of space travel physics.

This type of execution-by-airlock is itself a common trope in space-based science fiction, a modern version of the pirate story cliché of walking the plank. Popular science fiction television series like *Babylon 5, Battlestar Galactica*, and *The Expanse* portray shipboard airlocks as a mechanism of execution for treason and mutiny (Eisner 2019; Ferrara 1994; Rymer 2005), while in the novel series (and subsequent television adaptation) *The 100*, the last remnants of humanity, living on an orbital space station above the destroyed Earth, eject residents convicted of any crime out of the airlock (Morgan 2013). Even the anarchist lunar society of Heinlein's (1966) *The Moon is a Harsh Mistress* uses the airlock to dispose of men acting against the community's norms, such as behaving as a "bad-tempered straw boss" (19) or "touching a fem [woman] without her consent" (49). Rather than a formal, state-sanctioned execution, these incidents are ironically called "accidents", and the perceived threat of the airlock helped maintain order—or "good manners", as the narrator terms it—among the residents.

Capital punishment remains controversial on Earth, so why is it so popular in our fiction, even for protagonists portrayed as the "good guys"? Perhaps the cold equations of space and the emotional distance between ourselves and future (fictional) space settlements distract us from the ethical concerns that many activists point to with the death penalty on Earth today. Of course, even proponents of capital punishment today generally do not argue for it as a response to *every* crime, so we will need to find alternative criminal justice solutions in space regardless of future settlers' positions on the death penalty.

Science fiction can also be a vehicle for perpetuating harmful mythologies about humanity's history—and future—of colonization. *Star Trek*'s designation of space as "the final frontier" calls to mind the frontier of European settlement as it expanded across North America; the science fiction series was originally pitched as a "*Wagon Train* to the stars", referencing a long-running Western television series that aired in the U.S. in the 1950s and 1960s. Space settlement is more commonly referred to as "space colonization" among advocates, and political arguments for American funding of space activities in particular uses colonization metaphors like "America's manifest destiny in the stars" (Koren 2020). Recent years have seen a growing pushback against colonization rhetoric for space, noting that the history of colonization on Earth—including in North America—was rife with genocide and environmental destruction, and that the harmful legacies of colonization affect a significant fraction of the human population today (e.g., Nesvold 2023; Rubenstein 2022; Schwartz et al. 2023; Treviño 2020).

But for the purposes of this chapter, consider the overlap between the execution-by-airlock trope and the concept of "frontier justice" in fiction set in the American Old West. This kind of extrajudicial capital punishment is often portrayed as an unfortunate necessity to maintain order at the edge of "civilization", given the absence of a more formal and "civilized" justice system. Like the cold equations of orbital physics, the remoteness and harsh conditions of frontier towns requires a harsher response to criminal behaviour that threatens this order, and thus the survival of these communities. Setting aside the question of whether extrajudicial capital punishment actually decreased crime and increased safety in real-world American settlements during the expansion across North America, note that the concept of "frontier justice" was used to justify the lynching of Black men well into the twentieth century, long after the frontier was settled. The narrative long outlived the setting it was created for. As artist Ken Gonzales-Day (2006: 10) notes in his book *Lynching in the West: 1850–1935*: "The irony is that while terms like 'frontier justice or popular tribunal' still invoke images of cattle rustlers and stagecoach robbers meeting justice on a lawless frontier, they also mask a history of racial violence in a region that was not only culturally diverse but still is".

This dangerous concept of "frontier justice"—dangerous not least because it can and has been applied disproportionately to marginalized communities—still exists in our culture, as exemplified in the frequent presence of airlocks in today's science fiction. If we do not make an effort to design more just and ethical criminal justice systems for future space settlements, and instead assume that the first settlers will simply bring modern and forward-thinking systems with them from Earth, we risk them bringing this kind of cultural assumption with them as well, and convicted space residents may someday be harmed by the "cold equations" of our poor planning as much as by the uncaring vacuum of space.

Conclusion

Science fiction writers and space settlement planners are not, for the most part, experts in criminology or criminal justice advocacy. It is therefore unsurprising when, through this ignorance, they are drawn to the simplest possible solutions for the incredibly complex problems of criminal justice when they consider humanity's future in space: either the creation of space prisons (thus reproducing our most common current criminal justice models from Earth in space) or execution-by-airlock. However, both of these approaches will not only be impractical in the space environment, they also ignore the work of criminologists and activists who have spent generations working for better, more effective, and more ethical justice systems here on Earth.

Science fiction—and, by extension, proposing designs for actual future communities in space—provides an opportunity for us to develop more radical, non-carceral approaches to criminal justice than we have today. Because the criminal justice systems of humanity's first space settlements will be established in locations that, by definition, have no previously existing justice systems already in place, space settlement planning allows us to consider approaches that would be difficult to implement within our current systems on Earth. Residents of early space settlements will not be fighting for reform in the face of tradition and inertia; they will be creating new systems from scratch.

These unusual circumstances have led some space settlement advocates to argue that these isolated, brand-new human communities in space will act as "independent social laboratories", providing a unique opportunity for "experiments in society-building" (O'Neill 1977: 202). This argument applies to criminal justice as much as it does to political science or economics: we could, as a species, intentionally implement a variety of different criminal justice systems in different space settlements and then measure and compare the outcomes. This is, of course, the kind of proposal that one is more likely to hear from a researcher (like physicist Gerard K. O'Neill) whose work does not involve human subjects, and who therefore does not frequently need to consider the ethics of performing research on humans. Unlike adrienne maree brown's "exploring ground" of science fiction, failed experiments in these laboratories in space will have real-world costs. Residents of space settlements will be our descendants, not our social science guinea pigs. While future criminologists will undoubtedly eventually study how these communities respond to criminal behaviour, we should try to give these societies the best possible start by carefully and deliberately applying what we have learned on Earth to design effective and ethical off-Earth criminal justice systems.

We must also be cognizant of the risks of the temporal and spatial distance between us and the hypothetical residents of future space settlements, and this can translate into a psychological distance that can make it too easy for us to ignore the human consequences of criminal justice as we speculate about possible solutions. Objectivity is crucial in any field of research, including both criminology and space science, but the social scientists in these conversations must work to ensure that we do not lose sight of the ethical considerations of this work, given that most of the physical scientists advocating for space settlement have less formal ethical training. With all of these considerations in place, we have the opportunity to harness the public's excitement for space to develop a better vision for humanity's future together, in space and on Earth.

References

American Civil Liberties Union (2022) *Captive Labor: Exploitation of Incarcerated Workers.* From: https://www.aclu.org/report/captive-labor-exploitation-incarcerated-workers.

Beaumont G (1997) *Star Trek: Deep Space Nine.* Season 5, Episode 14. "In Purgatory's Shadow." Aired February 10, 1997.

Bole C (1996) *Star Trek: Voyager.* Season 2, Episode 16. "Meld." Aired February 5, 1996, on UPN.

brown a & Imarisha W (eds) (2015) *Octavia's Brood: Science Fiction Stories from Social Justice Movements.* AK Press and the Institute for Anarchist Studies.

Cockell C (2016) Exoconfac – The Extraterrestrial Containment Facility: An Essay on the Design Philosophy for a Prison to Contain Criminals in Settlements Beyond the Earth. *Journal of the British Interplanetary Society* 69: 27–30.

Davis A (2003) *Are Prisons Obsolete?*New York: Seven Stories Press.

Edgemon T & Clay-Warner J (2018) Inmate Mental Health and the Pains of Imprisonment. *Society and Mental Health* 9 (1): 33–50.

Eisner B (2019) *The Expanse.* Season 4, episode 10. "Cibola Burn." Released on December 12, 2019 on Amazon Video.

Ferrara L (1994) *Babylon* 5. Season 1, episode 21. "The Quality of Mercy." Aired August 17, 1994, on Prime Time Entertainment Network.

Fincher D (1992) *Alien* 3. USA: 20th Century Fox.

Godwin T (1954) The Cold Equations. *Astounding Science Fiction.* August 1954.

Gonzales-Day K (2006) *Lynching in the West: 1850–1935.* Durham, NC: Duke University Press.

Gordon S (1992). *Fortress.* USA: Miramax and Australia: Roadshow Film Distributors.

Grassian S (2006) Psychiatric Effects of Solitary Confinement. *Washington University Journal of Law & Policy* 22: 325–383.

Haynes T (2022) *Andor.* Season 1, episode 10. "One Way Out." Released November 9, 2022, on Disney+.

Heinlein R (1966) *The Moon is a Harsh Mistress.* New York: G. P. Putnam's Sons.

Imarisha W (2015). Rewriting the Future: Using Science Fiction to Re-Envision Justice. *Bitch Media.* From: https://www.bitchmedia.org/article/rewriting-the-future-prison-abolition-science-fiction.

Kain E (2022) Andor's Brilliant Prison Trilogy Takes 'Star Wars' to New Heights. *Forbes.* From: https://www.forbes.com/sites/erikkain/2022/11/13/andors-brilliant-prison-trilogy-takes-star-wars-to-new-heights/?sh=5312467919a5.

Kolbe W (1998) *Star Trek: Voyager.* Season 5, episode 9. "Thirty Days." Aired December 9, 1998, on UPN.

Koren M (2020) No One Should 'Colonize' Space. *The Atlantic.* September 17, 2020. From: https://www.theatlantic.com/science/archive/2020/09/manifest-destiny-trump-space-exploration/612439/.

Kroeker A (2004) *Battlestar Galactica.* Season 1, episode 3. "Bastille Day." Aired November 1, 2004, on Sky1.

Kutner L (1968) A World Outer Space Prison: A Proposal. *Denver Law Review* 45 (5): 702–718.

Landau L (1996) *Star Trek: Voyager.* Season 3, episode 3. "The Chute." Aired September 18, 1996 on UPN.

Le Guin U (1974) *The Dispossessed.* New York: Harper & Row.

Leger S & Mather J (2012) *Lockout.* Brussels: EuropaCorp.

Lucas G (1977) *Star Wars.* USA: 20th Century-Fox.

Meyer N (1991) *Star Trek VI: The Undiscovered Country.* USA: Paramount Pictures.

Morgan K (2013) *The* 100. New York: Little, Brown Books for Young Readers.

Morris N & Izenberg J (2023) Mental Health and Legal Implications of Access to the Outdoors during Incarceration. *Journal American Academy of Psychiatry Law* 53: 103–111.

Nesvold E (2023) *Off-Earth: Ethical Questions and Quandaries for Living in Outer Space.* Boston: MIT Press.

Novisky M & Peralta R (2020) Gladiator School: Returning Citizens' Experiences with Secondary Violence Exposure in Prison. *Victims & Offenders* 15 (5): 594–618.
O'Neill G (1977) *The High Frontier: Human Colonies in Space.* New York: William Morrow.
Pettit B & Gutierrez C (2018) Mass Incarceration and Racial Inequality. *American Journal of Economics and Sociology* 77 (3–4):1153–1182.
Piercy M (1976) *Woman on the Edge of Time.* New York: Knopf.
Rubenstein M-J (2022) *Astrotopia: The Dangerous Religion of the Corporate Space Race.* Chicago: University of Chicago Press.
Rymer M (2005) *Battlestar Galactica.* Season 2, episode 1. "Scattered." Aired July 15, 2005, on Sci Fi Channel.
Schaub M (2018) Antarctica Scientist Allegedly Stabs Colleague for Spoiling the Endings of Books. *Los Angeles Times.* From: https://www.latimes.com/books/la-et-jc-anarctica-stabbing-books-20181030-story.html.
Schwartz J, Billings L & Nesvold E (2023) *Reclaiming Space: Progressive and Multicultural Visions of Space Exploration.* Oxford: Oxford University Press.
Soble L, Stroud K & Weinstein M (2020) Eating Behind Bars: Ending the Hidden Punishment of Food in Prison. *Impact Justice.* From: https://impactjustice.org/impact/food-in-prison/#report.
Starhawk (1993) *The Fifth Sacred Thing.* New York: Bantam Books.
Thomas P (2022) Abolition Science Fiction. From: https://abolitionscifi.org/.
Treviño N (2020) *The Cosmos is Not Finished.* Dissertation. University of Western Ontario.
Twohy D (2004) *The Chronicles of Riddick.* USA: Universal Pictures.
Wetzel A (2014) *Alienating Punishment: Prisons in Science Fiction.* Dissertation. University of Washington.
Zebrowski G (1999) *Brute Orbits.* New York: HarperPrism.

11 Defining Space Debris Policy

A Perspective from Rawls' Veil of Ignorance

Sabine Winters

Introduction

The launch of Sputnik I in 1957 marked the beginning of the era of human-generated orbital debris and presented a significant challenge for space exploration and utilization. Space debris, which is defined as non-functional man-made objects that orbit Earth, particularly in low Earth orbit (LEO), has proliferated since then, with millions of pieces posing a threat to ongoing space activities (Pelton 2013). Efforts to regulate space debris involve a combination of international treaties and soft law instruments, such as the Inter-Agency Space Debris Coordination Committee (IADC) Guidelines, which aim to foster international consensus on mitigation (IADC 2024).

Defining terms related to space debris is essential for clarity and effective regulation, yet ambiguity in definitions poses challenges, particularly in addressing issues like military activities in space and liability for environmental damage. Political considerations, including geopolitical interests, influence how space debris is defined and regulated by spacefaring states, highlighting the need for stronger international cooperation and clearer legal frameworks in space governance.

However, the complexity of defining space debris reflects broader political, legal, and strategic interests in space exploration, which complicates efforts to achieve consensus on regulation (Meegan 2021). Challenges in space debris governance include the need for robust legal frameworks, clear definitions and international cooperation. These definitions should first focus on reconciling the competitive instincts inherent among states by providing incentives for greater collaboration. Second, they should impose stricter normative agreements on the limitation of space debris. Reaching effective agreements on the space debris issue requires a vision of extended timelines that often surpass immediate political and personal horizons and outcomes that may be uncertain and abstract.

In the *Theory of Justice* (1971), John Rawls (1921–2002) introduced the *original position*. The original position is a concept employed to determine principles of justice that individuals would agree to under conditions of fairness and equality. In this thought experiment, individuals are imagined to be

DOI: 10.4324/9781003437178-12

behind a 'veil of ignorance', meaning they are unaware of their own characteristics, such as their social status, wealth, or talents, as well as their personal preferences or circumstances. In this state of uncertainty, they would, according to Rawls, choose principles that safeguard the interests of all members of society, regardless of their personal attributes (Rawls 1999).

Applying Rawls' veil of ignorance to space governance, offers a framework for establishing norms and addressing concerns of intergenerational justice regarding space debris accumulation. However, there are practical limitations to implementing Rawls' theory in the context of space governance. Disparities in power and resources among spacefaring nations challenge the assumption of equality inherent in the original position. Nevertheless, the original position can serve as a tool for elucidating underlying values, fostering public debate, and formulating normative soft law.

This chapter proposes an application of Rawls' concept of the original position to examine notions of fairness and justice in the context of crafting a sustainable policy on space debris. It delves into the challenges posed by the soft law guidelines established by the IADC, with a particular focus on the ambiguity of certain definitions and difficulties in achieving voluntary compliance with agreements based on these definitions. The aim of this chapter is to advocate for the use of the veil of ignorance as a constructive approach to providing feedback on non-binding international space guidelines and definitions. Rather than introducing another normative framework atop existing guidelines, the goal of applying the veil of ignorance is to uncover underlying values, shape concepts and establish principles related to space debris governance through impartial and collaborative efforts.

Space Debris Context

The launch of Sputnik I in 1957 not only led to the formation of human-generated orbital debris, which consisted of the initial rocket stage and the satellite itself, but also underscored the importance of monitoring such objects in space. Since Sputnik I, man-made space debris has only increased. Researchers have found that millions of pieces of debris are circulating in various Earth orbits, predominantly in LEO, but also in medium Earth orbit and geosynchronous Earth orbit (Pelton 2013). Statistical models estimate that there are over 30,000 space debris objects larger than 10 cm and approximately 900,000 objects larger than 1 cm in orbit. Smaller fragments are potentially dangerous because they can disable spacecraft or damage sensitive instruments, while larger fragments may result in complete spacecraft fragmentation upon collision, especially in LEO, which can generate hundreds to thousands of additional orbiting fragments (IADC 2024). This cascade effect, known as the Kessler syndrome, poses a significant threat over the next decades and centuries, although the likelihood of immediate danger is low (Barry 2022). However, the 650 km to 1,000 km altitude region of LEO faces the highest orbital densities and collision velocities, with collision probabilities for satellites estimated at around 0.8% annually, a risk that increases with time (Barry 2022). Thus, the proliferation of human-made space debris in orbit has emerged as a significant and pressing concern.

In recent years, there has been exponential growth in natural debris generation, which is intensified by the rise in launches, particularly of planned large constellations of satellites. Predictive models suggest that there will be a significant increase in space debris over the next few centuries (ESA 2023). As the space industry expands, the likelihood of incidents related to the accumulation of space debris and atmospheric pollution in LEO is also expected to increase (ESA 2023: 117–118). Growing concern over space debris has garnered global attention and emerged as a prominent challenge. However, despite the fact that space agencies such as the European Space Agency (ESA) and National Aeronautics and Space Administration (NASA) have devised effective monitoring and modelling approaches that can be applied to debris of varying sizes across different orbital zones, it is noteworthy that the practical resolution of the escalating issue of accumulating space debris mostly remains unaddressed (Leonard & Williams 2023). While there are global efforts to ensure compliance to space debris mitigation, the number of objects larger than 10 cm in LEO is projected to double over the next 50 years (IADC 2024).

Soft Law

The primary framework of contemporary outer space law comprises five international treaties established by the United Nations Committee on the Peaceful Uses of Outer Space (COPUOS). COPUOS is the main United Nations (UN) body dealing with space-related matters and was established by the General Assembly in 1959 (Larsen 2018). While not legally binding, COPUOS guidelines represent best practices endorsed by the international community. Within the framework, there are several legally binding space law treaties and agreements that have been developed and overseen by member states. One of the treaties under the purview of COPUOS is the Outer Space Treaty (OST). Adopted in 1967, the OST is the non-legally binding cornerstone document governing outer space activities (Yang 2023).

Apart from COPUOS, there are a number of treaties that convey space principles adopted by the UN General Assembly, but that are not legally binding. These so called 'soft law' instruments emerged to govern space activities, including recommendations, guidelines, codes of conduct and other statements of expected behaviour (Beard 2016). Established in 1993, the Inter-Agency Space Debris Coordination Committee (IADC) composed a set of guidelines that forms a representative document of international soft law related to space debris regulation. Although the IADC Guidelines have limitations in terms of their enforceability and binding force, they aim to create a fundamental consensus on space debris mitigation among all countries (IADC 2024). In 2021, the latest iteration of the IACD Space Debris

Mitigation Guidelines was released; it represents a significant contribution to addressing the accumulation of space debris.

The IACD guidelines are appealing for voluntary and proactive contributions to implementing the guidelines in space debris policy, especially the long-term systematic space activities and requirements surrounding definitions related to space debris (Yang 2023). A fundamental challenge faced by space powers and other nations in relation to space debris revolves around the need for robust legal frameworks that promote international consensus and provide clear definitions for effectively tackling the challenges of mitigating and removing space debris (Meegan 2021). Because tackling the space debris challenges involves multiple countries and stakeholders, space policy requires extensive international cooperation. The IACD Guidelines and codes of conduct lack law enforcement mechanisms, but play a crucial role in establishing international norms and serve as a foundation for regulating future space endeavours (Ferreira-Snyman 2013).

While soft law instruments are typically characterized by reduced obligations and imprecise language due to their non-binding nature (Beard 2016), they still hold significant potential to influence state behaviour. Indeed, soft law tools such as guidelines, resolutions and principles are crucial for setting norms and regulating activities in outer space (Beard 2016). However, soft law documents often rely on imprecise or ambiguous language due to their non-binding nature. Therefore, developing definitions would enable greater clarity and understanding of the intended scope and application of the principles or guidelines outlined in the soft law instrument. By clearly defining terms and concepts, soft law documents can effectively communicate expectations and guide behaviour among stakeholders. In turn, this clarity may contribute to the implementation and effectiveness of law instruments by reducing ambiguity and promoting consistent interpretation and application. This approach would result in the emergence of 'political normativity' among participants and broader stakeholders (Bantekas & Akestoridi 2023). The term 'normative' does not solely indicate 'ought to be' but rather denotes norms and their significance within the political sphere. Normative theory not only provides a solid foundation for comprehending existing norms but also enables the critique of these norms and their origins (Pietrzyk-Reeves 2017).

Emerging challenges with space debris necessitate the ongoing consideration of future-proof solutions and stronger international collaboration in which each space project has its own Space Debris Mitigation Plan formulated based on IADC guideline systems. To effectively utilize the IADC Guidelines to address space debris, the global community must unify its stance and pinpoint mutual interests across all nations. To reach a consensus on space debris policy, including obligations, it begins with addressing the different interpretations of definitions, transcending the alignment of soft law with stakeholders' individual interests.

Definitions of Terms

Terms such as 'space debris' do not merely have a technical or scientific definition but involve complex negotiations and considerations shaped by political, legal and strategic interests. The lack of more clearly defined and enforceable definitions significantly increases the risk of a detrimental conflict in space (Ligor & Matthews 2022). For example, all activities that take place in outer space are intrinsically linked to earthbound activities and can therefore intersect with national security concerns (Ferreira-Snyman 2013). However, international space governance faces significant challenges in creating a safer and more secure space environment.

These obstacles include an increase in the number of participants in space activities; the scale and diversity of stakeholders plays a pivotal role in ensuring the stability and infrastructure needed to uphold shared spaces. The values, interests and goals related to space debris and held by stakeholders greatly vary, which entails challenges in reaching a form of consensus (Barry 2022). For example, ambiguity in the definitions of terms such as 'space debris' and 'sustainability' can carry significant political implications. Moreover, crucial treaty terms such as 'debris', 'space object', 'interference' and 'contamination' lack universally agreed-upon and legally binding definitions. Such ambiguity enables stakeholders to assert that actions such as destructive anti-satellite tests, which generate numerous hazardous debris fragments, do not technically violate the OST (Ligor & Matthews 2022). While states have pledged to solely employ space for 'peaceful purposes', the ambiguity of this term has historically led to conflicting interpretations. The lack of clear explanation of the terms of the definition 'peaceful purposes' allows space utilization 'for military purposes as long as they are non-aggressive', leading to an increase in diverse military activities in space and exacerbating issues such as space debris. One example is China's anti-satellite weapon test in 2007. Nicholas Johnson of NASA's Orbital Debris Program Office described this event as the most severe satellite fragmentation of the past 50 years, and numerous pieces of debris are expected to remain in orbit for more than a century (Beard 2016).

From a political perspective, spacefaring states may seek to define 'space debris' in ways that enable the protection of their assets in orbit, most likely also influenced by geopolitical considerations and relationships (Ligor & Matthews 2022). This is because spacefaring states may argue that activities such as satellite deployment, or space exploration missions, do not specifically constitute 'interference' or 'contamination' according to the current treaty language (Ligor & Matthews 2022). In terms of liability, Article VII of the OST and the Convention on International Liability for Damage Caused by Space Objects stipulate that both launching states and contributing states bear international responsibility for any damage caused by the entities that they launch into outer space (Yang 2023). Nevertheless, both of the aforementioned documents limit the definition of 'damage' to harm inflicted upon state parties or private individuals and do not explicitly include harm to the environment (Yang 2023).

Public Debate

Given the threat of the growing space debris problem, there is a crucial need for a framework of revised definitions that addresses sustainability issues in space governance. These definitions should first focus on reconciling the competitive instincts inherent among states by providing incentives for greater collaboration. Second, they should impose stricter normative agreements on the limitation of space debris. Reaching effective agreements on the space debris issue requires a vision of extended timelines that often surpass immediate political and personal horizons and outcomes that may be uncertain and abstract. However,

> in determining whether to adhere to international soft law, nations are guided by their own interests and the alignment of the soft law with their current and long-term objectives. To fully leverage the role of the IADC Guidelines in space debris mitigation, the international community must establish a consensus and identify common interests among all countries. This approach, on one hand, enhances the effective compliance of all countries with the IADC Guidelines and, on the other hand, supports the formulation of new rules of international law.
>
> (Yang 2023: 78–79)

Moreover, achieving an international consensus on what is considered reasonable in the realm of space debris policy, encompasses a broader array of stakeholders, such as environmental institutions, academia, experts, policymakers and non-governmental organizations. To benefit the consensus, it is of importance that all relevant members (stakeholders) of the community have access to the discussion (Wall 2022).

A debate that includes policymakers as well as the very practitioners (both public and private) who work with the technicalities of space debris problems, provides valuable input in policymaking through their experiences, translating technical expertise into political norms (Levit 2005). Definitions formulated through public debate promote mutual cooperation and restraint. These definitions serve as standards by which behaviours are guided through mutual agreement on what is just and reasonable (Levit 2005). Soft law, although not legally binding, relies on mutual understanding among participants. This understanding, nurtured through public debate, can lead to a consensus based on the reasonableness of stakeholders. Rawls' veil of ignorance can guide the shaping of definitions by fostering an attitude of objectivity and appealing to stakeholders' reasonability.

Theory of Justice

Since the release of Rawls' *A Theory of Justice* in 1971, the field of political philosophy and (by extension) normative political theory, has experienced significant growth and re-established itself as a crucial avenue for research in political science. This resurgence acknowledges the pivotal role that values play as the essence of political systems and structures, serving as intermediaries in both prescriptive and descriptive analyses of politics (Pietrzyk-Reeves 2017).

Rawls' theory aimed to establish the fundamental principles that should guide societal institutions. To derive these principles, Rawls proposed a thought experiment named the original position, which entails a process based on impartial agreement and in which impartiality is ensured through a 'veil of ignorance'. In essence, Rawls' approach involves the discussion of principles without knowledge of one's own circumstances. The veil ensures that parties involved in the agreement do not have access to any specific information about their own circumstances, advantages, or disadvantages in society. This means they remain unaware of their personal characteristics, such as gender, social status or economic position. In this state of uncertainty, they would, according to Rawls, choose principles that safeguard the interests of all members of society, regardless of their personal attributes (Rawls 1999).

The first principle, known as the *principle of equal basic liberties*, asserts that each individual possesses an equal right to the most extensive basic liberties in alignment with a similar system of liberty for all (Davies 2020). This principle underscores fundamental rights such as freedom of speech, assembly and personal property, which ensure that all individuals, regardless of their social or economic standing, enjoy equal access to essential liberties. The second principle, the *difference principle*, addresses social and economic inequalities. Rawls suggested that these inequalities should be structured to benefit the least advantaged members of society (Davies 2020). While acknowledging the existence of disparities in wealth, income and other social goods, Rawls advocated for a system in which such inequalities contribute to the well-being of the most disadvantaged individuals, thus promoting a more just and equitable society. The concept of *reflective equilibrium* in justice as fairness entails an ongoing process of reflection and adjustment of principles (Davies 2020).

Essentially, Rawls' veil of ignorance establishes objectives and criteria that societies should aspire to achieve, even if their immediate realization is challenging (Pietrzyk-Reeves 2017). Rawls' model is often seen as a way to reach consensus by considering the perspective of an impartial spectator and prompting the parties involved to contemplate what ought to be. Everyone who occupies the common standpoint shares this motive. 'Justifications from the impartial standpoint, are therefore, shared justifications' (Wall 2022: 4). Through the veil of ignorance, the original position exemplifies the normative beliefs and aspirations that typically guide thinking about justice within a democratic society.

Reasonability

Making a moral judgment entails consideration of another person's perspective and, in doing so, acknowledgment of this perspective to the extent that it could genuinely influence one's own viewpoint. Thus, reasonability requires impartiality and a certain kind of 'objectivity' and expresses itself in the notion of equity (Sibley 1953). Adopting a reasonable perspective implies a responsibility to consider the judgments of others within a framework devoid of instrumental, self-interested reasoning. Rawls contended that 'the reasonable is public in a way that the rational is not' (Rawls 1993: 53, in Rasmussen 2003: 102). He further suggested that 'it is by the reasonable that we enter as equals the public world of others and stand ready to propose, or to accept, as the case may be, fair terms of social cooperation with them' (Rawls 1993: 53, in Rasmussen 2003: 102). The introduction of 'reasonableness' in Rawls' work reflects a practical recognition of the diversity inherent to modern culture and its fundamental plurality. Furthermore, it allows morality to be viewed not merely as a constraint on self-interest but also more importantly as an independent normative criterion that gives way to implicit obligation regarding the political and economic interests of others.

What is considered reasonable is influenced by various types of evidence, the significance and strength of evidence, the ambiguity of concepts and their interpretation, the impact of individual experiences, diverse normative considerations and conflicting values (Rasmussen 2003). In finding a consensus in defining guidelines and definitions, it does not necessarily mean that the stakeholder does not understand the measures. The justification of a guideline, or the definition of a term, must meet the standard of reasonableness. Through the transparency of public debate, where all relevant members of the community are included in the consideration of forming the definition, there is an appeal to the notion of acceptability (Wall 2022). Acceptability relies on the reasonableness of the involved parties rather than their pursuit of personal advantage. In this context, rational endorsement entails accepting a political decision that may not be optimal but is nonetheless tolerable (Wall 2022).

Rawls acknowledged that, as a consequence of conflicting views of what is reasonable, expecting universal agreement in a public political culture is unrealistic. Moreover, in discussions regarding nonfundamental political matters, finding consensus in public reason might not always be feasible or beneficial (Wall 2022). However, he suggested that a robust and functional democracy relies on citizens' ability to navigate these challenges with a sense of fairness and respect for diverse viewpoints. This view underscores the idea that tolerance in a democratic society involves the acknowledgement and management of the inherent difficulties and disagreements that arise from the diversity of perspectives within a community (Rawls 1993). Thus, reasonability recognizes the concerns of others by impartially representing their interests and responsibilities. What is 'reasonable' must not be solely understood as a moral agreement but also as a political obligation.

Applying the Veil

Guidelines as proposed by the IADC in outer space activities are essential for establishing norms and regulating behaviour in space, especially when legally binding agreements are challenging to achieve. Rawls' veil of ignorance can serve as a valuable tool for creating a foundation for progress and facilitating cooperation among spacefaring and non-spacefaring nations. A feasible way to conduct the original position by Rawls, would be through a method (e.g., a working group) that presents feedback to IACD, with a focus on narrowing down the ambiguity of definitions in the already formed principles.

Primarily, the working group aims to clarify the present understanding of the damage caused by space debris in orbit, making it both meaningful and actionable for stakeholders. The context of justice as fairness emphasizes the fair distribution of societal benefits and burdens. This perspective acknowledges that fairness does not always require a strict quid pro quo or reciprocity in every interaction. If the underlying rules and structures that govern space activities are deemed just through the veil of ignorance, then differences in outcomes between various entities are not inherently indicative of injustice within the system. In particular, Rawls' difference principle suggests that social and economic inequalities should benefit the least advantaged.

Behind the veil of ignorance, developing an understanding of orbital damage is an impartial process that is free from bias towards any specific stakeholder's interests. This ensures that the understanding reached is objective and comprehensive and prioritizes the common good over individual or group interests. The promotion of normative reasonable behaviour aims to establish passive rules that benefit the entire stakeholder community rather than specific individuals or organizations. These norms can be further discussed in conversations in which financial, technological and data-sharing benefits and responsibilities are contextualized for stakeholders and norms are reaffirmed with continued enforcement. Decisions about enforcement should prioritize the preservation of the commons and the well-being of all stakeholders. Any changes resulting from the removal of negative externalities and the establishment of new norms should be evaluated based on their overall impact on the stakeholder community, with the goal of ensuring sustainability and fairness for all involved.

Intergenerational Justice

While the original position, as presented in *A Theory of Justice*, was primarily formulated within the context of domestic societies and aimed to address issues of justice within a nation-state, Rawls (1971) acknowledged the relevance of his theory to the global context, particularly in later works and discussions. He extended his ideas of justice to an international scale in his later work, *The Law of Peoples* (Rawls 1993). In this work, he introduces the concept of 'peoples' as political entities in acknowledgement of the pluralistic nature of the international system. While Rawls did not propose a comprehensive global theory akin to his domestic theory of justice, he outlined principles for just international relations.

Essentially, the veil of ignorance prevents individuals from exploiting specific circumstances to formulate principles, norms or institutions that would benefit their own interests (Rawls 1999: 17). Instead, all are equally represented because each person must make choices under the same conditions of perfect informational symmetry (Klaser et al. 2021). Rawls emphasized the impracticality of including all human generations in the decision-making process and preferred a model grounded in present realities. Despite this constraint, Rawls' social contract theory holds promise for just intergenerational approaches and offers a framework for deriving intertemporal norms from intragenerational considerations to present a viable solution to the challenge of resource allocation across generations (Klaser et al. 2021).

It is important to bear in mind that space debris is an intergenerational problem that will only worsen over the upcoming decade. The space community, therefore, should be invested in selecting a governance framework for space debris that is not only just but also stable over time. To achieve this, a blindfold must be applied to not only strip stakeholders of context but also to set aside present-generation privilege. This approach would allow us to establish intertemporal norms from within our own generation and enable perspectives to be considered across different time periods (Klaser et al. 2021; Rawls 1999). If present stakeholders are veiled from knowing the specific moment of history to which they belong, they can agree on and comply with fair principles which regulate the accumulation and mitigation of space debris (Rawls 1999).

Practical Limitations

There are doubts about the 'practicality' of a normative theory. Firstly, the original position assumes that individuals who are behind the veil begin from a position of equality. Moreover, Rawls made a number of assumptions about the thought experiment. For instance, he posited that a 'society' (or the structure of an organization) must be arranged in such a way that it is:

> well-ordered when it is not only designed to advance the good of its members but when it is also effectively regulated by a public conception of justice. That is, it is a society in which (1) everyone accepts and knows that the others accept the same principles of justice, and (2) the basic social institutions generally satisfy and are generally known to satisfy these principles.
>
> (Rawls 1971: 4–5)

However, nations, space-related entities and other stakeholders have vastly unequal resources, technological capabilities, and historical context. Thus, the equality assumption may not accurately reflect the disparities in power and influence among spacefaring nations.

Similarly, delving into the notion of a democratic society in the realm of space activities entails an understanding of the dynamics of social cooperation among spacefaring entities and institutions and the principles of societal reciprocity and fairness (von Platz 2017). If matters of basic justice and the basic rights of citizens are still points of discussion viewed worldwide, then in what way could the original position serve for international and intergenerational complex issues such as space debris policy? Already, the utilization of the term 'democracy' within the domain of space exploration and spacefaring poses a unique challenge. To democratize a system implies subjecting the system to democratic oversight. Thus, contemplating the role of space within a democratic global framework also means the recognition of space as a shared global resource and a critical reflection on the interpretation of 'democracy' and power structures.

Space is currently the stage for political power displays; thus, it continues to evolve into a highly militarized, contested and dangerously insecure domain in which many states view each other's activities with great suspicion and seek to counter growing, perceived threats (Beard 2016). Furthermore, in space debris policy, in which social, technological and economic values play a significant role, the challenge lies in identifying universal values. Setting aside all values, historical context and aims to arrive at normative principles is striving for a utopia. It is important to regard the thought experiment as an ideal for public debate, fostering public reason through discussions that expose inequality and establish norms for a just and fair space debris policy system.

The original position is a methodology for reflection that prioritizes the articulation of principles (Jaggar 2009) and can be employed to discuss ambiguities related to the guiding principles proposed by the IADC. The original position is not intended to introduce extra elements into the discussion on space debris but rather to closely examine the definitions and interpretation of definitions by space institutes such as the ESA and NASA, as well as commercial spacefaring institutes like SpaceX and spacefaring nations. In doing so, the original position serves as a method for elucidating the underlying values of stakeholders. Moreover, the original position is a tool for formulating normative, non-binding soft law. These tools are often used in areas where formal legal agreements are difficult to reach or where flexibility and adaptability are necessary, for example, because of rapid technological developments or the involvement of a variety of stakeholders with vast differences in values and objectives (Beard 2016).

Conclusion

International agreements such as the IADC emphasize that space is not subject to national appropriation, but geopolitical motivations play an important role in agreeing on law-binding definitions. Public debate plays a crucial role in shaping definitions, fostering mutual cooperation and restraint and establishing standards for guiding behaviour based on a shared agreement on what is just and reasonable. Soft law is not legally binding. Instead, it relies on mutual understanding among participants, fostered through public debate, to build consensus rooted in the stakeholders' reasonableness, thus emphasizing this process.

Additionally, Rawls' veil of ignorance can contribute to shaping definitions by encouraging objectivity and appealing to stakeholders' sense of reasonableness.

Clashing values require a forward-looking policy to promote justice for future generations of space users and a sustainable relationship with the space environment and its implication for Earth inhabitants. The most fitting conception of justice for managing space debris in a spacefaring society is one that effectively addresses and embodies these multifaceted values and objectives in well-formulated definitions. Above all, the value of political justification through public reason should be understood as an ideal. Something to strive for. The original position inherently conveys a dedication to a form of impartiality necessary to guarantee that the principles we choose reflect commitments to fairness and equal responsibility. In essence, the original position serves as a soft law tool to help think about commitments regarding space debris and provides clarity on existing power structures and the interpretation of definitions.

A Rawlsian approach to issues such as space debris and the sustainable management of spacefaring activities calls for a reflective and morally responsible attitude towards the future of space exploration. In practical terms, reasonability rooted in public reason questions existing power structures and contributes to understanding of political justification. Public reason improves, and makes better in one respect, the political order that realizes it. To exercise normative authority over space community, the authority needs to be justified on a consensus based on rationality and acceptability.

Rawls' concept of the veil of ignorance emphasizes the need to formulate rules for a just and fair spacefaring policy, without knowledge of one's future position or interests. The current state of space exploration, which is marked by uncertainties about debris responsibility and a lack of rules for a sustainable future in space, presents an opportunity to create equitable regulations. Because of the current uncertainties and weak organization in space debris policy, there is an opportunity to apply the original position in practice rather than merely as a thought experiment.

References

Bantekas I & Akestoridi K (2023) Sustainable Development Goals, Between Politics and Soft Law: The Emergence of 'Political Normativity' in International Law. *Emory International Law Review* 37(4): 499–560.

Barry K (2022) Space Debris Mitigation and Remediation: Historical Best Practices and Lessons Learned for Economically Preserving and Utilizing Common Areas. *New Space* 10(3): 246–258.

Beard JM (2016) Soft Law's Failure on the Horizon: The International Code of Conduct for Outer Space Activities. From: https://digitalcommons.unl.edu/spacelaw/87/.

Davies B (2020) John Rawls and the 'Veil of Ignorance'. From: https://open.library.okstate.edu/introphilosophy/chapter/john-rawls-and-the-veil-of-ignorance/.

ESA (European Space Agency) (2023) *ESA's Space Environment Report 2023*. From: https://www.esa.int/Space_Safety/ESA_s_Space_Environment_Report_2023/.

Ferreira-Snyman A (2013) The Environmental Responsibility of States for Space Debris and the Implications for Developing Countries in Africa. *The Comparative and International Law Journal of Southern Africa* 46(1): 19–51.

IADC (Inter-Agency Space Debris Coordination Committee) (2024) *IADC Report on the Status of the Space Debris Environment*. From: https://www.iadc-home.org/documents_public/view/id/248#u/.

Jaggar AM (2009) L'Imagination Au Pouvoir: Comparing John Rawls's Method of Ideal Theory with Iris Marion Young's Method of Critical Theory. In Tessman L (ed.) *Feminist Ethics and Social and Political Philosophy: Theorizing the Non-Ideal*. New York: Springer, pp. 59–66.

Klaser K, Sacconi L & Faillo M (2021) John Rawls and Compliance to Climate Change Agreements: Insights from a Laboratory Experiment. *International Environmental Agreements: Politics, Law and Economics* 21(3): 531–551.

Larsen PB (2018) Solving the Space Debris Crisis. *Journal of Air Law and Commerce* 83(3): 475–519.

Leonard R and Williams ID (2023) Viability of a Circular Economy for Space Debris. *Waste Management*155: 19–28.

Levit J (2005) A Bottom-up Approach to International Lawmaking: The Tale of Three Trade Finance Instruments. *Yale Journal of International Law*, 1 Jan. 2005.

Ligor D & Matthews LJ (2022) What a 1970s Philosophical Concept Can Teach Us about Space Governance. From: https//:www.fastcompany.com/90749631/what-a-philosophical-concept-from-the-1970s-can-teach-us-about-outer-space-governance/.

Meegan M (2021) The Pressing Need for Legal Certainty for Space Operators to Mitigate and Remove Space Debris. In Flohrer E, Lemmens S & Schmitz F (eds) *Proceedings of the 8th European Conference on Space Debris* 8(1). https://conference.sdo.esoc.esa.int/proceedings/sdc8/paper/200.

Pelton JN (2013) *Space Debris and Other Threats from Outer Space*. New York: Springer.

Pietrzyk-Reeves D (2017) Normative Political Theory. *Teoria Polityki*1: 173–185.

Rasmussen DM (2003) Reasonability, Normativity, and the Cosmopolitan Imagination: Arendt, Korsgaard, and Rawls. *Continental Philosophy Review* 36: 97–112.

Rawls J (1971) *A Theory of Justice*. Cambridge: Harvard University Press.

Rawls J (1993) The Law of Peoples. *Critical Inquiry* (20)1: 36–68.

Rawls J (1999) *A Theory of Justice*. Cambridge: Harvard University Press.

Sibley WM (1953) The Rational Versus the Reasonable. *The Philosophical Review* 62 (4): 554–560.

von Platz J (2017) The Veil of Ignorance in Rawlsian Theory. In Moghaddam FM (ed.) *The SAGE Encyclopedia of Political Behavior*. Thousand Oaks: Sage Publications, pp. 889–892.

Wall S (2022) Public Reason and Contemporary Political Theory. *Oxford Research Encyclopedia of Politics*.

Yang J (2023) Study on the Legal Regime for Space Debris Mitigation — Taking the Inter-Agency Space Debris Coordination Committee Space Debris Mitigation Guidelines as an Example. *Studies in Law and Justice* 2(3): 74–81.

12 Atmospheric Justice

Visualizing Atmospheric Harm by the Global Space Exploration Industry using Treadmill of Production Theory

Jack Lampkin and José Luis Carpio-Domínguez

Introduction

Pollution from the global space industry contributes negatively toward *ecological disorganization* through the production of ecological withdrawals and ecological additions; both central components to Schnaiberg's (1980) Treadmill of Production (ToP) theory. Lynch et al. (2019: 378) define ecological disorganization as 'disruptions in the normal functioning of the ecosystem in ways that prohibit its regeneration/reproduction, causing increasing ecosystem instability'. Green criminologists have written extensively on social and ecological disorganization because it helps to explain and contextualize why environmental harm is so problematic (Lynch et al. 2013). In this way, Stretesky et al. (2014) use Schnaiberg's (1980) work as a basis to explain a variety of environmental harms. They suggest that ecological disorganization is created through ecological *withdrawals* and *additions*. Withdrawals largely include the extractive industries which are synonymous with environmental destruction. Examples of withdrawals include crude oil abstraction (Ruggiero & South 2013), fracking and natural gas extraction technologies (Lampkin 2023), the mining of coal and other natural elements that exist below the Earth's surface (Stretesky et al. 2014), and the excessive use of fresh water resources for industrial processes (McClanahan 2014). Fundamentally, withdrawals are anthropogenic practices that lead to the abstraction and removal of naturally existing matter.

On the other hand, ecological additions are processes that result in adding emissions and pollutions into the natural environment. Examples include oil spills, vehicle exhaust fume emissions, non-biodegradable waste accumulation, the burning of fossil fuels (such as oil, coal, and natural gas), and the carbon released from forests as a result of deforestation. Succinctly, 'ecological disorganization is the sum of the deleterious effects of two connected economic-ecological processes: ecological withdrawals that disrupt ecosystems by removing resources from the environment and ecological additions that disrupt ecosystems by adding pollution to environments' (Lynch et al. 2019: 380).

This chapter will explore ecological disorganization as it applies to global space industries.

DOI: 10.4324/9781003437178-13

Ecological Additions and the Global Space Industry

Emissions from Launching Spacecraft

The physics of launching human-made objects into outer space is complex and tests the abilities of modern science. It is beyond the scope of this chapter to unravel the dynamics of rocket launching. However, this section will provide an analysis of the impact rocket launching emissions have on Earth's atmosphere. As a result, the area of focus is on the ecological additions created by space launching.

Part of launching rockets involves using fossil fuels because they can generate the requisite thrust and power required to send an object into (and through) Earth's orbit, known as escape velocity (NASA 2015). This is complicated by the fact rockets are always fighting to break away from the gravitational pull of the Earth (NASA 2015). Fossil fuels used in this process vary over time and between missions, but often involve the use of hydrazine, kerosene, and hydrogen peroxide (Berners-Lee 2020), amongst other propellants. Ozone-depleting substances are also emitted from rockets through launching, such as 'nitrous oxide, hydrogen chlorine and aluminium oxide' (Klinger 2019: 13). These substances can destroy ozone molecules before degrading (Voigt et al. 2013), leading to global warming (Environmental Protection Agency 2022). This is important from a critical climate change perspective because the ozone layer acts as a shield, screening the Earth from the Sun's ultraviolet rays (UV) (Anwar et al. 2015). Therefore, if the ozone is depleted through ecological additions, more UV can penetrate the ozone layer.

Space flight through rocket launching is different from traditional terrestrial aviation because rockets are launched through different, higher, layers of Earth's atmosphere. Conversely, aviation is restricted to heights that are much closer to Earth, because the object is to fly from one place on Earth to another. There is no requirement to fly any higher than necessary for the purpose of the objective. Space flight, on the other hand, must penetrate Earth's atmosphere in order to reach its objective – targeted destinations in Earth orbit or in outer space. As Ross and Vedda (2018: 9) explain:

> Rocket emissions inherently impact the stratosphere in a way that no other industrial activity does. This is a fundamental aspect of placing payloads into space using chemical propulsion. The different types of propulsion systems affect the stratosphere in different ways. This means that the various global launch organizations, national or commercial, have different impacts on the global atmosphere.

The problem is that rocket launching emits fuel by-products into the immediate area of the launch site which impacts air quality (*localized* emissions), but also into parts of Earth's atmosphere that are not usually used by humans, such as the stratosphere and Earth's fragile ozone layer (*stratospheric* emissions). Both localized and stratospheric emissions are types of ecological additions that negatively impact air quality and atmospheric health. In this sense, rocket launching is unique as it is the only anthropogenic activity that emits waste products directly into Earth's upper atmospheres (Klinger 2019) from that location, rather than creating emissions only from the Earth's surface (or lower atmosphere, as in the case of aviation). As Ross and Vedda (2018: 3) denote:

> Particles injected into the stratosphere absorb and reflect solar energy, changing the flow of radiation in the atmosphere, heating the stratosphere and cooling the surface, respectively. This radiative forcing has the effect of changing the Earth's albedo and so the amount of solar energy injected into the atmosphere. These thermal changes also deplete the ozone layer.

Aviation as a sector has gained a notorious reputation for being a large greenhouse emitter (White 2020). Contributing c.2.5% of global CO_2 emissions annually from air travel alone, the aviation industry has recently been described 'as one of the most energy-consuming and pollution-intensive sectors' (Yang & O'Connell 2020: 1), disrupting the global carbon cycle and contributing heavily toward global warming. Currently, the emissions impacts of the space industry are relatively negligible in terms of the physical quantity of emissions, compared to aviation. For instance, Sirieys et al. (2022: 147) denote that, 'overall, present day rocket Radiative Forcing (RF)[1] is small, approximately 0.008% of global, industry-wide cardon dioxide RF'. However, Ross and Toohey (2019) suggest that this deficit is changing as state and commercial space flight continues to accelerate. 'With 114 launches in 2018, the number of launches has been growing at an average rate of about 8% per year for the past decade' (Ross & Toohey 2019: 1). Such forecasted growth is problematic in terms of ecological additions because the impact of rocket launching on Earth's atmosphere is a scantly researched phenomenon. This is because the number of annual launches from the space industry was in decline since the first space age (1950s–1970s) until around the millennium, from 157 launches in 1967, to just 42 in 2005 (Ross & Toohey 2019). However, projected space industry launches for the coming decades are expected to rise exponentially in the wake of increased commercial interest in outer space. Amazon alone is anticipated to launch 3,000 new satellites by 2029 (O'Callaghan 2020). Elon Musk's SpaceX company also has permission for 12,000 new satellites with only a 5-year lifespan (McDowell 2020). Sirieys et al. (2022: 147) predict that it would take an annual rocket launching increase up to 1,000 launches per year (up from the current rate of 150–200 launches in 2022), before rockets produce 'similar amounts of positive RF as subsonic aviation'. Consequently, it may only be a matter of years before rocket launching emissions catch up with aviation emissions.

Ultimately, the increased projections for space launches are challenging for two reasons. Firstly, launches create emissions which are a form of ecological addition under Schnaiberg's (1980) concept of ecological disorganization. Secondly, due to the decrease in launches between 1967 and 2005, very little research was conducted into the impact of rocket launching emissions on Earth's atmospheric layers. Consequently, 'the literature is sparse and the present state of understanding of rocket emissions is weak' (Ross and Vedda 2018: 3). For example, the impact of black carbon (BC) is complex and little studied (Kravitz et al. 2012). Ross and Toohey (2019) suggest that BC is injected directly into the stratosphere through rocket launching, and such particles are known to absorb solar radiation.

Consequently, it is clear that more investment in atmospheric research and the stratospheric impacts of rocket launching is needed in order to understand the climatic impacts prior to a severe increase in the annual global number of rocket launches, as is forecasted in the coming years. However, there are other climate change related impacts from outer space activity that are not solely restricted to rocket launching, such as emissions and other by-products produced from the space industry supply chain. These are also classed as ecological additions under ToP theory because they contribute negatively towards climate change.

Emissions from the Supply Chain

An antithesis exists when considering the environmental impact of the space industry. On the one hand, there is undoubtedly a positive environmental impact resulting from the technological capabilities of satellites. While there are environmental costs associated with launching a satellite into Earth orbit, satellites also provide us with the ability to assess environmental harm and degradation. For example, satellites enable us to monitor droughts (Kogan & Guo 2011), predict extreme weather (like flooding), and monitor the impacts of climate change. Consequently, satellites are fundamental in our ability to accurately track changes in the Earth's weather and climate systems.

However, not all satellites are used for these purposes. They are also used in contemporary society for communications which provide 'us with global television, radio, business, internet, and telephone services' (Welti 2012: 29). Therefore, forecasted increases in satellite numbers in the coming years for commercial, rather than environmental monitoring purposes, poses questions over the necessity and environmental impact of regularly launching satellites into Earth orbit. As a result, it could be argued that satellite usage for *human wants* such as technological innovations (i.e. to communicate via social media channels) is beginning to outweigh satellite usage for *human needs* (i.e. to predict and respond to extreme weather conditions that may have a detrimental impact on human life).

As with rocket launching, there is very little by the way of research into the environmental consequences of satellite launching. Furthermore, there is even less research that considers the overall environmental impact of manufacturing rockets and satellites, including emissions produced from essential activities in the global space industry supply chain. There have been, though, some recent advances in this area (Martinez 2021). Some space organizations (such as the European Space Agency) 'have identified the life cycle assessment methodology (LCA)...to measure and then support the reduction of their environmental impact' (Maury et al. 2020: 122). In essence, an LCA involves tracking the 'potential environmental impacts of a product system throughout its life cycle' (Maury et al. 2020: 123). Regarding the launching of a rocket or satellite, this would consider the environmental impact of the long process that begins with the idea or purchase of a contract, through manufacturing, to the actual launch of the finished product. An LCA may even go further than this by assessing the impact of the launched object even after its operational life.

The legal standpoint for assessing the overall environmental impact of space operations is convoluted but exists in both domestic and international law. For example, individual nations have enacted their own legislation to govern the environmental impact of space manufacturing processes. For instance, the 'French Space Operations Act provides a national legislative framework concerning all the space operations [conducted within French territories] which aims to make operators responsible for the prevention of risks to people, property, public health and the environment' (Maury et al. 2020: 122). Nations may also be bound by the principles of other governing bodies which impact their domestic legislation. The European Union, for example, designs environmental measures through the use of directives 'which are transposed into domestic law by means of statutory instruments' (Wolf & Stanley 2014: 63).

An example of governing bodies attempting to regulate the environmental impact of space related activities can be seen in guidelines formulated by the Committee on the Peaceful Uses of Outer Space (COPUOS). Founded in 1958 by the United Nations Secretariat, the remit of COPUOS is now to assist 'any United Nations Member States to establish legal and regulatory frameworks to govern space activities and strengthen the capacity of developing countries' (United Nations Office for Outer Space Affairs 2021: 1). In 2019, COPUOS produced guidelines for the 'Long-Term Sustainability of Outer Space Activities', (LTSOSA) applicable for countries operating within the United Nations (Martinez 2021: 98). Guideline 27 of LTSOSA intends to 'promote and support research into, and the development of ways to support, sustainable exploration and use of outer space' (Martinez 2018: 16).

Although the LTSOSA produced by COPUOS was a decade in the making and claims to focus on *sustainability*, the concentration of the guidelines is mainly restricted to management of orbital space debris accumulation, rather than environmental sustainability and space-related activity. Where environmental issues are considered, the guidance is extremely weak. For example, while COPOUS recognizes the importance of liaising with the International Maritime Organization regarding 'the effects on the marine environment of waste jettisoned from commercial spaceflight activities', they suggest that it is 'the responsibility of member states to liaise and coordinate nationally' with such issues (United Nations 2019: 39). As such, devolving vague guidance on important environmental issues related to anthropogenic space activities arguably devalues the importance and substance of such guidelines.

A further problem with the development of guidelines and merely *suggesting* they should be incorporated into domestic law, is that they are 'voluntary and not legally binding under international law' (Martinez 2021: 102). Therefore, whilst guidelines may have noble intentions, there is no requirement for any member state to incorporate them into legislation. This sentiment is highlighted by the COPOUS sustainability guidelines which denote that:

> States and international intergovernmental organizations should voluntarily take measures, through their own national or other applicable mechanisms, to ensure that the guidelines are implemented to the greatest extent feasible and practicable, in accordance with their respective needs, conditions and capabilities, and with their existing obligations under applicable international law, including the provisions of applicable United Nations treaties and principles on outer space.
>
> (United Nations 2019: 52)

Consequently, COPOUS commitments to sustainability in relation to outer space activities do not have the sufficient political and legal weight to ensure the guidelines are abided by and enforced equally and across all spacefaring nations. As such, nations may take advantage of a lack of binding international law and commit actions that are environmentally detrimental.

Many industries herald technological innovation as a solution to induced environmental problems. However, some ecological withdrawals and additions have already been created before technological innovations related to environmental harms are implemented. This demonstrates the *reactive* nature of many contemporary industrial processes, including those activities associated with rocket launching, where the environment is considered after environmental harm has occurred. Therefore, in order to prevent such harm occurring in the first place, the space industry must consider the environmental impact of their operations *before* they take place. As Wooten and Tang (2018: 1019) suggest:

> as more private firms enter the space industry, there is a need to understand the interactions between a larger group of stakeholders. Specifically, these stakeholders need to ensure that firms take environmental sustainability, social responsibility, and financial responsibility into consideration when they develop new products and services.

So far this chapter has considered the ecological additions associated with both making and launching spacecraft. However, there are two further concerns relating to the creation of ecological additions. These include the pollution generated by the reentry of space objects into Earth's atmosphere, and the accumulation of space objects too large to burn up upon reentry.

The Reentry of Space Objects

At current rates of reentry, the atmospheric impact for most objects that re-enter Earth's atmosphere is negligible (Schildknecht 2007). However, such reentry does cause burnup whereby harmful gases are emitted into Earth's atmosphere. Consequently, reentry can be considered a form of ecological addition. Furthermore, some objects re-entering the atmosphere consist of radioactive materials which are considerably more harmful, and are much more ecologically destructive, than non-radioactive material.

Human-made objects in space might contain radioactive material for a variety of reasons, including for scientific instruments or to provide energy for heating or power generation systems (Wyss et al. 2011). The most common use of radioactive material, however, is simply for fuel. Whilst the type and properties of such fuel are known,

> the effect on human health of radioactive material returning from space has not been addressed in studies yet…(and furthermore), it is known that radioactive material can spiral down as far as the sea-level, even a long time after its release.
>
> (Ko et al. 2019: 6)

Ultimately, this means that the radioactive material returning from outer space is certainly an ecological addition, but it has the potential to cause ecological damage in three different ways – damage to human health, damage to the atmosphere and damage to non-human life (particularly in aquatic regions that are targeted for reentry).

During the Cold War, numerous satellites were constructed using radioactive components and 'today, there is an estimated 1,500 kilograms of radioactive material in orbit, which mostly rests in LEO' (Low Earth Orbit) (Kaineg 2020: 282). There have already been incidents where these satellites have re-entered the atmosphere, causing radioactive debris to be scattered across Earth's surface. Most notably this happened in 1978 when the Soviet Cosmos-954 satellite crashed into the Northwest Territories of Canada populated predominantly by indigenous communities (Power & Keeling 2018). Ultimately, however, the risk of this type of ecological addition will 'increase as the Cold War era satellites continue to age and reentry to Earth becomes more likely' (Kaineg 2020: 282).

The use of radioactive material for outer space endeavours is an essential part of space infrastructure (International Atomic Energy Agency 2020). They have enabled humans to begin to use outer space, while simultaneously presenting a risk to humans, the Earth's atmosphere and the Earth's ecology (through contamination potential). Consequently, the decision to adopt radioactive materials is a point of contention whereby the benefits of using such material are weighed up against the costs of doing so. Wyss et al. (2011) provide a clear rationale on the debate regarding the use of radioactive materials which they suggest have

> enabled humankind to pursue scientific research and gain insights into our universe that would have been impossible or impractical by other means. However, these same materials that have proven so beneficial in outer space can pose a significant hazard to human health and the environment if an accident causes them to be dispersed. For this reason, the risks associated with radioactive materials must be carefully managed, and a risk assessment is performed as a key step in the mission evaluation and launch approval process.

While such debates have occurred in the scientific community, and risk assessments do exist to attempt to overcome potential risks, there has been a clear lack of engagement in such debates from the social sciences and criminological community. Social scientists and green criminologists can be criticised for focussing too heavily on terrestrial events that have caused radioactive and other hazardous materials to contaminate the Earth environment, such as the Chernobyl and Fukushima nuclear power plant disasters, or the Gulf of Mexico oil spill in 2010 (Hall 2018). Although these events have created tremendous amounts of environmental harm through ecological withdrawals and additions and are certainly worthy of criminological study, the harms associated with outer space hazardous material have received scant attention, despite having the potential to be extremely ecologically damaging.

The lack of criminological research and critique relating to outer space and hazardous materials is probably due to one of three reasons, or a combination of them all. Firstly, there is a lack of knowledge of the harmful impacts of space debris (particularly those containing radioactive substances), due to a lack of research in the natural sciences (Ko et al. 2019) and social sciences. Secondly, anthropogenic space use and exploration has only emerged in the last 70 years and, although annual launches into space are predicted to rise sharply in the coming decades (Ross & Toohey 2019), the ecological impacts of terrestrial problems (such as from coal, oil or natural gas extraction, for example) have perhaps been perceived as more immediate and impactful than outer space related environmental harms. Finally, the fact that space debris exists far away from the gaze of most humans, has perhaps made social scientists less inclined to research such issues due to a lack of both knowledge and urgency. Even for those radioactive debris pieces that do survive reentry into Earth's atmosphere, most splash-land in the sea out of human view. Consequently, it can be argued that there exists a *dark-figure* of astro-environmental harm (Lampkin 2021), and a knowledge void concerning astro-environmental issues. Therefore, ecological harm stemming from the reentry of space debris needs further green criminological scrutiny in order to better understand the phenomenon. The same applies to debris that accumulates in marine environments.

Matter Accumulating at Point Nemo

Most objects careering towards Earth are small enough to burn up in Earth's atmosphere. However, a small quantity of large objects manage to withstand Earth's atmosphere remaining intact through reentry. On a positive note (from an environmental perspective), the reentry of large objects means the upper atmospheres avoid the contamination associated with smaller objects that completely burn-up upon reentry. However, large objects are targeted at oceanic environments to avoid them crash-landing into land masses occupied by humans. This presents some environmental hazards to marine environments through the creation of ecological additions. One might imagine at this point that the company who created the large object would be responsible for safely managing and disposing of the returning debris in an environmentally efficient way. But this is not the case.

The law of the sea dictates that the national jurisdiction of oceanic environments extends only 200 nautical miles from land. Seas further than 200 nautical miles are considered Areas Beyond National Jurisdiction (ABNJ) which equates to the majority (58%) of the ocean surface (O'Leary & Roberts 2017). Whilst, then, a state can regulate its activity and biodiversity impacts within its own national waters, ABNJ fall under the remit of international law. Here, there are parallels between international law pertaining to international waters, and those relating to appropriating *land* that exits in outer space (for instance, states attempting to appropriate planets or Moons). Previous international cooperation in both areas has led to the state of *global commons*, whereby space (or ABNJ in the case of marine law) are considered to be the property of human kind generally, rather than an individual state or company specifically. Whereas there have been some recent developments and United Nations involvement in ABNJ, there still exists 'a lack of regulation that threatens marine biodiversity' in oceanic areas beyond national jurisdiction (Farris 2021: 158). Consequently, space matter accumulating in ABNJ (such as at Point Nemo) is beyond the legal scope of a state or company due to the lack of domestic law pertaining to ABNJ and the failure of international law to effectively regulate and enforce binding commitments in such areas.

Ultimately, such legal failings have produced and will continue to produce repercussions for marine environments. ABNJ are not sufficiently protected by environmental laws, which has led to space matter collecting in areas such as Point Nemo. Although large reentry events are rare, they do contribute to environmental harm in marine environments. This section has suggested that this is problematic due to the absence of effective legal mechanisms to clean up such pollution. From an anthropocentric perspective, creating such pollution has consequences not just for the non-human life that occupies ABNJ, but for human life as well. As Warner (2018: 32) observes:

> The oceans, covering approximately 72 percent of the Earth's surface, are the repository for a vast array of species, habitats, and ecosystems, which make up global marine biodiversity. Marine biodiversity is critical to human survival, providing a key source of food through fisheries and aquaculture.

Consequently, it is clear that the ecological additions highlighted in this chapter so far fall neatly under the remit of green criminology – the study of environmental harms. However, ecological withdrawals also have the potential to create environmental harms and are therefore of comparable importance to green criminology.

Ecological Withdrawals and the Global Space Industry

The space industry suffers the same fate as any other industry when it comes to creating a product whereby natural resources must be exploited and manufactured in the creation of goods and services, such as materials used to create rockets, satellites, buildings and other associated infrastructure. Ecological withdrawals (that is, taking from the natural environment for anthropogenic purposes), is a term that can be used to describe this state of affairs (Lampkin & McClanahan 2023). Whilst there is nothing particularly unique about ecological withdrawals within the space industry compared to similar extractions in the buildings, chemical or engineering industries, exploiting and manufacturing plastics, fuels, metals and the like, produces a negative environmental footprint. Although there may be some elements of re-use and recycling of goods, it is undeniable (at the time of writing) that the Earth environment and biodiversity would be significantly improved if a space industry did not exist. This is not, however, to suggest that the space industry will always have a negative environmental impact. The later *ways forward* section of this chapter will go on to explore how technological innovations may go some way to reducing the sector's environmental footprint. Similarly, it may be possible in the future to take from the extraterrestrial environment in order to provide for Earth's environment, through the extraction of water, fuel or materials that are scarce on Earth but abundant elsewhere in the solar system. However, one preliminary question of importance for green criminology concerns whether it is morally acceptable to exploit the extraterrestrial world for anthropogenic purposes.

Ways Forward: Technological Innovation and Doing More with Less

In the context of human existence, serious interaction with outer space (i.e. rocket launching and space travel) are all very recent developments. It was not until the Second World War, for instance, that Wernher von Braun developed the first long distance V2 rocket (Laney 2008). Today, rocket launching is commonplace and indeed the main mechanism for propelling humans, satellites or other space-related infrastructure into outer space.

This is not to say that rocket launching is the only way of accessing places outside of Earth's atmosphere and gravitational pull. Bolonkin (2005), for example, provides a comprehensive analysis of a range of other propulsion systems that may grant such access, such as space elevator transport systems, cable accelerators, the potential for utilizing solar winds, as well as various other non-rocket transport systems. Such technological advancements have the ability to drastically improve the cost and environmental implications that are currently barriers for traditional rocket launching systems. These innovations are a welcome development and demonstrate the technical progression possible in the space industry and associated sectors.

However, is doing more (going further) with less (more environmental processes) enough to reduce, prevent, or eradicate terrestrial and extra-terrestrial environmental harm? Is it enough to react to environmentally harmful practices after harm has occurred, rather than preventing such harm from occurring in the first place?

To providc an early answer to such questions we can look at the past to inform the present. Following the Second World War, capitalism trumped socialism and fascism as the dominant world order. But capitalism itself is fraught with its own problems and inequalities that pave the way for environmental harm and destruction (Porritt 2007). The motor industry is a perfect example of capitalism in action, where automobiles were developed originally for the wealthy, but eventually for the middle and lower classes as cars became more affordable. Cars originally ran on fossil fuels, namely petrol and diesel, which were initially in plentiful supply and affordable *en masse*. However, burning these non-renewable products over a long period of time contributed to mass pollution through exhaust emissions, impacting climate change. Today, there are other technologies available to power motor and other mechanically propelled vehicles. Biofuels and electric cars are now available (albeit largely still unaffordable for the masses compared to fossil fuel powered transport). Consequently, while technology now exists that is much more environmentally friendly, this technology is not completely devoid of any environmental harm. Electric cars, for example, must still be made with natural resources, and powered by electricity which is not usually free of fossil fuels in its production.

It could be argued, particularly at this fairly nascent stage of space exploration, that space innovations may well lead to similar outcomes as the above example of the motor industry. Whilst technological innovations go some way in offsetting the total environmental damage caused by the industry, some environmental harm has to pursue in order for the space industry to operate. However, with the increased interest in outer space in the 21st century, and the forecasted increase in numbers of annual rocket launches (Ross & Toohey 2019), it is likely that the environmental gains made by technological innovation are likely to be outweighed by significant growth in the industry. This pattern was the same in the motor industry, and is typical of industrial expansion under capitalism and according to

Treadmill of Production theory. As Stretesky et al. (2014: 3) state, 'the nature of capitalism does not allow technology to solve the problem of green crimes related to ecological destruction and disorganisation'. Environmental harm can instead be viewed as a negative externality of economic growth, industrial expansion and capitalism generally.

Conclusion

Emissions from global space industries contribute towards global warming. Rocket launching is unique in that it emits pollutants both on the ground during manoeuvre and take-off, and while in-flight directly into Earth's atmosphere. Although some innovative solutions have been proposed for mitigating the climatic impacts of rocket emissions pollutions, the industry still relies heavily on fossil fuels to power spacecraft.

While green criminologists have made numerous theoretical and empirical advances in researching environmental harms and crimes, few have considered space industries. Consequently, this chapter has made a contribution to foundational literature in this area utilizing Treadmill of Production theory. However, due to the greenhouse gas impact of space industries, more green criminological research needs to be conducted in order to understand air pollution impacts and atmospheric harm more broadly. If states and corporations continue to contribute heavily to irreversible climatic change through ecological additions and withdrawals, then further climate injustices will occur as global warming exacerbates. This will include the warming of the polar regions and oceans, leading to sea-level rise, the subsequent loss of ecosystems and coral reefs, an increased prevalence of wildfires, and an upsurge of environmental refugees seeking more hospitable areas to survive. We must address climate change now in order to prevent its escalation. This includes analysing all industries that contribute to global greengas emissions, including the space industry.

Note

1 A metric called Radiative Forcing is used to measure the climatic impact of emissions, with a positive RF effect leading to atmospheric warming, and a negative RF effect leading to atmospheric cooling (Budd 2017).

References

Anwar F, Chaudhry FN, Nazeer S, Zaman N and Azam S (2015) Causes of Ozone Layer Depletion and its Effects on Human: Review, *Atmospheric and Climate Sciences*, 6(1): 129–134. https://doi.org/10.4236/acs.2016.61011

Berners-Lee M (2020) *How Bad are Bananas? The Carbon Footprint of Everything*. London: Profile Books Ltd. 2nd Edition.

Bolonkin AA (2005) *Non-Rocket Space Launch and Flight*. London: Elsevier.

Budd T (2017) Environmental Impacts and Mitigation. In: Budd L and Ison S (eds) *Air Transport Management*. pp. 311–334. Oxon: Routledge. https://doi.org/10.4324/9780429299445.

Environmental Protection Agency (2022) Nitrous Oxide Emissions. https://espanol.epa.gov/la-energia-y-el-medioambiente/emisiones-de-oxido-nitroso.

Farris CJ (2021) Progress Towards the New International Legal Framework for Protecting Biodiversity in Areas Beyond National Jurisdiction, *North Carolina Journal of International Law*, 46(1/4): 157–180. https://scholarship.law.unc.edu/ncilj/vol46/iss1/4.

Hall M (2018) Environmental Victimisation: Corporate Villainy or State Connivance? *Radical Criminology*. ISSN: 1929–7904. http://eprints.lincoln.ac.uk/id/eprint/15918/.

International Atomic Energy Agency (2020) Ensuring Safety on Earth from Nuclear Sources in Space. https://www.iaea.org/newscenter/news/ensuring-safety-on-earth-from-nuclear-sources-in-space.

Kaineg S (2020) The Growing Problem of Space Debris, *Hastings Environmental Law Journal*, 26 (5): 277–288. https://repository.uchastings.edu/hastings_environmental_law_journal/vol26/iss2/5.

Klinger JM (2019) Environmental Geopolitics and Outer Space, *Geopolitics*: 1–38. https://doi.org/10.1080/14650045.2019.1590340.

Ko N, Bouslama R, Betten T and Pfeuffer S (2019) Fundamentals for the Development of Impact Categories for the Application of LCA in Space – Space Debris, *Matériaux & Techniques*, 107 (106): 1–9. http://dx.doi.org/10.1051/mattech/2019011.

Kogan F and Guo W (2011) Early Detection and Monitoring Droughts from NOAA Environmental Satellites. In: Kogan F, Powell A and Fedorov O (eds) *Use of Satellite and In-Situ Data to Improve Sustainability*. Dordrecht: Springer.

Kravitz B, Robock, A, Shindell, DT and Miller MA (2012) 'Sensitivity of Stratospheric Geoengineering with Black Carbon to Aerosol Size and Altitude of Injection,' *Journal of Geophysical Research: Atmospheres*, 117(D09203): 1–22. https://doi.org/10.1029/2011JD017341.

Lampkin JA (2021) Mapping the Terrain of an Astro-Green Criminology: A Case for Extending the Green Criminological Lens Outside of Planet Earth, *Astropolitics: The International Journal of Space Politics and Policy*, 18 (3): 259–273. https://doi.org/10.1080/14777622.2020.1972542.

Lampkin JA (2023) The Harms and Crimes of Fracking. In: Pontell H (ed.) *Oxford Research Encyclopedia of Criminology and Criminal Justice*. New York and Oxford: Oxford University Press. https://doi.org/10.1093/acrefore/9780190264079.013.768.

Lampkin JA and McClanahan B (2023) Astronomical Withdrawals: A Green Criminological Examination of Extreme Energy Mining on Extra-terrestrial Objects, *Crime, Law and Social Change*: 1–20. https://doi.org/10.1007/s10611-023-10123-9.

Laney M (2008) Wernher Von Braun and Arthur Rudolph: Negotiating the Past in Huntsville. In: Schulze M, Skidmore JM, John DG, Liebscher G and Siebel-Achenbach S (eds) *German Diasporic Experiences: Identity, Migration, and Loss*. Canada: Wilfrid Laurier University Press.

Lynch MJ, Long MA, Barrett KL and Stretesky PB (2013) Is it a Crime to Produce Ecological Disorganization? Why Green Criminology and Political Economy Matter in the Analysis of Global Ecological Harms, *British Journal of Criminology*, 53 (6): 997–1016. https://doi.org/10.1093/bjc/azt051.

Lynch MJ, Long MA, Stretesky PB and Barrett KL (2019) Measuring the Ecological Impact of the Wealthy: Excessive Consumption, Ecological Disorganization, Green

Crime, and Justice, *Social Currents*, 6 (4): 377–395. https://doi.org/10.1177%2F2329496519847491.

Martinez P (2018) Development of an International Compendium of Guidelines for the Long-Term Sustainability of Outer Space Activities, *Space Policy*, 43: 13–17. https://doi.org/10.1016/j.spacepol.2018.01.002.

Martinez P (2021) The UN COPUOS Guidelines for the Long-Term Sustainability of Outer Space Activities, Journal of Space Safety Engineering, 8 (1): 98–107. https://doi.org/10.1016/j.jsse.2021.02.003.

Maury T, Loubet P, Serrano SM, Gallice A and Sonnemann G (2020) Application of Environmental Life Cycle Assessment (LCA) Within the Space Sector: A State of the Art, *Acta Astronautica*, 170: 122–135. https://doi.org/10.1016/j.actaastro.2020.01.035.

McClanahan B (2014) Green and Grey: Water Justice, Criminalization, and Resistance, *Critical Criminology*, 22 (3): 403–418. http://dx.doi.org/10.1007/s10612-014-9241-8.

McDowell JC (2020) The Low Earth Orbit Satellite Population and Impacts of the SpaceX Starlink Constellation, *The Astrophysical Journal Letters*, 892 (L36): 1–18. https://doi.org/10.3847/2041-8213/ab8016.

NASA (National Aeronautics and Space Administration) (2015) Escape Velocity. https://www.nasa.gov/audience/forstudents/k-4/dictionary/Escape_Velocity.html.

O'Callaghan J (2020) Amazon is Going to add 3,000 More Satellites into Earth's Orbit – and People are Not Happy. Produced for: *Forbes*. https://www.forbes.com/sites/jonathanocallaghan/2020/07/31/amazon-is-going-to-add-3000-more-satellites-into-earths-orbit–and-people-are-not-happy/.

O'Leary BC and Roberts CM (2017) The Structuring Role of Marine Life in Open Ocean Habitat: Importance to International Policy, *Frontiers in Marine Science*, 4 (268): 1–7. https://doi.org/10.3389/fmars.2017.00268.

Porritt J (2007) *Capitalism: As if the World Matters*. London: Earthscan.

Power E and Keeling A (2018) Cleaning up Cosmos: Satellite Debris, Radioactive Risk, and the Politics of Knowledge in Operation Morning Light, *Northern Review* (48): 81–109. https://doi.org/10.22584/nr48.2018.004.

Ross MN and Toohey DW (2019) The Coming Surge of Rocket Emissions. Produced for: Eos. https://eos.org/features/the-coming-surge-of-rocket-emissions.

Ross MN and Vedda J (2018) April 2018: The Policy and Science of Rocket Emissions. Produced for: The Aerospace Corporation. https://aerospace.org/sites/default/files/2018-05/RocketEmissions_0.pdf.

Ruggiero V and South N (2013) Toxic State–Corporate Crimes, Neo-Liberalism and Green Criminology: The Hazards and Legacies of the Oil, Chemical and Mineral Industries, *International Journal for Crime, Justice and Social Democracy*, 2 (2): 12–26. http://dx.doi.org/10.5204/ijcjsd.v2i2.115.

Schildknecht T (2007) Optical Surveys for Space Debris. *The Astronomy and Astrophysics Review* 14: 41–111.

Schnaiberg A (1980) *The Environment: From Surplus to Scarcity*. New York: Oxford University Press. http://hdl.handle.net/10822/785551.

Sirieys E, Gentgen C, Jain A, Milton J and De Weck O (2022) Space Sustainability Isn't Just About Space Debris: On the Atmospheric Impact of Space Launches, MIT Science Policy Review (3): 143–151. https://doi.org/10.38105/spr.whfig18hta.

Stretesky P, Long MA and Lynch MJ (2014) *The Treadmill of Crime: Political Economy and Green Criminology*. New York: Routledge. https://doi.org/10.4324/9780203077092.

United Nations (2019) *Report of the Committee on the Peaceful Uses of Outer Space: Sixty-Second Session (12–21 June 2019)*. https://www.unoosa.org/res/oosadoc/data/documents/2019/a/a7420_0_html/V1906077.pdf.

United Nations Office for Outer Space Affairs (2021) About Us. https://www.unoosa.org/oosa/en/aboutus/index.html.

Voigt C, Schumann U, Graf K and Gottschaldt KD (2013) Impact of Rocket Exhaust Plumes on Atmospheric Composition and Climate: An Overview, *Progress in Propulsion Physics*, 4: 657–670. http://dx.doi.org/10.1051/eucass/201304657.

Warner, R (2018) Oceans in Transition: Incorporating Climate-Change Impacts into Environmental Impact Assessment for Marine Areas Beyond National Jurisdiction, *Ecology Law Quarterly*, 45: 31–52. https://www.jstor.org/stable/26568786.

Welti CR (2012) *Satellite Basics for Everyone: An Illustrated Guide to Satellites for Non-Technical and Technical People*. Bloomington: iUniverse.

Wolf S and Stanley N (2014) *Wolf and Stanley on Environmental Law*. Oxon: Routledge. 6th Edition.

Wooten JO and Tang CS (2018) Operations in Space: Exploring a New Industry, Decision Sciences, 49 (6): 999–1023. https://doi.org/10.1111/deci.12312.

White R (2020) *Climate Change Criminology*. Bristol, UK: Bristol University Press.

Wyss GD, Polansky GF, Mitchiner KW, Allahdadi F and Bakhtiyarov SI (2011) Modeling the Behavior of Radioactive Materials in Combined-Phenomenology Environments. US Department of Energy. Office of Scientific and Technical Information. https://www.osti.gov/servlets/purl/1118555.

Yang H and O'Connell JF (2020) Short-Term Carbon Emissions Forecast for Aviation Industry in Shanghai, *Journal of Cleaner Production*, 275 (122734): 1–22. https://doi.org/10.1016/j.jclepro.2020.122734.

13 ‘Spaceport of Call’

Developing a Geopolitical-Criminological Perspective on Spaceport Crime and Policing

Sarah Poss, Yarin Eski and Jack Lampkin

Introduction: Setting the Sci-Fi Scene

Running since 2019, *For All Mankind* is a sci-fi series that imagines an alternate history where the space race continued after the Soviets landed on the Moon first (Moore, Wolpert & Nedivi 2019). The show explores a world where Soviet success prompts the US to intensify its space efforts. Episode 4 of season 4, titled ‘House Divided’, opens with a scene of a Martian base, called *Happy Valley* (Edwards 2023). Happy Valley is where all spacefaring nations, including the US, Soviet Union, India, North Korea and others, have their settlements living and working together constructively. It is an example of peaceful international cooperation on another planet. Or it was, until two Soviet cosmonauts got into an argument about domestic political developments back on Earth. This led to a scuffle between the two that resulted in one cosmonaut puncturing the other’s spacesuit. The interpersonal violence on Mars led to geopolitical tension back on Earth, and to a trial for assault and crimes against ‘a patriot of the Soviet Union’, after which the perpetrating cosmonaut most likely awaited a show trial and disproportionate punishment. An American astronaut in charge at Happy Valley tells NASA that she will not allow a subordinate to be shipped off to ‘some gulag’. However, the hands of Happy Valley leadership are tied by the terms of Martian governance, as defined by the M7 treaty. The final decision seems to rest with US President Al Gore.

This fictional scene illustrates how interpersonal conflict off-Earth interacts with geopolitical tension on Earth. The storyline touches on the sensitive interplay between the small presence of people and their livelihoods in space, and macrolevel (geo)politics on Earth. Space is where the rules of the game at the individual and group-level in confined spaces of space stations and lunar/Martian bases directly influence the game of geopolitical contestation on Earth.

Off-Earth spaceports, like Happy Valley, are therefore uniquely influential nodes where microlevel personal interactions in small space-settling communities that are unsafe, insecure, harmful and criminal, bear serious macrolevel consequences millions of miles away on Earth. They can be, much as ports

DOI: 10.4324/9781003437178-14

and airports on Earth, characterised as potential centres of malpractice, serving as crucial arenas for geopolitical-criminological investigation due to the flows of people, goods and concerns associated with continued space exploration. Given how the fields of criminology and political science have largely neglected the challenges of spaceborne conflict, crime and policing overall (Eski 2023; Lampkin & White 2023), this chapter aims to identify specific avenues for studying spaceport crime (control) and policing from a geopolitical-criminological perspective. By contextualising anticipated key crime and insecurities within spaceports and the potential frameworks addressing these concerns, this chapter is an initiator to enable interdisciplinary research grounded on intersecting criminological and geopolitical theoretical and methodological debates. This begins with the contextual question, what is a spaceport?

What is a Spaceport?

To understand what a *spaceport* is, we first consider the fundamentals of ports as transport nodes. The term 'port' has different meanings: harbours, havens, docks or marinas. It is generally agreed that its linguistic roots are in Old English and Latin, and involves a place that comprises both landside and waterside areas that serve as a place for boats to load or unload, start or end voyages and where charges may be levied on boats using the facilities (OED Online 2014). Certain ports, such as international airports, can also serve as border crossings supervised by customs and immigration (Nichols & Williams 2009). Both as a concept and phenomenon, the port has evolved over 5,000 years, from Byblos to Rotterdam to Shanghai, exemplifying the shifts in trade systems and technologies (Verhetsel & Sel 2009), which correspond to the changing epicentres of power over time.

In recent history, globalisation, technological advancements and neoliberalism have reshaped ports, replacing traditional labour with automation and altering demand patterns (Branch 2007). Global ports, as interdependent nodes in the global supply chain, occupy a crucial position in maritime transport and its security. Still, despite being historically recognised as crime-prone spaces, port security has received less attention than ship safety, with port security measures gaining prominence only after the 9/11 attacks (Booth & Altenbrun 2002). Port crime, criminalisation and policing are, to an extent, (geo)politically shaped (Sergi & Reid 2021).

Conceptualising spaceports includes considering whether the location from where objects are launched into space is by definition a spaceport or if it entails more than having launching capacity and capability. The development of spaceports in the 20th century illuminates these conceptual questions, among others. The attempt to define spaceports begins with a historical overview of their development in the 20th century. Roberts (2019: 1) states the following on spaceports:

> Over 60 years ago, the Soviet Union used a derivative of its R-7 rocket—often called the world's first intercontinental ballistic missile (ICBM)—to launch an artificial satellite into orbit, marking the first orbital space launch from the spaceport now known as the Baikonur Cosmodrome. Since then, launch vehicles have reached orbit from 27 spaceports around the world. With the rate of space launches projected to grow exponentially in the coming years, spaceports will become an increasingly important and potentially limiting factor in the global space industry.

A spaceport, like a seaport or airport, serves as a transport hub for launching and receiving spacecraft. Originally associated with orbital and interplanetary missions, the term now extends to sites for sub-orbital flights, space stations and potential lunar bases. Rocket launch sites, encompassing safety areas and tracking stations, are integral to spaceport infrastructure. Earth-based infrastructure can include storage and production facilities for propellants as well as runways for aircraft that support operations. In particular, major, busy spaceports such as Cape Canaveral Air Force Station (US) and the Baikonur Cosmodrome (Kazakhstan) feature multiple launch complexes tailored to diverse launch vehicle types (see Pellegrino 2023).

In general, spaceports on Earth are located in geo-strategically favourable locations, often near the equator for optimal launch conditions. Originally tied to ballistic missile research, some spaceports were kept secret due to political risks. So, although spaceports ideally need equatorial locations with safe drop zones and weather protection, their global placement rather suggests political factors influence their locations significantly more than geographic considerations (Roberts 2019).

Spaceports also exist off-Earth. There, they can include large space stations, like the International Space Station (ISS) and future settlements on the Moon (e.g., Artemis Lunar Base Camp) and on Mars (NASA 2024). Spaceports on and off-Earth will together form an interplanetary nodal network of space transport hubs, much like seaports do nowadays (Eski 2016). This nodal network of spaceports forms and enables spacefaring routes which transcend national, international, global and other planetary boundaries and characteristics; it ties together locations on and off-Earth that are inherently differently structured, and connect fundamentally different atmospheres, including differences in microgravity, radiation and chemical compositions (e.g., no oxygen) (see Atomi 2015). Off-Earth spaceports – often because of these existential differences – comprise places and spaces that are air-tight and very small, requiring confined living conditions with limited nutrition to survive the harsh conditions of outer space (Williams et al. 2009).

Spaceport Crime and Policing

Spaceport Crime

Being away from Earth, living and working at a spaceport is heavily embedded in space logistics, far more than working at a seaport entails, for example (see Cowen 2014). Hence, spaceport crime and policing manifest most prevalently at the junction of Earth into cosmic travel, people and goods, particularly in the docking zones of off-Earth space stations and other planetary bases, where spacecraft navigation remains the primary mode of space transportation. Moreover, space includes extensive (actually infinite) regions that operate beyond singular jurisdictions in which a culture of impunity may linger; limited enforceability of Earth-bound, predominantly national laws could lead to more crime than on Earth. Hence, experiences of lawlessness could be in and of themselves criminogenic at spaceports, which is already filtering through in the following 'Governing Law' section of Starlink's Terms of Service:

> For Services provided on Mars, or in transit to Mars via Starship or other colonization spacecraft, the parties recognize Mars as a free planet, and that no Earth-based government holds authority or sovereignty over Martian activities. Accordingly, disputes will be settled through self-governing principles established in good faith at the time of Martian settlement.
>
> (Starlink 2024 – online source)

Therefore, spaceports off-Earth will be located in lawless spaces while they function as nodes that are essential for safe, secure and lawful space exploration where heightened control and preventive measures against potential threats, malpractices and crime would be assumed to be in place.

Currently, there is no specific definition of the term 'space crime'. Definitions are provided by several authors (see Lampkin & White 2023; Sachdeva 2023), understanding space crime as place-dependent, relating to locations in outer space, celestial objects such as asteroids, moons, and planets; and Earth-bound but space-related activities, including environmental crimes. The term 'space crime' is a complex phenomenon, not least because there is no legal criminal jurisdiction in outer space. This means that behaviours conducted off-Earth are regulated by just a handful of 'soft' international laws. While there are a plethora of domestic laws relating to outer space, these only apply to persons representing the state from which domestic space laws derive. With contemporary expansion and interest in outer space, some are now calling for more stringent, universal and ultimately – meaningful – space laws and jurisprudential approaches to address future space crimes (Butler 2017; Lampkin & White 2023). Without more purposeful legal recognition, unwanted behaviours in outer space simply cannot, arguably, be automatically treated as space crimes. Therefore, as Lampkin and White (2023) suggest, zemiological notions of 'harm' and 'deviance' may be more appropriate lexical choices.

This convoluted legal 'system' indicates there is uncertainty surrounding 'what will be defined as a crime, who will do the defining, and what the social and ecological result will be once crimes in space are given legal definition' (Lampkin & White 2023: 22). If the development of laws follows traditional mechanisms, outer space law will be heavily influenced by nation-states and corporations who have a vested interest in exploring, extracting, or otherwise – utilising – outer space to satisfy national or commercial interests. As on Earth, law could be used to facilitate the exploitation of outer space environments.

Spaceports could create space crimes in several different ways. For instance, a spaceport could be a place where a satellite is launched illegally. A good example of this is anti-satellite tests (ASAT) where a nation-state facilitates the intentional destruction of a defunct satellite to test military capability. This happened recently, in 2021, when Russia conducted a successful ASAT test on its defunct Kosmos-1408 satellite, months before the first attempt on a Ukrainian invasion (Bugos 2021). Consequently, this is a good example of how 'space crime is defined initially by place' (Lampkin & White 2023: 9).

Spaceports can also contribute to space crime *indirectly* through facilitating the launch of the people and infrastructure required to commit a crime in outer space. Without spaceports it would be difficult to imagine that any space crime could take place, as they are quintessential to the logistical development of space. Spaceport crime thus comprises numerous harms, crimes and transgressions within outer space by powerful entities, as well as a whole range of interpersonal crimes, and actions that adversely affect the space environment.

The distinction between *direct* and *indirect* spaceport crimes is important because it is relevant to discussions of legal jurisdiction. Indirect spaceport-facilitated crimes suffer the same jurisdictional difficulties as described earlier, such as the lack of national sovereignty attached to outer space locales. Direct spaceport-facilitated crimes, on the other hand, avoid these jurisdictional woes because they physically exist within a nation-state and are owned by a state or company. Consequently, activities conducted there are subjected both to relevant international laws as well as domestic laws. This means that it would be possible to prosecute someone for the violation of a law that has taken place at, or was directly facilitated by, a spaceport. Consequently, activities at spaceports are only ever likely to experience prosecution if they impact another nation-state, or some person or company that exists outside of the spaceport itself.

Spaceport Policing

In space, and on other planets, there is no internationally agreed upon universal criminal justice system or governance apparatus in place that is legally and practically responsible for enforcing legal compliance and policing (see Li 2020). Next to policing crimes on Earthbound spaceports, of specific interest here is policing crimes at spaceports in orbit or deep space, and policing potential future crimes at spaceports on celestial bodies like the Moon or Mars (Lampkin & White 2023).

Human activity on celestial bodies, while mostly speculative, presents a new frontier requiring critical examination of law enforcement practices as they expand into space. Policing of spaceports in Earth's orbit (and in deep space) and on off-Earth celestial bodies are of particular interest due to their distance from Earth and the practical, policy and legal challenges involved. For example, who will be policing aboard a spacecraft and how to prosecute a space crime and where (Seshagiri 2005)? Which courts will be available to try and penalise space crime (Szocik et al. 2016)? Is this realisable via a digital at-a-distance court (see Swierczynski 2022) on Earth? Would that constitute a fair trial of a space crime? Should we explore classic suggestions of space-faring being served best by individuals capable of self-policing and -punishment (Taubenfeld 1961–1962)?

So far, only the Canadian government have proposed a law to prosecute crimes committed by Canadian astronauts, specifically on the Moon (Thrower 2022). Other countries and space agencies have not followed up yet. Still, space crime policing is essential for spacefaring crews' well-being, knowing what is acceptable human behaviour and which available responses exist, which should 'be clearly defined before a spacecraft leaves Earth' (Seshagiri 2005: 478). There is a strongly felt need for space policing and appropriate mechanisms for promulgation, enforcement and administration 'to manage the space environment to control and deter criminal action' (Doyle 1990: 130).

It is important that space policing carries international legitimacy, with space policing authorities being acknowledged. This is extremely difficult. Russia and the US already do not recognise the ICC and its authority, so when it comes down to policing space (state) crimes (Vinjamuri 2016), it may very well have a similar stance. The same goes for policing corporate crimes in space, given how large space-going companies seem to treat space as a legal vacuum, for example, the aforementioned SpaceX that will settle disputes through self-governing principles (StarLink 2024).

As said, a particular policing issue regarding space crime is the absence of national borders in space and this raises uncertainties about the enforcement of legal compliance (Li 2020). The virtually limitless nature of space and the lack of defined borders creates legal uncertainties, potentially contributing to criminogenic situations. Astro-mining companies, for instance, exploit cosmic bodies' resources under the assumption of their infinite availability, potentially justifying unregulated space ecocrime (Lampkin & McClanahan 2023). These factors contribute to the complexity of space policing of crimes in the vast, limitless expanse of space, making it worth considering the presence and role of spaceports (as political places) in these complexities. As such, and much like maritime ports once were considered beacons of some sort of (policing) stability in an overall lawless domain (Eski 2011), spaceports could have an initiating role to fulfil regarding the first serious steps towards space policing. Put differently, spaceport policing *is* space policing. And policing is political (Reiner 2010).

Order in the Port! The Geopolitics of Spaceports

The capacity to expand, secure and maintain economic activity beyond borders is a function of power, which is established through ports. The construction and control of ports along new routes represents a challenge to preexisting geographies and the ideas and capabilities that underpin that order (Noorali, Flint & Ahmadi 2022: 2–3). Contested geographies underscore attempts to secure access to the world economy, which make ports and their administration a geopolitical consideration in maritime security strategies (Germond 2015).

Ports are sites that exhibit the process and outcomes of geopolitical contestation due to the goods and services traded, at what cost, by and for whom. The site selection, financing, construction, operation, ownership, tax status, tariffs, legal framework, among other aspects of expanding ports, are also manifestations of geopolitical competition. The hegemonic contestation to connect markets in East Asia to Europe and beyond by way of the *Belt and Road Initiative*, and competing infrastructure projects, illustrates the importance of infrastructure development in hegemonic contestation (Flint & Zhu 2019). To conform or consent to an imposed trade order reinforces the notion that transport geographies are not neutral, but instead are crafted imaginaries guided by rules and backed by force (Lin 2019). Rulemaking and the enforcement thereof is shaped by subjective notions of danger that reflect power dynamics (Campbell 1992; Tuathail 2005).

As key nodes in trade networks, ports are governed according to dangers and crimes perceived as threats to order. Port policing and security measures are among the preventative efforts to uphold an unabated circulation of goods and capital and to limit activities that are disruptive or illegal (Germond 2015). What is considered disruptive and illegal is a manifestation of power (Campbell 1992). For example, the securitisation of sea (and airports) following the September 11th attacks represents an attempt to assert a monopoly on violence for the preservation of economic, and increasingly environmental, interests (Germond 2015). The attacks on the shipping route through the Gulf of Aden in late-2023 and early-2024 and the US military response (Al Jazeera 2024) illustrate how a society of (oil) consumption dependent on the flow of resources results in enduring, violent contestation over the use of extra-territorial space (Campbell 2007).

Whether expansion into outer space follows a similar course is a matter of contention. Contestation unfolds among Astro-capitalists (Dunnett 2019), state actors and the EU (Pekkanen 2020; Peter 2007; Rementeria 2022; Riddervold 2023; Wang 2009), indigenous groups (Finnegan 2022), the private space sector (Shammas 2019; Weinzierl 2018) and media (Damjanov 2015), among others. When, or if, off-world spaceports become part of existing trade networks, the policing of the spaceport crime can reflect the power dynamics and struggles that are important to subject to critical geopolitical analysis.

Similar to ports, and like geopolitical analyses, spaceports will form at the intersection of politics, power and geography. For example, maritime ports serve as strategic launch points for projecting material and ideological power beyond a state's territory. The expansion of trade and transit routes therefore has implications for world order (Noorali et al. 2022), from the Silk Road of old to the 21st-century Belt and Road Initiative, which now includes a Space Information Corridor focusing on satellite and other space infrastructure to rival GPS, for example (BRI Forum 2023). Attempts to manifest and sustain a social construction of a place connected by extra-territorial nodes – such as China's efforts creating a space corridor enhancing connectivity among participating states, or the ports along the maritime 'Belt' from Myanmar to Sudan dubbed the String of Pearls (Brewster 2017; Khurana 2008) – are examples of geopolitical contestation. Attributing meaning, identity and order to connected places, such as spaceports, represents a challenge among competing discourses from rival states over conflictual notions of danger and what or whom should be secured (Campbell 1992).

Therefore, human expansion into outer space is not only about the physical presence of humans or humanmade objects, but also about the construction of meaning, identity and order out of the off-world expansion of economic activity. As humanity gets closer to expanding trade networks into outer space, there is an opportunity to critically examine the ongoing contestation over the production and ordering of outer space and how and whether expansions into space challenge and/or reproduce predominant dynamics of world order. Therefore, spaceports can be considered as geopolitical objects and manifestations of power, shaped by and shaping space law and governance in which the logics of expansion play a vital role.

As such, seen from classical and critical approaches to geopolitics, spaceports (and their securitisation) are contested geopolitical objects in transit and trade routes. Dolman's (2001) *Astropolitik* comes to mind, which places MacKinder's classical geopolitical dictum about controlling the Eurasian heartland into a new context: 'who controls low-Earth orbit controls near-Earth space. Who controls near-Earth space dominates Terra. Who dominates Terra determines the destiny of humankind' (Dolman 2001: 6–7). Critical geopolitical scholarship oriented toward outer space considers the production of outer space and problematises the classical, 'problem-solving' theories which explain and extend state-centred logic into outer space (Klinger 2019). However, a future-oriented critical analysis of space trade infrastructure, namely spaceports, has yet to be included in the emerging field of critical astropolitics.

Thus, the administration of space is a result of contestation between ways of establishing order and meaning in a delineated area, what Tuathail (2005: 1) refers to as 'an active writing of the Earth by an expanding, centralizing, imperial state'. Geographical knowledge is therefore not neutral, and instead utilised for producing (writing) and managing territory in an expression of *geo-power*, whereby in practice the interconnected forces of industrialisation and imperial expansion in the 19th century were mapped across spaces constructed in line with state interests (Tuathail 2005: 5–6). After centuries of inventing the Americas and grappling with the confrontation with the Christian Self of the Old World and the Indigenous Other of the New World, European exploration of the Americas ultimately fostered a profound realignment toward the modern identity due to confrontations with the Other that challenged engrained metanarratives of the time (Campbell 1992: 97). The maritime ports along transit routes are active sites of confrontation with the Other and contestation over geographies, making them not only pivotal for international trade but also the production of geographies. The creation of the Other in the context of outer space has a similar potential to recalibrate existing Earth-bound distinctions between selves and others, which can be studied as activities in space increase and as identities related to those space activities take shape, especially at spaceports.

Conclusion: Toward a Geopolitical-Criminology of Spaceport Crime and Policing

Political scientists using criminology and criminologists using political science: both groups of scientists operate at the interplay between political processes and crime (control)-related issues (see Bigo 2016; Holley et al. 2018; Loader & Percy 2012; Newman 2016; Patio 2020; Whitehead 2016). At this geopolitical-criminological intersection, it is examined how order and institutions shape crime policy and policing. Focus is laid upon various aspects, including the structure of the criminal justice system and policing politics, legislative bodies crafting laws and the link between laws and political systems. Arguably, a politico-criminological perspective on macro-crime, policing and policy phenomena makes sense, but why so for spaceport crime and policing?

First, given the many (anticipated) challenges of spaceport crimes and policing, criminology needs to develop theoretical, methodological and empirical research to understand the complex dynamics that can be involved in space as the final frontier to be filled with vital (transport) infrastructures of which spaceports are part.

Thinking about space crime as a crime-warp phenomenon (Behan 1990; Bennett 1987), space*port* crime forms an interesting subcategory, as well as how to police spaceport crimes and insecurities, not too different from maritime and airport crime and policing (Blackwood 2015; Eski 2016). The difference though is that a criminology of spaceport crime and policing, is not *after-the-fact* research on maritime port and airport crime and policing, but instead, a criminology that is pro-actively working toward a *before-the-act* critical and empirical space criminology (Eski 2024a/b, 2023). A criminology of spaceport crime and policing is in that sense a criminology of crime-warps and policing-warps: it is future-oriented.

Professionals and organisations tasked with spaceport security are confronted with multifaceted, interplanetary challenges within these cosmic hubs.

They are specifically interesting for criminology, hence, a future-oriented criminology on spaceport crime and policing (Eski, Clamp & Pemberton 2024) ought to study spaceport policing and security personnel, as well as space-logistical staff, next to states and international bodies in charge of spaceports. Also, astronauts and space tourists are crucial actors to consider for criminological research. Focusing on these and other actors, criminologists would need to dissect the dimensions of risk, insecurity, security and crime in spaceport social realms, especially as on-Earth and off-Earth security of spaceports are fully overlooked phenomena in public and academic discussions. However, they form critical intersections where policing, security and safety measures and the governance of space activities converge. In sum, a criminology of spaceports and their small communities is about the micro- and meso-aspects and interactions at a spaceport, on and off-Earth.

Second, a geopolitical-criminological analysis is not just about criminology, but also about (geo)politics. This indicates that next to the micro- and meso-orientations in criminologies of spaceport crime and policing, it is key to examine the manifestation of contemporary geopolitical tensions, which shape policing and security ideologies at home and on the cosmic frontier, which may be analysed narrowly by focusing on spaceports.

The incorporation of outer space into trade routes and infrastructure reflects the well-trodden grounds of geopolitics due to the geopolitical implications of the expansion of trade and transit routes. States project power beyond territorial boundaries in part by shaping the norms and institutions by which international trade is conducted and secured. The manner of incorporating trade routes and infrastructure in space into the world economy reflects contestation among hegemonic and counter-hegemonic forces over the attribution of meaning and order to a place, and the rules of a given order shape the policing of geopolitical objects such as ports. The expansion into outer space coincides with the production of new geographies, of which transit infrastructure and its policing is a part, which is possible to study through the lens of critical geopolitics.

There is noticeable ambition to develop strategic foresight (Wilkinson 2017) and futures literacy (Miller 2015) to make innovative policy that is future-proof. It is therefore prudent to understand the geographies of outer space as they are constructed. The expansion of trade networks and nodes such as spaceports represents a slice of the increasing activity in space, one that faces governance challenges that include the policing of off-world activity. The study of spaceports and their governance as a topic in critical geopolitics and criminology could proceed in myriad ways, particularly in discerning contested understandings of the construction of cosmic geographies by the constellation of relevant space-oriented actors. The role of off-Earth spaceports and their administration, as nodes in outer space trade routes, can reveal how and if the past manner of imperial expansion on-Earth is reproduced and/or challenged in outer space, especially through spaceport crime, criminalisation and policing. These challenges or reproductions may come from (counter) hegemonic (and associated social forces) such as states, the EU, institutions such as the UN, billionaire Astrocapitalists, the private space sector more broadly, environmentalists and indigenous groups, to name a few. This approach may also include the responses to cosmic geographies put forward by the multitude of actors. Any emancipatory potential for humanity in the expansion into outer space would be served by understanding its construction, for whom it serves, and how the expansion of trade networks and spaceport nodes proceeds, especially given the historical role of trade/transit in the exercise and experience of power.

References

Al Jazeera (2024, February 22) US Military Targets Houthi Antiship Cruise Missiles in Preemptive Strikes. From: https://www.aljazeera.com/news/2024/2/22/us-military-targets-houthi-antiship-cruise-missiles-in-pre-emptive-strikes.

Atomi Y (2015) Gravitational Effects on Human Physiology. *High Pressure Bioscience: Basic Concepts, Applications and Frontiers*, 627–659.

Behan CJ (1990) Responding to Change and Managing Crimewarps. *American Journal of Police*, 9: 171.

Bennett G (1987) *Crimewarps: The Future of Crime in America*. Garden City, NY: Anchor Books.

Bigo D (2016) Rethinking Security at the Crossroad of International Relations and Criminology. *British Journal of Criminology*, 56(6): 1068–1086.

Blackwood L (2015) Policing Airport Spaces: The Muslim Experience of Scrutiny. *Policing: A Journal of Policy and Practice*, 9(3): 255–264.

Booth F and Altenbrun L (2002) Maritime and Port Security, Piracy, and Stowaways: Renewed Concerns over Old Problems. *University of San Francisco Maritime Law Journal*, 15(1): 1–48.

Branch AE (2007) *Elements of Shipping*. Routledge.

Brewster D (2017) Silk Roads and Strings of Pearls: The Strategic Geography of China's New Pathways in the Indian Ocean. *Geopolitics*, 2(22): 269–291.

Bugos S (2021) Russian ASAT Test Creates Massive Debris. *Arms Control Today*, 51 (10): 30–31.

Butler DA (2017) Who Owns the Moon, Mars, and Other Celestial Bodies: Lunar Jurisprudence in Corpus Juris Spatialis. *Journal of Air Law & Commerce*, 82: 505–518.

Campbell D (1992) *Writing Security: United States Foreign Policy and the Politics of Identity*. University of Minnesota Press.

Campbell D (2007) *The Biopolitics of Security: Oil, Empire, and the Sports Utility Vehicle*. US: Palgrave Macmillan.

Cowen D (2014) *The Deadly Life of Logistics: Mapping Violence in Global Trade*. University of Minnesota Press.

Damjanov K (2015) The Matter of Media in Outer Space: Technologies of Cosmobiopolitics. *Environment and Planning D: Society and Space*, 33(5): 888–906.

Dolman EC (2001) *Astropolitik: Classical Geopolitics in the Space Age*. Routledge.

Doyle SE (1990) International Space Plans and Politics: Future Roles of International Organizations. *Journal of Space Law*, 18 (2): 123–138.

Dunnett OM (2019) Geographies of Outer Space: Progress and New Opportunities. *Progress in Human Geography*, 43(2): 314–336.

Edwards R (2023) 'For All Mankind' Season 4, Episode 4, Review: The Shove that Shook Two Worlds. From: https://www.space.com/for-all-mankind-season-4-episode-4-review.

Eski Y (2011) 'Port of Call': Towards a Criminology of Port Security. *Criminology & Criminal Justice*, 11(5): 415–431.

Eski Y (2016) *Policing, Port Security and Crime Control: An Ethnography of the Port Securityscape*. London: Routledge.

Eski Y (2023) Crime Among the Stars: Space Criminology and the Domain of Limitlessness. *De Criminoloog*15 From: https://research.vu.nl/ws/portalfiles/portal/233723169/Nieuwsbrief28_def.pdf.

Eski Y (2024a) Space Crime, Control and Space Criminology. From: https://space.n2k.com/podcasts/t-minus/215.

Eski Y (2024b) How to Deal with Space Crime? From: https://www.nporadio1.nl/fragmenten/spraakmakers/af44620e-f8d4-4d3a-b196-f9b9cd3a9826/2024-02-07-hoe-moeten-we-omgaan-met-ruimtecriminaliteit.

Eski Y, Clamp K & Pemberton A (2024) A Future-Oriented Response to Crime: Exploring Space Crime & Criminology. Papers presented at the Criminal Justice Research Centre, University of Nottingham, 13 March 2024, Nottingham.

Finnegan C (2022) Indigenous Interests in Outer Space: Addressing the Conflict of Increasing Satellite Numbers with Indigenous Astronomy Practices. *Laws*, 11 (2): 1–19.

Flint C & Zhu C (2019) The Geopolitics of Connectivity, Cooperation, and Hegemonic Competition: The Belt and Road Initiative. *Geoforum*, 99: 95–101.

Forum (2023) The Belt and Road Initiative: A Key Pillar of the Global Community of Shared Future. Retrieved from http://www.beltandroadforum.org/english/n101/2023/1010/c124-895.html.

Germond B (2015) The Geopolitical Dimension of Maritime Security. *Marine Policy*, 137–142.

Holley C, Shearing C, Harrington C, Kennedy A, & Mutongwizo T (2018) Environmental Security and the Anthropocene: Law, Criminology, and International Relations. *Annual Review of Law and Social Science*, 14: 185–203.

Khurana GS (2008) China's 'String of Pearls' in the Indian Ocean and its Security Implications. *Strategic Analysis*, 32(1): 1–39.

Klinger J (2019) Environmental Geopolitics and Outer Space. *Geopolitics*, 26 (3): 666–703.

Lampkin JA & McClanahan BW (2023) Astronomical Withdrawals: A Green Criminological Examination of Extreme Energy Mining on Extraterrestrial Objects. *Crime, Law and Social Change*, 1–20.

Lampkin JA & White R (2023) *Space Criminology: Analysing Human Relationships with Outer Space*. Springer Nature.

Li AS (2020) Ruling Outer Space: Defining the Boundary and Determining Jurisdictional ity. *Oklahoma Law Review*, 73: 711.

Lin W (2019) Transport Geography and Geopolitics: Visions, Rules and Militarism in China's Belt and Road Initiative and Beyond. *Journal of Transport Geography*: 1–8.

Loader I & Percy S (2012) Bringing the 'outside'in and the 'inside'out: Crossing the criminology/IR divide. *Global Crime*, 13(4): 213–218.

Miller R (2015) Learning, the Future, and Complexity. An Essay on the Emergence of Futures Literacy. *European Journal of Education*, 50(4): 513–523.

Moore RD, Wolpert M & Nedivi B (2019–present) *For All Mankind*. United States: Apple TV+.

NASA (2024) Mars. From: https://www.nasa.gov/humans-in-space/humans-to-mars/.

Newman E (2016) Human Security: Reconciling Critical Aspirations with Political 'Realities'. *British Journal of Criminology*, 56(6): 1165–1183.
Nichols CR and Williams RG (2009) *Encyclopaedia of Marine Science*. New York: Infobase Publishing.
Noorali H, Flint C, & Ahmadi S (2022) Port Power: Towards a New Geopolitical World Order. *Journal of Transport Geography*, 1–12.
OED Online (2014) Port, n.1. Oxford University Press. Available at: www.oed.com/.
Patio GR (2020) The Ineludible Employment of Political Science and Criminology to Guarantee the Integral Security of the Population in the Iberoamerican Countries. *Vox Juris*, 38: 223.
Pekkanen S (2020) China, Japan, and the Governance of Space: Prospects for Competition and Cooperation. *International Relations of the Asia-Pacific*, 21(1): 37–64.
Pellegrino S (2023) Spaceports: The Most Active Are Surprisingly Close to Urban Areas. From: https://citymonitor.ai/environment/infrastructure/spaceports-the-most-active-are-surprisingly-close-to-urban-areas.
Peter N (2007) The EU's Emergent Space Diplomacy. *Space Policy*, 23(2): 97–107.
Reiner R (2010) *The Politics of the Police*. Oxford: Oxford University Press.
Rementeria S (2022) Power Dynamics in the Age of Space Commercialisation. *Space Policy*, 60: 1–13.
Riddervold M (2023) The European Union's Space Diplomacy: Contributing to Peaceful Co-operation? *The Hague Journal of Diplomacy*, 18(2–3): 317–350.
Roberts TG (2019) *Spaceport of the World*. Washington DC: CSIS.
Sachdeva GS (2023) *Crimes in Outer Space: Perspectives from Law and Justice*. Cham: Springer Nature.
Sergi A & Reid A (2021) *Ports, Crime and Security: Governing and Policing Seaports in a Changing World*. Policy Press.
Seshagiri L (2005) Spaceship Sheriffs and Cosmonaut Cops: Criminal Law in Outer Space. *Dalhousie Law Journal*, 28(2): 473–500.
Shammas VL (2019) One Giant Leap for Capitalistkind: Private Enterprise in Outer Space. *Palgrave Communications*, 5(1): 1–9.
Starlink (2024) Terms of Service. From: https://www.starlink.com/legal/documents/DOC-1020-91087-64.
Swierczynski M (2022) Critical Evaluation of New Council of Europe Guidelines Concerning Digital Courts. *Review of European and Comparative Law*, 48: 133–156.
Szocik K, Lysenko-Ryba K, Banaś S, & Mazur S (2016) Political and Legal Challenges in a Mars Colony. *Space Policy*, 38: 27–29.
Taubenfeld HJ (1961–1962) Regime for Outer Space. *Northwestern University Law Review* 56 (1): 129–167.
Thrower A (2022) Canada Warns it Will Prosecute Astronauts for Crimes they Commit on the Moon. From: mirror.co.uk/news/world-news/canada-warns-prosecute-astronauts-crimes-26857101.
Tuathail G (2005) *Critical Geopolitics: The Politics of Writing Global Space*. London: Routledge.
Verhetsel A & Sel S (2009) World Maritime Cities: From Which Cities do Container Shipping Companies make Decisions? *Transport Policy*, 16: 240–250.
Vinjamuri L (2016) The International Criminal Court and the Paradox of Authority. *Law and Contemporary Problems*, 79(1): 275–287.
Wang S (2009) The Making of New 'Space': Cases of Transatlantic Astropolitics. *Geopolitics*, 14(3): 433–461.

Weinzierl M (2018) Space, the Final Economic Frontier. *Journal of Economic Perspectives*, 32(2): 173–192.

Whitehead P (2016) *Transforming Probation: Social Theories and the Criminal Justice System*. Policy Press.

Wilkinson A (2017) Strategic Foresight Primer. European Political Strategy Centre. Retrieved from https://cor.europa.eu/Documents/Migrated/Events/EPSC_strategic_foresight_primer.pdf.

Williams D, Kuipers A, Mukai C & Thirsk R (2009) Acclimation During Space Flight: Effects on Human Physiology. *Canadian Medical Association Journal*, 180(13): 1317–1323.

14 Towards International Criminalisation of Orbital Debris Pollution

Eman Aboelkhair

Introduction

Not only did humankind pollute the environment on the Earth, but the effects of its actions extended to outer space and contaminated the environment there. We did not care about the future dimensions and the damage caused by this assault on the natural environment, which was granted to all humanity to utilise and harness its resources for humanity's general development and prosperity.

Space pollution usually refers to space debris or waste in Earth's orbit. Space debris includes objects that accumulate in space orbits. Space debris also includes inactive spacecraft, spent rocket bodies, materials released during space operations, fragments of disintegration of satellites, as well as fragments of explosions and collisions, and microparticles, which orbit around the Earth without control (McDonald 2022). Everything launched into outer space may become space debris. Space debris poses a significant threat to current and future active spacecraft. This threat is evident in the risk of debris colliding with active space objects and possibly interference with radio frequencies used in space. Without a doubt, this negatively affects space communications, navigation systems, space monitoring, and other space technology-based services (Flury & Contant 2001).

Despite the seriousness of the space pollution problem with space debris, international space treaties have not sufficiently protected the space environment. Even more complicated, the definition of space debris has not yet been addressed in any of the international instruments despite being one of the most significant issues threatening the space environment and its sustainability (Viikari 2015); leading international efforts to move towards the adoption of non-binding space debris mitigation guidelines. There have also been repeated calls to criminalise massive violations of the environment to provide legal protection to the natural environment under international criminal law.

Space Pollution Dimensions

Space activities are far more than environmentally friendly activities. Space launches cause various forms of environmental damage, including producing dust, emissions, and industrial waste that adversely affect the environment.

DOI: 10.4324/9781003437178-15

Rocket engine discharges could also cause atmospheric degradation and ozone depletion. Additionally, space activities may result in back-contamination, where pollutants transported by spaceships return to Earth. This pollution poses a risk of introducing alien pollutants to the environment upon their return. These environmental threats seem less critical than those posed by space debris. Hence, this research will focus on the problem of space debris stuck in orbits around the Earth (Raju 2019).

The Threat of Space Debris

Space debris has become a global issue that threatens the sustainability of space activities. The danger of space debris resulting from even the smallest particles scattered in space depends on the enormous speed at which the particles move and, consequently, the speed and strength of collisions in orbits. A particle not more than 1 cm can completely disrupt a functioning satellite. Even if the collision does not result in significant losses, it leads to more debris. The fragmentation of large pieces into smaller ones may increase the risk of further collisions with other space objects, resulting in many manoeuvres to avoid collisions (Adushkin et al. 2020). Satellites have become in need of manoeuvring regularly to avoid collisions. The proliferation of space debris in orbits around the Earth threatens space assets due to congestion and the proximity of debris to other space objects in a specific orbital region (Viikari 2015).

Additionally, space debris poses a clear threat to humans in space. Astronauts carrying out activities outside their vehicles are directly and most seriously exposed to space debris since the spacesuit cannot afford as much protection against space debris as the spacecraft's protective shields. The situation worsens if radioactive contamination occurs due to leakage of radioactive materials. Hence, space debris threatens crewed space activities (Tan 2000). The danger of space debris, moreover, lies in its ability to penetrate the atmosphere and its collision with the surface of the Earth, which in turn massively endangers human life as well as the Earth's environment, particularly when it comes to satellites that contain nuclear power sources.

Sustainability Challenges in Space Activities

Earth's orbits allow us to study our planet, send communications, and help provide many infrastructure services that have become indispensable in our daily lives. Using them sustainably is necessary, allowing future generations to benefit from their resources (National Academies of Sciences 2019).

Although the scope of space is vast, the near-Earth space environment, where most active satellites are based, is limited, i.e., limited in terms of where satellites can settle. Additionally, it is important for states to take into account that radio frequencies are limited natural resources (Stephens & de Zwart 2017, Rule 140). Every year, more and more satellites are launched.

Currently, there are nearly 9,000 active satellites in space orbits owned by countries and commercial entities (European Space Agency 2023). Satellite launch capability is being developed daily; for example, telecommunication companies plan to envelop the Earth by launching large groups of satellites, known as "Mega Constellations", to make access to broadband wireless internet a practical reality globally. If successful, up to 50,000 more satellites will be in orbit around the Earth (Millwood 2023). This means an increased risk of collisions between space objects and further manoeuvres to avoid collisions to ensure the safety of such satellites.

As the amount of space debris resulting from space activities continues to increase, the costs of collisions could eventually stifle the industry (Hays Parks 1990). Based on MASTER-8 data and prospects for 2023 (Hulme 2004), estimates of statistical models indicate a large number of debris objects in space orbits. This number is estimated as follows: about 36,500 space debris objects larger than 10 cm, 1 million space debris objects ranging from 1 cm to 10 cm, and about 130 million space debris objects larger than 1 mm and up to 1 cm. These large figures show the scale of the challenge we face in space orbit debris management, where not all space debris objects are tracked and classified (Hurova 2023). Approximately 35,000 space debris objects are in orbits around Earth, tracked by the United States Space Command's Space Surveillance Network (SSN) (European Space Agency 2023).

While recognising the limited nature of outer space resources, it is necessary to preserve the space environment and make rational and effective use of its resources to ensure that countries have equitable access to those orbits and frequencies, taking into account the particular circumstances and needs of developing countries and the geographical situation of certain countries (Jakhu & Freeland 2022).

The Legal Status Quo of Protecting the Space Environment

The Outer Space Treaties laid a foundation for international liability for damages caused by space activities. Generally, they indicate that the launching state is always liable for any damage caused by their space objects (McDonald 2022). However, space treaties have been criticised for the absence of a clear standard for establishing liability, particularly regarding damage caused by space debris of unknown origin. In addition, the current system does not provide a fair compensation mechanism for damages. What makes the situation even more complicated are the broad definitions contained in the texts of space treaties. That is why there is a need to prepare precise and unified definitions of terms such as "space object", "the launching state", "launch", and "damage", in order to establish liability of the state causing the damage (Viikari 2008). This makes the provisions of space treaties so vague as to render them useless, especially concerning the problem of space pollution and the damage caused by it.

Establishing liability for damage in space is highly complex because it is impossible to identify the origin of most space debris objects. Despite the technological development of states' capabilities, countries have tended to track space debris objects of 10 cm in size and do not track smaller objects. Even if governments decide to use the technology needed to track smaller objects, this does not necessarily mean being able to identify "owners" of small debris objects in order to establish liability for damage resulting from them (Viikari 2008). Therefore, the author recommends developing a new technique for identifying space objects, which could be called "space footprint". Here, each space object should be assigned a unique fingerprint specific to that object, registered with the launching state and the UN Secretariat. This idea parallels the idea of a genetic footprint; just as every cell in a human body has a genetic footprint that this object accounts for, every micro-particle or fragmentation of space debris should have a space footprint that expresses the space object as a whole. Hence, it is possible to identify the entity that owns the space object, no matter how small its size, and thus identify international liability for damages resulting from any collision between space objects as a vital step towards effectively addressing the problem of space debris.

International law does not explicitly provide specific rights or obligations for protecting the space environment. Hence, it will be necessary to resort to the general principles and rules of international law that generally aim to preserve the natural environment, as states and international organisations have to take the necessary measures to maintain a safe and sustainable space environment and avoid irresponsible acts that cause space pollution. However, the general protection afforded by general international law needs to be improved for the space environment, being a very fragile environment with unique characteristics and a weak ability to repair itself and recover from contamination and other consequences resulting from space activities (Moltz 2011).

The Need for International Criminalisation of Space Debris Pollution

Although the person and protecting his/her rights are the goal of every national and international social order, in international law, humanity manifests itself and prevails over all other individual interests, as humanity represents the mass population in the entire international community. The principles of international law ensure equal human treatment to all members of the international society since they criminalise acts that undermine the supreme goal of safeguarding humanity.

International law recognises the correlations between environmental rights and human rights. The importance of preserving the natural environment and its impact on achieving security and stability was emphasised earlier in 1972 in the Stockholm Declaration, a historic legal instrument described as "the birth of international environmental law", which states that "defending and improving the human environment for present and future generations has become an inevitable goal for humankind; a goal to be achieved in harmony with the established and fundamental objectives of peace and economic and social development at the global and local levels" (United Nations, no date).

In 2002, the International Law Commission issued the New Delhi Declaration of Principles of International Law Relating to Sustainable Development, which stipulates the duty of states to ensure the sustainable use of natural resources (New Delhi Declaration 2002). In July 2022, the United Nations General Assembly adopted a historic resolution, declaring the right to a clean, healthy, and sustainable environment as a universal human right linked to other rights and current international law. This step represents a landmark achievement in the efforts to safeguard the human right to a clean and sustainable environment and the other human rights that depend on it (Cima 2022).

The Legal Basis for Criminalising Environmental Pollution Under International Criminal Law

In international law, several interests emerge on which the international community is based, and international criminal law ensures that they are protected according to their relevance to peace and security worldwide. There is no doubt that maintaining international peace and security is a supreme goal for all civilised countries to pursue in response to the teachings of divine religions and philosophers' shouts and for their mutual benefit, and violating this noble international goal is at the forefront of international criminalisation. If protected interests in the international sphere are violated, international crime arises to protect those supreme international interests and maintain international public order.

While all international crimes are characterised by a violation of an international obligation necessary to safeguard the fundamental interests of the international community, specific interests are at the forefront, including those relating to the maintenance of peace, the right of peoples to self-determination, and the protection of human rights, and those involving the preservation of the environment in which human beings live (Yarnold 1994). Environment and sustainability issues have undoubtedly become integral to international public order. Many international conventions on protecting the environment and preserving natural resources have become part of civilised countries' obligations (Louka 2006).

Moreover, it is widely recognised that customary international law requires that an individual's actions do not cause massive environmental destruction to constitute a threat to fundamental human rights. Consequently, the deliberate targeting of the environment can and should establish criminal liability, and international legal instruments should lead to criminal accountability for those individuals who destroy the environment. Massive violations of the environment that shock the conscience of all humanity should constitute an international crime that brings criminal liability for its perpetrators (Freeland 2015).

The Impact of Space Debris Pollution on International Peace and Security

In 1992, the United Nations Security Council (1992) concluded that "[t]he absence of war and military conflicts amongst states does not in itself ensure international peace and security. The non-military sources of instability in the economic, social, humanitarian, and *ecological* fields have become threats to international peace and security. The United Nations membership as a whole needs to give the highest priority to the solution of these matters".

Undoubtedly, there is a correlation between environmental degradation and political and social conflicts. Activities that destroy the space environment can destabilise the international public order. The most significant of these activities is the use of anti-satellite (ASAT) weapons and their widespread and long-term effects, as well as the threat of weaponisation and the militarisation of outer space. The lack of transparency regarding the nature and actual purpose of space activities and objects stuck in the space environment, as well as the inability to track and determine the ownership or affiliation of most space debris objects, thereby establishing liability for damage, are the most important catalysts of international escalation, mistrust and tensions among spacefaring countries. They underscore the need to move towards the international criminalisation and prosecution of such serious acts (Steer 2020).

The Threat of Anti-Satellite Weapons

Anti-Satellite Weapons are a significant challenge facing the international space industry community. This issue poses a long-term threat to space sustainability based on its remarkable role in creating chaos in the outer space environment and the ensuing threats to international peace and security.

The dangers of using ASAT are reflected in their indiscriminate and uncontrollable effects, making it difficult to quantify the amount of space debris produced and the direction of its deployment. Space debris that lost control may cause collisions with space objects serving infrastructure systems, as well as the potential to cause jamming of satellite signals, which would render devices on Earth idle or operate wrongly, ultimately resulting in civilian or military casualties on Earth without discrimination. For example, losing signals of the Global Navigation Satellite System (GNSS) can pose a threat to systems on Earth that rely on those signals, such as airplanes that will be forced to land, smart bombs that may miss their targets, or self-driving vehicles that may stop suddenly during heavy traffic. For example, during a 2018 NATO military exercise, the Russian army jammed a GPS satellite, eventually causing serious danger to both military and commercial aircraft (Žilinskas & Marozas 2022).

Demonstrations of Power and the Race to Weaponise Outer Space

Unfortunately, international law still lacks explicit regulations or customs that determine the unlawfulness of the use of anti-satellite weapons. Notwithstanding the international community's knowledge of the quantities of space debris resulting from such activities and of the dangers that threaten the sustainability of the entire space environment, no common standards governing such weapons have been reached.

Four major spacefaring nations, the United States of America, China, Russia and India, have conducted anti-satellite weapons tests. Of concern, following India's test of anti-satellite weapons in 2019, India publicly stated that it had joined the Elite Club of major space states that had demonstrated this capability to test anti-satellite weapons in the past (Lauer 2022).

Accordingly, once a state proceeds to assert its unilateral authority to weaponise in outer space with the implicit threat of using such weapons, other states will react quickly in the same way to assert their unilateral authority. Such an arms race in space would increase international tensions and complicate international relations in a manner that threatens international peace and stability. Suppose this state of showing power and the race for weaponising outer space occurs. In that case, the "point of no return", referred to by US President Eisenhower in 1960, may be reached much sooner than expected; as President Eisenhower warned of the danger of an arms race in Space and the potential consequences for international security (Menon 1987).

Hence, the need for international criminalisation of the use or testing of anti-satellite weapons is demonstrated in order to prevent outer space from being a theatre for showing powers because unleashing anti-satellite weapons tests will motivate any state wishing to strengthen its defence capabilities to develop anti-satellite weapons. This would affect the safety and security of the entire international community. Therefore, it is necessary to criminalise it internationally. The major space powers have come to closely monitor each other's space activities, which are likely to threaten assets in space that are of high value not only economically but militarily as well.

Crimes of Space Debris: National, Transnational, and International Crimes

Based on the seriousness of the problem of space pollution in its various dimensions, which is essentially based on space debris, it is necessary to create a new category of crimes, namely "space debris crimes". Space debris crimes refer to any prohibited act in which space debris contributes to prejudice of a legally protected interest, which requires states to develop mechanisms to combat such crimes and to enact legislation providing for their criminalisation. Furthermore, space debris crimes require the creation of a raised awareness, among those entities active in the space field, of the illegality of these irresponsible practices that would create more space debris without justification. The author defines space debris crime here as "any unlawful act committed by a person with a considered will in which space debris is used or contributes to the violation of the rights of others protected under the law".

Therefore, many illegal acts that violate the rights of others can fall within the scope of space debris crimes. It may include collisions of space debris with other objects and property floating in space, leading to their destruction and thus disrupting the access of essential space services to stakeholders. It may also include cases of falling debris to Earth, causing serious damage to property and individuals. Practices that pollute the space environment create space debris and lead to its scattering in space, which can also violate others' rights to free access to space and the utilisation of space resources, thus falling within the framework of space debris crime.

With the emergence of space debris crimes, there will also be a need to classify them and clarify the possibility of them being included among national crimes or transnational (universal) crimes so that national courts have jurisdiction over them. National criminal legislation applies to them in those cases, or they may be included among international crimes if they meet the elements and conditions of international crime. Therefore, in those cases, international courts have jurisdiction over them.

In general, an international crime differs from a national crime in that an international crime affects the interests of the human community, disturbs its public order, and is therefore criminalised by international criminal law. It is a violation of an international obligation aimed at preserving the international community's fundamental interests and maintaining international peace and security (Omar 2006: 111), while national crime affects the interest of society and is criminalised by domestic criminal law. As for transnational crime, it is a criminal act committed in several states and punishable by all of them. Global crime is a national crime, but its criminal activity is practiced in several countries. Its danger is not limited to one country. Still, rather more than one country suffers from it, so international agreements are concluded to criminalise it. States cooperate to combat it, and through them (i.e. international agreements), states commit to criminalising it in their national criminal legislation and to cooperate in combating it.

First: Space Debris Crimes as National Crimes

Space debris crimes may fall within the framework of national crimes if the crime is committed in its entirety and in all its elements within the territory of a single state, and its effects and repercussions are limited within the borders of that state, so the national courts have jurisdiction over them. For example, the act of recovering a defunct spacecraft loaded with nuclear power sources and allowing it to fall into the territory of a state other than the launching country without permission from the relevant authorities of that state, resulting in massive losses, is a national crime. This crime will be subject to the national law applicable within that state, and the crime will also fall within the scope of its national jurisdiction.

Second: Space Debris Crimes as Transnational Crimes

Space debris crimes could be considered transnational crimes if the criminal behaviour harms the interests of more than one state that criminalises these acts in national legislation. For example, if the use of anti-satellite weapons and their resulting debris destroyed several spacecraft belonging to several states, which suffered significant losses, both human and economic, or if debris resulting from the use of these weapons impedes other countries from accessing a specific area in orbit around the Earth, in contravention of the internationally recognised principle of free access to, and use of space. In this case, the national law of any state affected by the crime applies, and states are committed, through international agreements, to criminalise it in their national legislation and to cooperate in holding liable, and punishing the perpetrators of, these crimes.

Third: Space Debris Crimes as International Crime

Space debris crimes can also be included in international crimes if they result in destabilising international security and stability and if they fulfil the conditions and elements of one of the four international crimes provided for in the Statute of the International Criminal Court. We would be facing a crime against humanity if the use of an anti-satellite weapon and the resulting debris caused the destruction of a crewed spacecraft and the killing of its astronaut crew if the act is committed as part of a widespread or systematic targeted attack. Moreover, causing widespread, long-term and severe damage to the space environment, such as creating clouds of space debris as a result of illegal space activities, such as the use of anti-satellite weapons, could be considered a war crime if the other objective conditions for the crime stipulated in Article 8 of the Rome Statute were met.

Protection of the Outer Space Environment Under International Criminal Law

The Rome Statute, in a single article, addresses the direct protection of the natural environment. Article 8(2)(b)(iv) refers clearly to the environment in the Rome Statute under the name “war crimes”. This article expressly criminalises the conduct of launching an intentional attack with the knowledge that such an attack will result in widespread, long-term, and severe damage to the natural environment that would be excessive in relation to the concrete and direct overall military advantage anticipated. Although the explicit reference to environmental damage as a war crime is a step forward and a praiseworthy development in international criminal law, meeting the conditions for applying the Article’s provisions is virtually impossible in practice (Freeland 2015: 210).

Restrictions on the Application of Article 8(2)(b)(iv) to Acts of Environmental Destruction

Article 8 of the Rome Statute established several objective conditions for the establishment of environmental war crimes as international crimes. The first is that criminal conduct must be committed during times of war or armed conflict. Criminal liability and international sanctions for environmental damage cannot be imposed outside the context of armed conflict (Freeland 2015: 204–210).

Article 8 also requires an aggravating standard of environmental damage resulting from a criminal act to fall within the scope of the war crime description. It required that the resulting damage be widespread, long-term, and severe. Noticeably, the conjunction "and" was used to connect the three terms in describing environmental damage, and the use of the disjunctive "or" was avoided, indicating that all these elements must coexist for the offence to be deemed a war crime, thereby imposing more stringent criteria for environmental harm to constitute the offence (Minkova 2023).

Article 8(2)(b)(iv) added an objective requirement of accountability for acts of environmental destruction in the context of armed conflict, requiring an attack to be contrary to the principle of proportionality (Rome Statute, art. 8.2(b)(iv)). It requires proving that there has been a clear imbalance or excess in the amount of environmental damage resulting when compared to the expected direct and tangible military gains. Even if the required level of environmental damage is achieved, a war crime will not be considered if this falls within what is acceptable in light of the expected military advantage (Hulme 2004). This makes environmental damage secondary to military gains, which means that this article considers environmental damage as an unintended by-product of war, even where environmental destruction is deliberate and direct (Freeland 2015).

The Relevance of *Jus in Bello* Principles to Protect the Outer Space Environment

Space is on the verge of becoming a battleground where spacefaring nations compete to enhance their military capabilities. Since the advent of space technology, information gathered from outer space has not only been used, for example, to plan military operations on Earth, but space resources are now used to direct and execute military operations directly and have become an essential part of the military equipment of the major powers (Bowen 2020).

Once the situation becomes a state of armed conflict, the *jus in bello* or international humanitarian law comes into force. Therefore, deliberate acts leading to the massive pollution of the space environment will be contrary to the law of war if they result in severe, widespread, and long-term damage and are disproportionate to anticipated military gains (Freeland 2006). If, during an armed conflict, anti-satellite weapons were tested as a threat of force or used in direct attacks on the satellites of states parties to the conflict, they were expected to violate all standards of environmental preservation, sustainability, and safety under international humanitarian law (Žilinskas & Marozas 2022).

The principles of distinction and proportionality in the law of war can be considered distinguished challenges facing the protection of the space environment during war (Hays Parks 1990); the principle of distinction addresses the need to distinguish between civilians or civil assets and military objectives, whereas the principle of proportionality means that the incidental damage inflicted on the civilian population or the natural environment during a military attack must not be excessive in comparison to the direct military advantage that would result (Wolfrum 2022).

The application of both the principle of distinction for military objectives and the principle of proportionality between military gains and environmental damage in the case of space pollution faces challenges, given the complex nature of space technology. With the common dual use of space technology, it has become challenging to determine the nature and purpose of space objects. Space systems and technologies are used not only to gather information but also to effectively control and coordinate military operations (Hurova 2023).

It is not easy to determine the nature and purpose of the space object in Earth orbit, which are often not precise, as a particular space object may serve military purposes. Hence, its military targeting is legitimate, and the same space object may also be intended for civilian purposes; therefore, its military targeting is unlawful. For example, the GPS positioning satellite provides a service to manage civil aviation traffic while at the same time providing a service to guide military weapons (Žilinskas & Marozas 2022). The ambiguity of military objectives thus prevails and is difficult to identify precisely, as well as the civilian victims targeted by military attacks, which entails a lack of clarity of responsibility for environmental damage occurring in the space environment.

Even if a space object is identified as a legitimate military target, we must consider the principle of proportionality, which may involve some uncertainty. For an environmental war crime to arise, a balanced assessment between military advantages and potential damage to the outer space environment is required, i.e., the damage resulting from a military attack on space must be excessive enough, compared with direct and anticipated military advantages. Consideration should be given to all incidental damage expected to be caused to civilians and civilian objects in space and on the surface of the Earth, whether directly or indirectly, including the targeting of dual-use space objects that become military targets and the potential harm they may cause to civilians by cutting off the vital services provided by these satellites. The danger posed by space debris and the potential for a cycle of successive collisions between such debris and other active space objects must also be considered (Žilinskas & Marozas 2022).

Conclusion

In conclusion, the provisions of international criminal law aimed at protecting the natural environment are primarily based on protecting human interests during times of war, providing indirect protection for the environment itself. In the absence of international legislation specifically on environmental crime, the space environment continues to be covered by limited criminal protection, beset by legal obstacles that would prevent adequate international liability for the severe destruction of the space environment and threaten its sustainability. Accordingly, for a genuine green shift in international criminal law, it would be necessary to establish a legal provision in the Rome Statute that would include a new international crime that recognised the seriousness of environmental damage per se, recognised the complexities of its commission and took into account the diverse situations in which it might arise.

References

Adushkin, VV, Aksenov, OY, Veniaminov, SS, Kozlov, SI & Tyurenkova, VV (2020) The small orbital debris population and its impact on space activities and ecological safety. *Acta Astronautica*, 176, 591–597.

Bowen, BE (2020) *War in Space: Strategy, Spacepower, Geopolitics.* Edinburgh: Edinburgh University Press.

Cima, E (2022) The right to a healthy environment: Reconceptualizing human rights in the face of climate change. *Review of European, Comparative & International Environmental Law*, 31(1), 38–49.

European Space Agency (2023) *ESA's Annual Space Environment Report*. https://www.sdo.esoc.esa.int/environment_report/Space_Environment_Report_latest.pdf (accessed 10 July 2024).

Flury, W & Contant, JM (2001) The updated IAA position paper on orbital debris. *Proceedings of the Third European Conference on Space Debris*, (2), 841–849.

Freeland, S (2006) The applicability of the jus in bello rules of international humanitarian law to the use of outer space. In *57th International Astronautical Congress* (pp. E6–4).

Freeland, S (2015) *Addressing the Intentional Destruction of the Environment During Warfare Under the Rome Statute of the International Criminal Court*. [Doctoral Thesis, Maastricht University].

Hays Parks, W (1990) Air war and the law of war. *Air Force Law Rev*, 32(1), 1–226.

Hulme, K (2004) *War Torn Environment: Interpreting the Legal Threshold* (Vol. 7). Brill.

Hurova, A (2023) Does dual use of satellites challenge the future space governance? *Advanced Space Law*, 12, 20.

Jakhu, R & Freeland, S (2022) *McGill Manual on International Law Applicable to Military Uses of Outer Space, Volume I: Rules*. Canada: McGill University.

Lauer, RS (2022) When states test their anti-satellite weapons. *Astropolitics: The International Journal of Space Politics and Policy*, 20(1), 1–26.

Louka, E (2006) *International Environmental Law: Fairness, Effectiveness, and World Order*. Cambridge: Cambridge University Press.

McDonald, S (2022) Earth and beyond: The creation of international environmental space law. *Golden Gate University Environmental Law Journal*, 14, 69.

Menon, PK (1987) Prevention of an arms race in outer space. *Military Law & Law of War Review*, 26, 225–326.

Millwood, S (2023) *The Urgent Need for Regulation of Satellite Mega-constellations in Outer Space.* Springer.

Minkova, LG (2023) The fifth international crime: Reflections on the definition of "ecocide", *Journal of Genocide Research*, 25(1), 62–83.

Moltz, J (2011) *The Politics of Space Security: Strategic Restraint and the Pursuit of National Interests.* United States: Stanford University Press.

National Academies of Sciences (2019) *Thriving on Our Changing Planet: A Decadal Strategy for Earth Observation from Space.* Washington DC: National Academies Press.

New Delhi Declaration of Principles of International Law Relating to Sustainable Development, ILA Resolution 3/2002, adopted at the 70th Conference of the International Law Association, New Delhi, India, April 2–6, 2002. https://www.ila-hq.org/en_GB/documents/conference-resolution-english-new-delhi-2002-3.

Omar, HH (2006) *Governors' Immunities and Trials for War Crimes, Aggression, Extermination and Crimes against Humanity "Trial of Saddam Hussein"*, Arab Renaissance House, First Edition.

Raju, N (2019) *The Regulation of Rocket Emissions.* PhD Thesis submitted to McGill University: Canada.

Steer, C (2020) *Why Outer Space Matters for National and International Security.* Center for Ethics and the Rule of Law: University of Pennsylvania.

Stephens, D & de Zwart, M (2017) The manual of international law applicable to military uses of outer space (MILAMOS). RUMLAE Research Paper (17–12), 2020–2046.

Tan, D (2000) Towards a new regime for the protection of outer space as the province of all Mankind. *Yale Journal of International Law*, 25, 145–194.

United Nations (no date) United Nations Conference on the Human Environment, 5–16 June 1972, Stockholm. https://www.un.org/en/conferences/environment/stockholm1972 (accessed 10 July 2024).

United Nations Security Council (1992) Note by the President of the Security Council (Presidential Statement S/23500). https://www.securitycouncilreport.org/atf/cf/%7B65BFCF9B-6D27-4E9C-8CD3-CF6E4FF96FF9%7D/PKO%20S%2023500.pdf.

Viikari, L (2015) Environmental aspects of space activities. In: van der Dunk, F (ed.) *Handbook of Space Law.* Cheltenham: Edward Elgar Publishing (pp. 717–768).

Viikari, L (2008) *The Environmental Element in Space Law: Assessing the Present and Charting the Future.* Brill Nijhoff.

Wolfrum, R (2022) Proportionality: Reconsidering the application of an established principle in international law. *International Law Studies*, 99(1), 677–696.

Yarnold, BM (1994) Doctrinal basis for the international criminalization process. *Temple International and Comparative Law Journal*, 8(1), 85–116.

Žilinskas, J & Marozas, T (2022) Weapons reviews for ASATs: Assessing distinction, proportionality, and effects on the natural environment of space. *Air and Space Law*, 47(2), 209–232.

15 The Legal Framework for Policing Outer Space

Giulio Calcara and Mika Launiala

Introduction

Police forces and services have traditionally been equipped and tasked by states to carry out their functions solely within their territorial boundaries. This is due to a vast array of practical and legal reasons, but mainly because the activity of policing is strictly connected to the administration of territorial jurisdiction. Unauthorised enforcement of the criminal law of a state over the territory of another state constitutes a serious violation of sovereignty (Boister 2018). Thus, whenever operating, domestic police bodies must be subservient to the limitations posed by the sovereignty of states. Nevertheless, instances of policing beyond national borders can take place in limited circumstances. For instance, if this is allowed by specific international instruments, such as conventions, or if this occurs in the context of specific international policing missions organised through the United Nations Police (UNPOL), for example, in the case of peace operations.

However, there are places in this world and beyond, which are legally outside the sovereign reach of any state. Places where states are barred from claiming sovereign rights include the high seas, as per the United Nations Convention on the Law on the Sea, and outer space, as per the Outer Space Treaty (von der Dunk 2002). When criminal acts take place in these areas, it can be arduous to identify the appropriate legal basis for states to assert jurisdiction and even more challenging to conduct policing activities, as there may be a plethora of overlapping domestic and international legal regimes. Crucially, outer space is an area where human activities are destined to increase over time with ventures such as research missions organised by states, space tourism, deployments of satellites, and in the future perhaps controversially, space mining. Additionally, projects are already being developed to allow for human settlement of outer space and celestial bodies. If some of these activities were to be disrupted or sabotaged through criminal activities, there could be serious repercussions.

DOI: 10.4324/9781003437178-16

For all these reasons, this chapter addresses fundamental questions regarding policing outer space, including:

1 What is the legal framework for policing crime in outer space?
2 Who is allowed to police crimes committed in outer space?

The Legal Frameworks for Policing

Within different jurisdictions, there are legal frameworks that regulate how policing operations are to be conducted. These frameworks regulate different types of activities, which can include the gathering of information, collecting testimonies, conducting surveillance, patrolling, performing arrests, and many others. The onus for conducting such activities can lie on a multitude of different actors including police institutions, the private sector, and different types of authorities (Brodeur 2010; den Boer 2018). Laws in matters of policing tend to differ significantly among states. Laws can bestow different legal powers, notably including the use of force and power of arrest, rights, privileges, and immunities to specific police institutions while limiting their scope, for instance, by posing strict conditions on how they should carry out their tasks. On a general level, it is possible to classify policing laws as either "permissive" or "restrictive". Studies highlight how varied these can be across the globe (den Boer 2018). Interestingly, research shows that laws in matters of policing can be numerous and at times contradicting even inside the territory of a single state (den Boer 2018). Naturally, such issues tend to be particularly evident in the context of federal states. Then, there are unfortunate situations in which police forces act mainly, if not solely, on behalf of illiberal regimes to deliberately harass or suppress political opponents. In these situations, laws tend to be ineffective, if not officially suspended, and it is possible to identify "legal black holes" (den Boer 2018). It should be noted that aside from the noticeable divergences in laws, approaches and strategies on how to conduct policing activities which differ among countries and even within domestic boundaries, they tend to change significantly over time (O'Neill 2010).

Spurred by the development of technology and the ease of travelling, there has been an increase in the number of international and transnational crimes since the second half of the 20th century. This has meant that domestic institutions have been compelled to shift their focus beyond domestic borders (Calcara 2019). However, police institutions are traditionally equipped to carry out their functions within their territorial jurisdiction. Extraterritorial law enforcement presents challenges of both a practical and legal nature. Most of all, domestic police institutions need to be subservient to the limitations posed by the sovereignty of other states. In the absence of specific permissive laws, international law does not allow for one state to exercise its power in the territory of another state, including conducting law enforcement activities. This rule was illustrated in the famous Lotus case judged by the Permanent Court of International Justice in 1927 and has been confirmed ever since in various conventions and international instruments (Boister 2018).

In practice, domestic police officers can only perform law enforcement activities beyond their national borders and in the territory of another state in very limited instances, which are strictly regulated by specific international instruments. Examples of such limited instances include conducting cross-border surveillance and transnational hot pursuit regulated in the Schengen acquis, and conducting law enforcement activities through the United Nations Police UNPOL in the limited context of peace operations (Calcara 2019; Hughes, Hunt and Curth-Bibb 2013). Thus, international police cooperation remains the main tool for domestic police institutions to deal with criminal offences that originate from the territory of other states (Boister 2018; Calcara 2019). Crucially, in the context of international police cooperation, the police institution of State A applies the laws of State A, while the police institution of State B applies the laws of State B. If the police institution of State A contacts the police institution of State B to initiate cooperation concerning an alleged criminal offence originating in State B, the police institution of State B ultimately decides whether and to what extent to cooperate according to the domestic laws of State B (Boister 2018). In the past, international police cooperation used to be conducted through diplomatic channels. Nowadays, domestic police institutions tend to communicate with their counterparts by using the platforms of regional or international entities, such as Europol or INTERPOL, which are not always devoid of legal issues (Calcara 2019; Lemon 2019; Zabyelina 2019).

Interestingly, there are areas outside the sovereign reach of any state. As previously mentioned, places on this planet and beyond where states are barred from claiming sovereign rights include the high seas, as per the United Nations Convention on the Law on the Sea (Calcara and Launiala 2023), and outer space, as per the Outer Space Treaty (von der Dunk 2002). Naturally, in case criminal acts take place in these areas, identifying the appropriate legal basis for states to assert jurisdiction or conduct policing activities can be challenging, as several domestic and international legal regimes may overlap. The following section will first focus on the laws and principles governing outer space and assess them on how, and according to what legal frameworks, policing activities can take place.

Outer Space Law

While the entire legal framework governing space, the *corpus juris spatialis*, is seen as comprising a large number of domestic and international laws, space law is ultimately a sub-branch of public international law, and it is largely shaped by treaties. Thus, the traditional sources and the system of public international law apply to space law as well (Steer 2017). This is also specifically highlighted in Article III of the Outer Space Treaty, where it is ruled that the states parties must carry out their activities involving the exploration and use of outer space, the Moon, and other celestial bodies, in accordance with international law and the UN Charter. For this reason, when analysing what legal framework and principles regulate policing activities taking place in outer space, it is crucial to consider treaties on outer space, customary laws, and principles of public international law. There are five core treaties that represent the pillars of international space law:

1. The Outer Space Treaty – The Treaty on Principles Governing the Activities of States in the Exploration and Use of Outer Space, including the Moon and Other Celestial Bodies, 27 January 1967, 610 UNTS 205.
2. The Rescue Agreement – Agreement on the Rescue of Astronauts, the Return of Astronauts and the Return of Objects Launched into Outer Space, 22 April 1968, 672 UNTS 119.
3. The Liability Convention – Convention on International Liability for Damage Caused by Space Objects, 29 March 1972 961 UNTS 187.
4. The Registration Convention – Convention on Registration of Objects Launched into Outer Space, 14 January 1975, 1023 UNTS 15.
5. The Moon Agreement – Agreement Governing the Activities of States on the Moon and Other Celestial Bodies, 18 December 1979, 1363 UNTS 3.

All the above treaties were developed during the 1960s and 1970s by the UN Committee on Peaceful Uses of Outer Space (UNCOPUOS).

The most influential is undoubtedly the Outer Space Treaty (OST), as it establishes the fundamental principles concerning the laws on outer space, and thus it is often indicated as the Constitution for space activities. Among the principles set by the treaties, it is possible to mention that outer space is regarded to be the province of all humankind, and that in outer space there is freedom of exploration and utilisation, and no state can claim sovereignty, i.e., the principle of non-appropriation. Additionally, states shall cooperate and provide mutual assistance in the context of exploration and use of outer space (Pečujlić 2023; United Nations Office for Outer Space Affairs 2023a).

Crucially, through the OST, space law has been shaped with some unique traits, if compared to traditional public international law. For instance, according to Article VI of the OST, the responsibility of states is not to be viewed as being limited to conducts attributable to states which constitute internationally wrongful acts directed against other states. Instead, all activities that qualify as national activities in outer space, even when performed by private entities and Non-Governmental Organisations, can trigger state responsibility. Further, in case space activities are performed by an international organisation, responsibility for such activities is to be shared among states involved in the organisation and the organisation itself. Then, in line with Article VII of the OST, states bear responsibility for any damage caused by space objects even when these have been handled by private entities (Hermida 2004; von der Dunk 2017). Signalling the widespread acceptance of this document, as of the 1st of January 2023, the OST has been ratified by 112 countries (United Nations Office for Outer Space Affairs 2023b).

Beyond the five core treaties created under the umbrella of UNCOPUOS, it should be noted that there are other treaties that directly or indirectly contribute to the regulation of outer space concerning specific areas of interest. Among those are the Partial Test Ban Treaty (Treaty Banning Nuclear Weapon Tests in the Atmosphere, in Outer Space and Under Water, 5 August 1963, 480 UNTS 43), the ITU Constitution (Constitution of the International Telecommunication Union, 22 December 1992, 1825 UNTS 331), and ITU Convention (Convention of the International Communication Union, 22 December 1992, 1825 UNTS 390) (Steer 2017). Many treaties and different types of agreements have also been created specifically for scientific purposes, such as the intergovernmental agreement on the International Space Station (ISS) and the convention instituting the European Space Agency (von der Dunk 2017).

As previously mentioned, besides what could be identified as international space law, there are national space laws, which are domestic regulations that have been adopted in certain countries. While international space law deals mainly with the activities and responsibilities of states and intergovernmental organisations, the domestic space laws have been created mainly to regulate private space activities, which have exponentially increased in recent times. The presence of domestic regulations is important as norms belonging to the sphere of public international laws are not necessarily directly applicable to private entities (Marboe 2017).

Policing Outer Space

The general focus of this chapter is directed towards an exploration of the principles and legal frameworks involved in the process of policing outer space. However, much of this discussion is hypothetical as lack of resources, technological issues, and several other practical limitations impede policing in outer space. For instance, patrolling or conducting investigations in outer space would be an extremely costly activity to undertake for domestic police institutions. Additionally, the processes involved in the policing of outer space pose challenges that at the moment appear almost insurmountable. Such challenges include lack of qualified personnel, difficulties in gathering evidence or preventing deterioration of crime scenes due to different conditions than on Earth, difficulties in procuring or transporting appropriate equipment, and many others (Lampkin and White 2023). Nevertheless, due to the increasing interests of public and private entities in pursuing exploration, and scientific and commercial endeavours in this area, it is still a worthy endeavour to illustrate what could be the legal basis for undertaking policing activities in outer space. Moreover, for half a century, humans have already inhabited space stations and as later discussed, there are already plans in place for colonising celestial bodies, such as Mars and the Moon. Furthermore, where human beings inhabit, there is always a chance for criminal activities to occur (Lampkin and White 2023).

This section is divided into three subsections. First, the only existing precedent of policing in outer space is examined. Second, the five principles of jurisdiction are viewed. Lastly, the adjudicative, prescriptive, and enforcement categories of jurisdiction in outer space are illustrated.

The Only Precedent – an Alleged Crime on Board the ISS

Before attention is directed to provide a systematic review of the legal framework and principles that could enable states to conduct policing activities over outer space, it is beneficial to assess and analyse existing precedents.

There are some precedents involving crimes which, while connected to space activities, have taken place on Earth. Crimes such as bribery, harassment, discrimination, and violations of health and safety standards have taken place on our planet among public and private entities involved in the space industry (Lampkin and White 2023). Some of these crimes are complex and of a transnational nature, thus affecting the territory or interests of two or more states. A relevant example is the Airbus bribery scandal, where the European company failed to prevent the payment of bribes as an unlawful way to secure contracts in Ghana, Indonesia, Malaysia, Sri Lanka, and Taiwan (Boakye, Siaw, and Sarpong 2022).

However, when the focus is shifted towards outer space, US astronaut Anne McClain is the first and only person accused, and later cleared, of committing a crime in outer space while on board the ISS in 2019. In this case, Summer Worden, the estranged spouse of McClain, reported to the Federal Trade Commission and NASA's Office of Inspector General that her bank account was accessed by McClain while she was an astronaut onboard the ISS. In doing so, Worden claims that McClain committed an improper access to her financial records and identity theft. On the other hand, McClain refuted any wrongdoing and justified her conduct based on her finances being intertwined with those of Worden, and pointed to the still extant authorisation by Worden to access the bank account. The following investigation cleared McClain of any criminal wrongdoing (Baker 2019; Williams 2021).

As the alleged misconduct took place inside the ISS, it should be noted that there were already specific agreements put in place to define applicable laws and address potential issues of jurisdiction, such as the previously mentioned 1975 Registration Convention and, more specifically, the ISS Agreement (Baker 2019; Williams 2021).

Focusing on the ISS Agreement, Article 22 of the agreement regulates issues of criminal jurisdiction aboard the International Space Station:

> Article 22 Criminal Jurisdiction
>
> In view of the unique and unprecedented nature of this particular international cooperation in space:

1 Canada, the European Partner States, Japan, Russia, and the United States may exercise criminal jurisdiction over personnel in or on any flight element who are their respective nationals.
2 In a case involving misconduct in orbit that: (a) affects the life or safety of a national of another Partner State or (b) occurs in or on or causes damage to the flight element of another Partner State, the Partner State whose national is the alleged perpetrator shall, at the request of any affected Partner State, consult with such State concerning their respective prosecutorial interests. An affected Partner State may, following such consultation, exercise criminal jurisdiction over the alleged perpetrator provided that, within 90 days of the date of such consultation or within such other period as may be mutually agreed, the Partner State whose national is the alleged perpetrator either:
3 concurs in such exercise of criminal jurisdiction, or
4 fails to provide assurances that it will submit the case to its competent authorities for the purpose of prosecution.

According to Article 22(1), the US was allowed to establish jurisdiction over the alleged criminal conduct of astronaut McClain. In general, this case appeared to be quite straightforward: it involved an alleged perpetrator who was a US national aboard the ISS, and the alleged victim was another US National, who was not aboard the ISS but presumably located in US territory when the alleged misconduct took place (Ireland-Piper 2019).

Interestingly, in Article 22(2), more specific rules are provided in case a misconduct endangers the lives of nationals of another state party, damages the flight element of another state party, or occurs in the flight element of another state party. Naturally, Article 22(2) did not apply to the case of McClain; however, it is worth being examined in detail. Evidently, in all the cases illustrated in Article 22(2), two or more states could be in a position to legitimately and concurrently claim jurisdiction over the same acts. Thus, this provision is meant to coordinate such occurrences and provide clear guidelines. In brief, if requested, the state party (state A) whose national has done one of the alleged misconducts has the duty to contact the affected states concerning their prosecutorial interests. In doing so, state A maintains the prerogative of activating jurisdiction over the acts of its national. However, the affected states may exercise jurisdiction in case of inactivity or permission of state A.

Furthermore, Article 22(4) mandates the states parties to cooperate over alleged misconducts taking place in orbit. This obligation is tempered by the fact that states can cooperate according to their national laws and regulations, which can hamper the process of legal assistance.

While the case of McClain remains an interesting precedent on how the relevant provisions of the ISS Agreement may apply, questions remain if in the future crimes take place outside the confines of the International Space Station. In such different and unpredictable contexts there are various domestic legal regimes in place and conflicts of jurisdiction might ensue. With more than 50 states actively involved in outer space, it is possible to reflect on potential crimes happening in the context of commercial or military activities, or simply tourism (Ireland-Piper 2019). In these types of cases, whenever clear regulations are not available, public international law principles may provide some guidance.

The Five Principles of Jurisdiction

Traditionally, in public international law, there are five principles according to which states can claim jurisdiction. These are the principles of territoriality, nationality, protection, passive personality, and universality. While not being under any obligations to do so, states are entitled to construe their laws, including their domestic criminal laws, to allow for the implementation of such principles. Interestingly, due to the way these principles work, often more than one state can simultaneously claim jurisdiction over the same acts. Thus, states can, and sometimes do, sign treaties that are meant to coordinate cases of jurisdictional conflicts (Klabbers 2021). A relevant example is the previously mentioned ISS Agreement that contains provisions on coordinating matters of criminal jurisdiction onboard the ISS.

The first two principles of jurisdiction, territoriality and nationality, are the most straightforward, while the others sometimes cause legal controversies if applied too broadly (Klabbers 2021). Next, a brief overview of all five principles is laid out.

First, according to the principle of territoriality, sometimes referred to as the territorial principle, states can pass legislation, adjudicate, and prosecute individuals who violate laws in their own territories. Unsurprisingly, to this day, this remains the most adopted type of criminal jurisdiction (Gallant 2022; Klabbers 2021).

Second, according to the principle of nationality, known also as active personality, states can affirm their authority and prosecute their nationals even when they act outside of their territories. Nationals can be interpreted as to include foreigners, if they are resident in the interested country. Evidently, matters are complicated in cases where individuals possess more than one citizenship (Ireland-Piper 2019).

Third, the principle of protection, or protective principle, allows for states to claim jurisdiction over acts that cause danger to themselves, even if they originate beyond their territories and are not executed by their nationals (Klabbers 2021).

Fourth, the passive personality principle posits that states can claim jurisdiction over acts that harm their nationals, even beyond their territories. It should be noted how this principle is shrouded in controversy, as it has been mainly applied by countries belonging to the civil law tradition. In contrast, common law countries have traditionally objected to the application of this principle. However, nowadays common law states have started to adopt this principle on matters of terrorism or whenever their nationals become targets of a crime because of their nationality (Gallant 2022).

Finally, the universality principle is predicated on the concept that certain crimes are so abhorrent that states are allowed to pass laws and prosecute individuals regardless of whether such crimes happen in their territories or are committed by their nationals (Klabbers 2021).

As the principles of jurisdiction have demonstrated, it is evident that states are expressly allowed to claim jurisdiction over criminal acts committed by their own nationals in outer space, according to the principle of nationality, or over acts that occurred in their own registered spacecraft in outer space, as these are assimilated to the territory of the state (Gallant 2022). The latter idea is also confirmed in Article VIII of the OST: "A State Party to the Treaty on whose registry an object launched into outer space is carried shall retain jurisdiction and control over such object, and over any personnel thereof, while in outer space or on a celestial body." Unfortunately, the treaty fails to define who qualifies as "personnel", which could then lead to confusion over jurisdiction if the term is to be interpreted restrictively. This is the case where individuals on board, who may or may not be considered as personnel, end up facing accusations of committing a crime. For instance, Ireland-Piper (2019) has highlighted how tourists on board registered vessels in outer space may not qualify as personnel. Nevertheless, even if the treaty is to be interpreted as restrictive in scope, the previously mentioned principle of territoriality could still be applied by extension on board registered spacecraft by the relevant state as a way to claim jurisdiction over all acts on board in case an alleged crime has taken place.

In specific contexts and if the right conditions arise, the other principles for claiming jurisdiction may also be invoked by interested states (Klabbers 2021). Naturally, due to the overlapping nature of these principles, conflicts of jurisdiction are destined to ensue. Interestingly, in the case of unregistered space vehicles committing acts of piracy in outer space, it could be safe to assume that by virtue of customary law, all states are allowed to claim jurisdiction. This is in line with how the universality principle is applied in the context of the high seas (Calcara and Launiala 2023; Ireland-Piper 2019).

Prescriptive, Adjudicative, and Enforcement Jurisdiction in Outer Space

Beyond the previously mentioned five principles, the concept of jurisdiction can be further subdivided into three distinct and alternative categories: prescriptive, adjudicative, and enforcement jurisdiction (Gaeta, Viñuales, and Zappalá 2020).

Prescriptive jurisdiction usually refers to the power of states to enact legislation, while adjudicative jurisdiction to the power of solving legal disputes and adjudication. Usually, prescriptive and adjudicative jurisdiction are both meant to be applied to individuals and entities present in the territory of a state. However, as previously discussed, it is even possible to apply them beyond the territory, especially if the circumstances align with the previously mentioned five principles of jurisdiction, particularly the last four. Thus, in this light, it is possible to explain the proliferation of domestic legislation currently applied to outer space, particularly concerning the activities of private entities. Interestingly, adjudicative jurisdiction is not only a prerogative of states. Additionally, it can refer to the powers of international courts and tribunals. In theory, this is relevant in the context of space law. It is possible to hypothesise that in the case of international crimes, such as war crimes, crimes against humanity, genocide, and aggression, the International Criminal Court (ICC) could be in a position to claim jurisdiction, even when these are committed in outer space (Ireland-Piper 2019). In particular, the ICC could claim jurisdiction if an international crime is committed by a national of a state party to the Rome Statute, or if it is committed in the territory of a state party or of a state who has accepted jurisdiction of the court; in the case of outer space, this could also mean on board the registered spacecraft of a state party to the Rome Statute. Additionally, the ICC could claim jurisdiction if the ICC prosecutor has been tasked by the UN Security Council. In the case of crimes of aggression, the UN Security Council may task the court to activate jurisdiction regardless of whether states parties of the Rome Statute are involved. Furthermore, the prosecutor of the ICC may initiate investigations *ex officio* or at the request of a state party to the Rome Statute (ICC 2023). In any of these instances, there are no express limitations concerning outer space.

Enforcement jurisdiction refers instead to the power of states to enforce their own rules. This power is seen as to include conducting all the policing activities illustrated in section II of this chapter, such as investigations and arrests. By virtue of a judgment of the Permanent Court of International Justice in the previously mentioned Lotus Case of 1927, it is still widely accepted that enforcement jurisdiction can only be performed by a state on its own territory, unless there are specific permissive rules agreed by all the other interested states (Ryngaert 2015).

The reasons why enforcement jurisdiction is to be strictly applied on a territorial basis are perhaps self-evident. Allowing a state to enact coercive measures beyond its territory could result in the enactment of such measures over the territory of other states, thus ultimately infringing upon the principle of sovereign equality (Ryngaert 2015). Evidently, enforcement jurisdiction is ultimately limited to a strictly territorial basis mainly because of the idea that beyond the territory of one state there are areas that belong to the sovereignty of other states.

Partly similar to concerns regarding the high seas, policing in outer space warrants additional considerations. Policing on an extraterritorial basis in outer space does not automatically violate the sovereignty of other states. Thus, it is clear that outer space challenges the ordinary construct of criminal jurisdiction as devised on planet Earth (Gallant 2021). For this reason, it seems possible to deduce that unless there are specific restrictive or clarificatory rules, such as the well-defined jurisdiction of a state on board its registered vessels, states are allowed to conduct different types of policing activities in outer space. In case conflicts of jurisdiction arise, the same few and unsatisfactory international rules on matters of criminal jurisdiction devised on Earth apply. Thus, principles and problems on Earth appear destined to be mirrored in outer space.

The Involvement of Military and Private Entities in Policing Outer Space

Once the extent according to which states are allowed to police outer space is ascertained, attention should be directed towards examining the entities that will conduct such activities. As previously discussed, domestic police services are best suited to operate on a territorial basis. For this reason, it is quite likely that different actors will be involved in carrying out policing activities in outer space, namely military and private entities. This, however, can pose legal and practical issues, if not ethical dilemmas.

Regarding the involvement of military bodies, there is already a well-known precedent (Grunert 2022). On the 20th of December 2019, the US established a new military branch named the United States Space Force. According to the law 10 U.S. Code § 9081 – The United States Space Force, the functions of the new military branch are: "The Space Force shall be organized, trained, and equipped to– (1) provide freedom of operation for the United States in, from, and to space; (2) conduct space operations; and (3) protect the interests of the United States in space". Evidently, some of the functions of the US Space Force include traditional policing activities.

Delegating policing tasks to military bodies is not a new nor an extraordinary phenomenon. On the contrary, it is possible to mention various examples also in recent times. For what concerns the air space, the North Atlantic Treaty Organisation (NATO) Air Policing is a mission focused on securing the air space of the members of the alliance (NATO 2023). More significantly, on the high seas, an area by definition beyond national boundaries and territorial waters, military bodies such as navies have been tasked by states to conduct policing activities on matters of maritime crime (Calcara and Launiala 2023). Thus, it is a foreseeable development that other countries will start the process of instituting military bodies tasked with policing outer space, mainly as a way to protect their own immediate interests and assets.

However, the use of military bodies to conduct policing activities can be a source of concern, as already experienced in the case of policing the high seas. Indeed, in this context, military bodies such as navies are not always properly equipped nor trained to perform those tasks and functions that have been historically delegated to domestic police and judicial institutions (Gottlieb 2011). Additionally, there are legitimate concerns about whether information and evidence gathered or produced by military bodies are subjected to the same procedural standards, legal safeguards, and scrutiny employed by police services and traditional judicial institutions (Calcara and Launiala 2023). At last, navies have been shown to routinely categorise all data gathered as classified information, even when this is not appropriate nor necessary. Therefore, police cooperation among institutions of different countries in the context of maritime crime has been significantly hampered (Calcara and Launiala 2023; Gottlieb 2013). This is problematic, as policing the high seas would certainly benefit from cooperation among nations. Cautionary warnings could thus be extended also to future practices of military bodies that will be involved in the policing of outer space.

Additionally, the pioneer scholar Taubenfeld (1961–1962) theorised how outer space may eventually become an area of competition and conflict. In particular, Taubenfeld speculated how ongoing tensions on Earth may reappear in different forms in outer space. The author was also sceptical about the willingness or capacity of states of self-policing their conducts in outer space. Thus, it is possible to reflect on how, if policing through the use of military bodies were to become common, significant risks may occur of increasing tensions among countries involved in this area. For instance, controversial situations could take place due to policing activities taking place in the context of conflicts of jurisdiction where states are unwilling to cooperate. Finally, it should be noted how authoritarian states have in the past shown to unscrupulously use their domestic policing and judicial entities to persecute dissidents or adversaries inside and outside their domestic borders (Calcara 2019). It cannot be excluded that similar wrongful actions may reoccur in the context of outer space, and how the consistent use of military entities could further exacerbate pre-existing political hostilities.

Regarding the future role of private entities in policing outer space, additional considerations are in place. It is well-documented that corporations and individuals are increasingly turning their interests towards outer space. While they are already pursuing activities such as space tourism and exploration for commercial purposes, they are also planning more ambitious endeavours, such as the human colonisation of celestial bodies. Concerning the latter, relevant examples include the project Blue Origin by Jeff Bezos and SpaceX by Elon Musk, which ultimately aim to colonise respectively the Moon and Mars (Lampkin and White 2023). Thus, authors such as Lampkin and White (2023) rightfully speculate that the aforementioned companies and individuals might play a dominant role in the colonisation processes of outer space. Additionally, they may almost exclusively rely on private entities to conduct all those policing activities traditionally reserved for domestic police bodies on Earth. Such private entities may provide different types of services from cybersecurity to private guards on board vessels or, in the more distant future, space colonies (Lampkin and White 2023). Evidently, most of the issues raised in the context of policing activities carried out by military bodies, also apply to policing activities carried out by private entities. Additionally, further concerns can be highlighted. Similar to police bodies, military bodies are ultimately institutions of states. Thus, both should act to pursue the public interests and be subjected to multifaceted forms of democratic control. The same cannot be said for private entities, who are employees of individuals or corporations, and yet they may be tasked to perform particularly sensitive acts, such as employing the use of force (Lampkin and White 2023).

Furthermore, acts of public officials, such as military or police bodies, can trigger state responsibility in cases of the commission of internationally wrongful acts (Klabbers 2021). This is extremely important in several circumstances, for instance in case a domestic police institution of a state violates the human rights of individuals, and the relevant state fails to prevent and/or correct the situation. On the other hand, acts of private entities, unless officially endorsed by states, do not usually trigger state responsibility. Obviously, in general, this may pose additional problems in providing avenues for assuring accountability in front of the international community. However, as previously mentioned, Article VI of the OST expands the concept of state responsibility to acts of private entities, if such acts qualify as national activities conducted in outer space. Thus, at least according to an extensive interpretation of Article VI, even acts of private entities in matters of policing may in theory trigger state responsibility.

Conclusion

In this chapter, the various principles and legal frameworks that allow states to conduct policing activities in outer space and over celestial bodies have been explored. In general, as international law is permissive in nature, states may be able to invoke jurisdiction over criminal acts occurring in outer space or over celestial bodies if the circumstances involving those criminal offences are so that they may invoke one of the five principles of jurisdiction. For certain, states are always able to police activities involving their nationals or occurring on their registered vessels.

It should be noted that this chapter has opted to analyse these issues in line with the current framework provided by public international law on Earth. As a pure matter of speculation, it might not be too hard to conceive that in the future, if human settlement of space becomes a reality and a widespread phenomenon, settlements may over time achieve independence and demand for a reassessment of the current laws and principles applying on Earth. After all, similar events have not only been envisioned in works of science fiction, but they have already taken place on Earth centuries ago.

References

Baker M (2019) NASA Astronaut Anne McClain Accused by Spouse of Crime in Space. From: https://www.nytimes.com/2019/08/23/us/astronaut-space-investigation.html.

Boakye D, Siaw D, and Sarpong D (2022) The Airbus Bribery Scandal: A Collective Myopia Perspective. *European Management Review* 19(4): 654–670.

Boister N (2018) *An Introduction to Transnational Criminal Law*. Oxford: Oxford University Press – second edition.

Brodeur JP (2010) *The Policing Web.* Oxford: Oxford University Press.

Calcara G (2019) Rethinking Legal Research on Matters of international Police Cooperation: Issues, Methods and raison d'être. *Liverpool Law Review* 40(2): 95–111.

Calcara G and Launiala M (2023) An Overview of INTERPOL's Involvement in Tackling Maritime Piracy: History, Developments, and Legal Issues. In Eski Y and Wright M (eds) *Maritime Crime & Policing*. New York: Routledge, pp. 141–159.

den Boer M (2018) Introduction to Comparative Policing from a Legal Perspective. In den Boer M (ed.) *Comparative Policing from a Legal Perspective.* Research Handbooks in Comparative Law series. Cheltenham: Edward Elgar Publishing, pp. 1–14.

Gaeta P, Viñuales JE, and Zappalá S (2020) *Cassese's International Law.* Oxford: Oxford University Press – third edition.

Gallant K (2022) *International Criminal Jurisdiction: Whose Laws Must We Obey?* New York: Oxford University Press.

Gottlieb Y (2011) Article 3 of INTERPOL's Constitution: Balancing International Police Cooperation with the Prohibition of Engaging in Political, Military, Religious, or Racial Activities. *Florida Journal of International Law* 23: 135–186.

Gottlieb Y (2013) Combating Maritime Piracy: Inter-disciplinary Cooperation and Information Sharing. *Case Western Reserve Journal of International Law* 46(1–2): 303–334.

Grunert J (2022) *The United States Space Force and the Future of American Space Policy: Legal and Policy Implications.* Leiden: Brill Nijhoff.

Hermida J (2004) *Legal Basis for a National Space Legislation.* New York: Kluwer Academic Publishers.

Hughes B, Hunt CT, and Curth-Bibb J (2013) *Forging New Conventional Wisdom Beyond International Policing: Learning from Complex Political Realities.* Leiden: Brill.

ICC (International Criminal Court) (2023) *How the Court Works.* From: https://www.icc-cpi.int/about/how-the-court-works.

Ireland-Piper D (2019) Star laws: what happens if you commit a crime in space? From: https://theconversation.com/star-laws-what-happens-if-you-commit-a-crime-in-space-122456.

Klabbers J (2021) *International Law.* Cambridge: Cambridge University Press – third edition.

NATO (2023) NATO air policing: securing allied airspace. From: https://www.nato.int/cps/en/natohq/topics_132685.htm (accessed 30 March 2023).

Lampkin J and White R (2023) *Space Criminology: Analysing Human Relationships with Outer Space.* Cham (Switzerland): Palgrave MacMillan.

Lemon E (2019) Weaponizing Interpol. *Journal of Democracy* 30(2): 15–29.

Marboe I (2017) National Space Law. In von der Dunk F and Tronchetti F (eds) *Handbook of Space Law.* Cheltenham: Edward Elgar, pp. 127–204.

O'Neill M (2010) The Police Response to Crime. In Shoham SG, Knepper P and Ket M (eds) *International Handbook of Criminology.* Boca Raton: CRC Press-Taylor and Francis Group, pp. 483–510.

Pečujlić AN (2023) *The Space Law Stalemate: Legal Mechanisms for Developing Norms.* New York: Routledge.

Ryngaert C (2015) *Jurisdiction in International Law.* Oxford: Oxford University Press – second edition.

Steer C (2017) Sources and law-making processes relating to space activities. In Jaku RS and Dempsey PS (eds) *Routledge Handbook of Space Law.* New York: Routledge, pp. 3–24.

Taubenfeld HJ (1961–1962) Regime for outer space. *Northwestern University Law Review* 56(1): 129–167.

United Nations Office for Outer Space Affairs (2023a) *Treaties & Principles: Treaty on Principles Governing the Activities of States in the Exploration and Use of Outer Space*, including the Moon and Other Celestial Bodies. From: https://www.unoosa.org/oosa/en/ourwork/spacelaw/treaties/introouterspacetreaty.html (accessed 12 September 2023).

United Nations Office for Outer Space Affairs (2023b) *Status of International Agreements Relating to Activities in Outer Space as of 1 January 2023*. From: https://www.unoosa.org/res/oosadoc/data/documents/2023/aac_105c_22023crp/aac_105c_22023crp_3_0_html/AC105_C2_2023_CRP03E.pdf (accessed 13 September 2023).

von der Dunk F (2002) Sovereignty and space: When and where shall the twain meet? In Kreijen G, Brus M, Duursma J, De Vos E and Dugard J (eds) *State, Sovereignty, and International Governance*. Oxford: Oxford University Press, pp. 462–481.

von der Dunk F (2017) International Space Law. In von der Dunk F and Tronchetti F (eds) *Handbook of Space Law*. Cheltenham: Edward Elgar, pp. 29–126.

Williams K (2021) Space crime continuum: Discussing implications of the first crime in space. *Boston University International Law Journal* 39(1): 79–108.

Zabyelina Y (2019) The role of major intergovernmental organizations and international agencies in combating transnational crime. In Natarajan M (ed.) *International and Transnational Crime and Justice*. Cambridge: Cambridge University Press, pp. 305–310.

16 Extraterritorial Criminal Jurisdiction in International Law and Environmental Space Crime

Alla Pozdnakova

Introduction

This chapter explores the role of international law rules of jurisdiction in addressing environmental crime in outer space. Although some of the issues and findings may be relevant for space-related crimes more generally, an examination of Astro-Green Criminology (AGC) and law perspectives on misconduct causing environmental harm is particularly interesting: as other chapters in this book also argue, environmental space crimes are not yet conceptualized and studied sufficiently in criminology and law (Eski 2023; Gorove 1972a/b; Lampkin and White 2023). Yet, while offences actually committed in outer space are rare (Sachdeva 2023), environmental harm from space activities is not. It is already occurring, both in 'near outer space' and on Earth, caused both by state and commercial actors. There is no doubt that environmental misconduct – albeit obviously different in outer space and on Earth in terms of both shape and consequences – will continue to grow as humanity advances into the depths of outer space.

This chapter also examines the concept of an environmental space crime in its criminological and legal dimensions. Environmental harm is seldom limited to one state's boundaries, and pollution of outer space by space debris is inherently global in terms of its impact (Lampkin and Wyatt 2023). The international community should take measures, including criminal law (Lampkin and White 2023), to stop particularly egregious forms of environmental misconduct in the space sector.

From an astro-green criminological standpoint, jurisdictional rules are only a means to accomplish the ultimate goal: prevention of environmental degradation caused by space activities. At the same time, due to the global scope of environmental harm in the space sector, an effective criminal justice system for the sector is dependent on the jurisdiction of states under international law to criminalize and punish environmental misconduct. The scope and reach of criminal jurisdiction in international law, and the inherent challenges of traditional approaches to extraterritorial jurisdiction, are also analysed. This leads us further in our search for improved solutions to extraterritorial jurisdiction over environmental space crime. Section 4

DOI: 10.4324/9781003437178-17

discusses the findings and argues that development of international space law principles of criminal jurisdiction in the outer space sector should be informed by insights from AGC research. Section 5 concludes.

The Concept of Environmental Space Crime at the Intersection of Astro-Green Criminology and Law

What Is Environmental Space Crime?

The question of what conduct amounts to an environmental space 'crime' depends on who exercises the power to define the terms 'crime', 'offender', 'victim' and 'harm'. In contemporary criminal law, this power is held by a state (legislative branch). In literature on green criminology, (causing) environmental harm can in itself be deemed a social and ecological crime, regardless of its legal status (White 2021). Research on AGC elaborates upon green criminological approaches to environmental harm caused by space activities, including those creating space debris in orbits (Lampkin and Wyatt 2023: 12). Environmental harm is being caused, not only in outer space, but also on the local and indigenous communities, ocean ecosystems and atmosphere (Byers and Byers 2017). However, the societal implications of such harm are little explored by policymakers and the environmental impact of space activities on Earth's ecosystems has not been thoroughly studied by scientists (Pozdnakova 2023).

One of the decisive factors driving forward unsustainable activities in space is our perception of outer space as a common area belonging to no-one, which everyone is free to explore – and exploit. This freedom is intrinsically linked to how we (humanity) use outer space and other common spaces and resources (Takemura 2019: 12). As Cesare Beccaria wrote in 1764 in his essay 'On Crimes and Punishment', '[n]o man ever freely sacrificed a portion of his personal liberty merely on behalf of the common good' (McLaughlin and Muncie 2013: 6). A commitment in the shape of an individual or collective *restraint* on our freedom to use outer space unsustainably may promote eco-justice and sustainability. Green and astro-green criminology may assist us in deciding what restraints may work effectively to prevent and punish environmental space crime, including feasibility of criminal or other (administrative, civil) law measures (Hall 2015: 67–75), and overcoming problems of criminal justice (e.g., difficulties of crime investigation) with regard to environmental space misconduct (Lampkin and White 2023).

Law and Legal Approaches to Space Crime

In democracies, the rule of law principle of legality (*nullum crimen sine lege*) prohibits punishing without a basis in law and requires criminal behaviour to be clearly defined in domestic criminal statutes (Simester 2021: 58). Thus, a state (via its domestic legislative and adjudicative powers) has a monopoly over such key questions as to the suitability of criminal law to protect the environment, thresholds for unacceptable environmental harm, how criminal proceedings are conducted, and what punishment is feasible.

By comparison to civil liability and administrative penalties, criminal liability should arguably be reserved for particularly egregious misconduct. Although, as Hall puts it, 'ascribing criminal blame for an environmentally destructive activity remains a challenging proposition' (2015: 61), many states have criminalized serious environmental offences (Hall 2015). As explored further in this chapter, international harmonization of environmental space crime may be indispensable for ensuring that such misconduct is regulated similarly by states. Environmental harm to outer space or to the Earth environment is, however, a relatively 'new' form of environmental misconduct, making prospects of quick international harmonization unlikely. Thus, as Lampkin and Wyatt point out, 'many environmentally destructive practices were – and still are – lawful' (2023: 3). By revealing environmental harms of some space activities, AGC approaches may, however, contribute to the development of environmental criminal law in the space sector.

In law, 'crime' against outer space or Earth environments is a much narrower concept than one we may have concluded from criminological reflection. For example, environmental crimes caused by *states* – the most prominent examples being use of Anti-Satellite (ASAT) weapons causing space debris, or dumping toxic jettisoned waste and end-of-life space crafts into oceans – are outside the ambit of national criminal law (Lampkin and Wyatt 2023), although such misconduct is of interest to criminology and may be punished under international law rules of state responsibility (Hall 2015).

National criminal law will not treat *any* environmental harm as illegal (and punishable), usually allowing certain levels of pollution (Gorove 1972b; Lampkin and Wyatt 2023). This means that *cumulative transboundary* environmental harm to the global commons is likely to escape the reach of criminal law. With punishment needing to be imposed on a particular person (natural or legal, as the case may be), i.e. a specified 'offender', it may be impossible to identify the perpetrators of the environmental harm to the global commons.

A state-centric perspective on environmental crime against outer space can be challenged by pluralistic approaches to law. For example, Santos (2020) argues that states do not have a monopoly on law-making. There are other legal communities and groups which are engaged in making their own 'laws'; most notably, transnational corporations as a new form of 'global legality' which 'create a global space that often conflicts with the national legal space' (Santos 2020: 514). While transnational corporations are unlikely to promote effective rules to prevent and punish environmental crime against global commons, one may think of the role of other 'legal communities' in defining and re-defining what constitutes a space (environmental) crime. This could include local and indigenous communities whose livelihood is negatively impacted by space pollution, or scientists endeavouring to identify and expose new sources and forms of environmental harm.

'Spaces', Offenders and Jurisdiction in International Law

The central issue of jurisdiction – which state is competent to exercise (criminal, administrative or civil) jurisdiction over offenders and misconduct – falls within the domain of public international law. State sovereignty over its territory and its subjects (natural and legal persons) is the alpha and omega of international law. By confining its jurisdiction to land, maritime and air space within its boundaries, each state avoids intervening into other states' sovereignty and territorial integrity. The criminal jurisdiction of a state – understood broadly as a state's power and prerogatives emanating from the state's sovereignty to regulate, control and punish persons under its jurisdiction – is generally exercised within the spatial limits of this sovereignty (Ryngaert 2015). Thus, a state will apply its criminal law to both nationals' and foreigners' misconduct under its territorial jurisdiction. Obviously, in cases of harmful behaviour against the global commons such as outer space, located outside any state's territorial boundaries, but probably affecting the whole international community, this territorial principle of jurisdiction may prove inadequate.

While the territoriality principle of jurisdiction is the main rule in international law, exercise of *extraterritorial* jurisdiction by states is also relatively common (Ryngaert 2015). International law on extraterritorial jurisdiction is examined in a significant amount of scholarship, supplemented by empirical observations on states' practical approaches to extraterritorial crimes on Earth (Ryngaert 2015). The following discussion will highlight central principles of international law on extraterritorial criminal jurisdiction and explore the solutions and challenges they present regarding environmental misconduct in outer space.

Principles of Extraterritorial Criminal Jurisdiction in International Law

Extraterritorial jurisdiction in international law may be broadly defined as a state's power (a right) to perform acts in the exercise of its sovereign prerogatives beyond its territory. The jurisdiction may generally be divided into legislative (prescriptive), executive (enforcement) and adjudicative (Ireland-Piper and Freeland 2020: 52). Obviously, exercise of *enforcement* (and adjudicative) jurisdiction by one state within another state's territory may only take place once assured that the latter state's sovereignty is not violated (i.e., in case of consent). Adoption by a state *of legislative acts* applicable extraterritorially is less controversial, but may nonetheless result in violation of another state's sovereign prerogatives, and the exercise of extraterritorial prescriptive jurisdiction is generally justified by a recognized international law principle of jurisdiction (Ryngaert 2015: 101).

This law has largely evolved through consistent practice of states and *opinio juris* (i.e., customary international law), despite some attempts at harmonization through treaties (Draft Convention on Jurisdiction with Respect to Crime 1935).

The UN Convention on the Law of the Sea (UNCLOS) and the Treaty on Principles Governing the Activities of States in the Exploration and Use of Outer Space, including the Moon and other Celestial Bodies (the Outer Space Treaty, OST) contain provisions on state jurisdiction for, respectively, oceans and outer space. In the contemporary literature on the law of the sea, supported by the UNCLOS approach, the dominant view is that the ship bears its flag state's *nationality* (Mukherjee 2020), i.e., nationality-based jurisdiction of the flag state applies to ships on the high seas. The principle of nationality (also known as the 'active personality principle') is well-established in general international law. As explored in the next section, outer space law largely applies a similar approach to jurisdiction, i.e., a state's jurisdiction over spacecraft of its registry and the personnel in orbit.

The crucial purpose of flag state jurisdiction is to preserve the flag state's freedom of navigation and protect its sovereignty against encroachment by other states on the high seas, in line with the aforementioned considerations of general international law on jurisdiction. The adoption of exclusive flag state jurisdiction in the law of the sea was prompted by the well-known *Lotus* case of 1927, in which the Permanent Court of Justice controversially ruled in favour of the affected (non-flag) state (Csernus 2023). However, recent international jurisprudence arguably strengthens the principle of extraterritorial jurisdiction held (exclusively) by the flag state, even in its prescriptive dimension (Csernus 2023).

Flag state jurisdiction may only be disregarded in exceptional cases of very serious crimes such as piracy (Ireland-Piper and Freeland 2020). The rationale for such an exception is that the misconduct committed is so serious that universal considerations (interests of the international community) outweigh flag state interests. Universal jurisdiction may also apply to other international crimes (Ireland-Piper and Freeland 2020).

However, while universal jurisdiction is reserved for exceptionally grave international crimes, in general international law, there are other alternatives to the nationality principle of extraterritorial jurisdiction, including the 'effects'-approach (or the objective territorial principle). This approach grants jurisdiction to a state regarding conduct causing effects within that state (Ryngaert 2015).

The suitability of such approach(es) to extraterritorial jurisdiction for ensuring effective (criminal) legal order for global commons is arguable. Firstly, competing claims for criminal jurisdiction by two or more states affected by the transboundary environmental harm may complicate the effective investigation and adjudication of the case (Ryngaert 2015: 142–143). Secondly, and even more problematic in cases of environmental crime against the global commons, no state may altogether claim or accept jurisdiction over misconduct, because no state has been sufficiently affected. In principle, the responsibility to exercise jurisdiction will lie with the nationality state of offenders, and the issue will revolve instead around this state's *willingness* or capacity to exercise it. In both cases, a likely outcome is impunity for perpetrators of the environmental harm.

International Legal Status of Outer Space and Extraterritorial Criminal Jurisdiction in Space Law

The principle of non-appropriation of outer space, including the Moon and other celestial bodies, laid down in Article II of the OST, is one of the fundamental principles of international space law (Lyall and Larsen 2018). Like the high seas, outer space is *res communis omnium* not subject to sovereignty claims or appropriation by use or occupation, or by any other means. The binding nature of the non-appropriation principle as a customary rule of international law is generally accepted by the international community (Tronchetti 2008). The principle is also incorporated and detailed out in the Agreement Governing the Activities of States on the Moon and Other Celestial Bodies (Moon Agreement).

The absence of any sovereignty does not mean that states are not allowed to exercise jurisdiction in outer space. On the contrary, Article VI of the OST imposes international responsibility on states parties for their national governmental and non-governmental space activities and requires 'the appropriate State' to ensure authorization and continuous supervision of non-governmental activities (Gerhard 2009). Article VII, detailed in the Convention on the International Liability for Damage Caused by Space Objects (Liability Convention), imposes strict liability on launching states to compensate for damage caused by the space objects launched from their territories and facilities. However, the liability regime established by this Article and the Liability Convention does not address criminal or other public law penalties for non-state actors involved in launches (Kerrest and Smith 2009).

Importantly for this discussion, the first sentence of Article VIII of the OST provides that '[e]ach State Party to the Treaty on *whose registry* an object launched into outer space is carried shall *retain jurisdiction and control over such object*, and over *any personnel thereof*, while in outer space or on a celestial body'. Registration requirements for space objects are detailed in the Convention on Registration of Objects Launched into Outer Space (Registration Convention), which establishes an international system of registration of space objects and requires states to maintain a register of objects they launch into space, as well as to inform the Secretary General of the United Nations of such launches (Agrawal 2023). The Convention does not contain further rules on states' jurisdiction over persons on board such space objects, and has been criticized as outdated and insufficient to tackle the issues of the contemporary space sector, including space debris (Agrawal 2023).

Article VIII of the OST only refers to jurisdiction of the state of registry over space objects and personnel onboard. In international space projects, involving several states, it is likely that the nationalities of persons on board will be different from the spacecraft's state of registry. This presents problems.

For instance, if such a person commits a criminal offence, which state will have criminal jurisdiction? Will it be the state of registry of the space object, the state of nationality of the offender, or, perhaps, a third state harmed by the offence (e.g, nationality state of the victim or state whose material interests were affected)? How can we address a jurisdictional gap in Article VIII of the OST with regard to crimes committed outside any spacecraft or space object (Gorove 1972a: 321–322; Ireland-Piper and Freeland 2020: 61)?

These questions may become quite problematic in a (relatively distant) future of crewed spaceflight in outer space. In practice, crime in outer space has so far been (allegedly) perpetrated only on board the International Space Station (Lampkin and White 2023). The *McClain*-case is an example of an (unfounded) accusation against an astronaut of a crime committed from the ISS Station on Earth (Sachdeva 2023). An incident of damage (a hole) in a Russian module of the Station is another example of misconduct committed and seeking to cause damage *in outer space* (Sachdeva 2023). This illustrates the need to develop rules on criminal jurisdiction in outer space further, preferably through treaty-based law.

The International Space Station Agreement is unique as a positive source of criminal law for space (Blount 2007: 312). Its provisions on criminal jurisdiction may arguably be transferrable to future outer space colonies (Hardenstein 2016). According to Article 1(1), the ISS Agreement seeks 'to establish a long-term international cooperative framework among the Partners, on the basis of genuine partnership, for the detailed design, development, operation, and utilization of a permanently inhabited civil ISS for peaceful purposes, in accordance with international law'. Article 2(1) further provides that '[t]he Space Station shall be developed, operated, and utilized in accordance with international law, including the Outer Space Treaty, the Rescue Agreement, the Liability Convention, and the Registration Convention'.

The ISS consists of several elements (modules) belonging to different partner states, with Article 5 regulating issues of registration, jurisdiction and control, in line with the Registration Convention and OST. Each partner is required to register, according to the Registration Convention, the flight elements (as listed by an annex it provides) as space objects (in Article 9). In line with Article VIII of the OST, each partner also retains jurisdiction and control over the elements it registers *and* over personnel in or on the Space Station who are its nationals (also Article 9). It is also specified that the exercise of such jurisdiction and control shall be subject to relevant provisions of the ISS Agreement and implementing arrangements, including relevant procedural mechanisms established therein.

Although the ISS Agreement does not seek to harmonize criminal laws of participating states, it contains relevant provisions on partner states' criminal jurisdiction over astronauts and misconduct on the ISS. Article 22 of the ISS Agreement establishes the relevant provisions on criminal jurisdiction and recognizes that the states parties (partner states) 'may exercise criminal jurisdiction over personnel in or on *any flight element* who are *their respective nationals*'. Although this provision also emphasizes 'the unique and unprecedented nature of this particular international cooperation', the approach taken is generally in line with the earlier discussed principle of extraterritorial jurisdiction based on nationality.

The ISS Agreement also addresses the issue of criminal jurisdiction when a national of one state partner carries out misconduct harming interests or nationals of another state partner. Article 22 specifies that this would be the misconduct affecting 'the life or safety of a national of another Partner State', or which 'occurs in or on or causes damage to the flight element of another Partner State'.

While following the generally recognized approaches to extraterritorial criminal jurisdiction, the ISS Agreement does more: it envisages a procedural framework for the smooth and effective exercise of such jurisdiction by a relevant state. The provisions of Article 22 envisage consultations between the partner states involved with regard to 'their respective prosecutorial interests'. The state of the perpetrator's nationality retains the right to exercise criminal jurisdiction over its national, provided it offers assurances to the aggrieved state that it will submit the case to its competent authorities for prosecution within the specified time (90 days from such consultation, or within such other mutually agreed period). However, if the nationality state fails to provide such assurances, or concurs in the exercise of criminal jurisdiction by the aggrieved state, the latter may exercise its jurisdiction over the alleged perpetrator. Article 22 of the ISS Agreement also deals with related questions of criminal proceedings: extradition and mutual assistance in case of alleged misconduct in orbit.

In essence, the ISS Agreement envisages practical solutions to tackle issues raised by the discretionary nature of state jurisdiction in international law, by recognizing – even if imperfectly – the aggrieved state's right to invoke the 'passive nationality' principle of jurisdiction (Ireland-Piper and Freeland 2020: 63).

The wording of Article 22 indicates that it applies only to sufficiently serious misconduct against nationals of another partner state (capable of 'affecting life or safety' or causing 'damage' to flight elements). Thus, jurisdiction for less serious offences is not regulated in this provision (although the fifth paragraph refers to the Code of Conduct which regulates the order and conduct of crew activities on the ISS). Misconduct affecting nationals of the same state is not regulated, since jurisdiction can be resolved on the basis of Article 9 mentioned earlier.

Finally, the ISS Agreement (including Article 22) does not regulate criminal jurisdiction over the misconduct in point for this chapter, i.e., violations affecting interests of third states and the international community, as well as persons, property or environment on Earth or in outer space. In essence, the Agreement only implements provisions of Article VIII of the OST, without expanding its scope. To resolve the question of criminal jurisdiction in cases beyond the reach of Article VIII and the Agreement, one must fall back on the principles of international space law and other relevant rules of international law examined earlier.

Protection of Outer Space Environment through Criminal Law: Appraisal and Discussion

Protection of the global commons will be more effective if states choose to wield their jurisdictional powers to criminalize and punish conduct causing environmental harm to outer space and Earth environments (Lampkin and White 2023). However, the research confirms that states are likely to refrain from exercising criminal jurisdiction with regard to environmental crimes against global commons on Earth (Ryngaert 2020). With regard to crimes committed outside any state's sovereign boundaries and not affecting any particular state, it can be problematic for a state to find a justification required by international law, such as connection through nationality or other sufficient interest. To a large extent, this reflects the anthropogenic view of the environment and the notion of environmental harm in law (Lampkin 2020), potentially leading to impunity and jeopardizing the preventive objectives of criminal law.

The same problems as described above are likely to arise for environmental space crimes. This section explores how international law principles of jurisdiction may be applied and developed to make environmental law principles effective, irrespective of where the environmental harm is caused.

The Ecosystem-Based Approach to Outer Space Environment

International law on extraterritorial criminal jurisdiction is preoccupied by boundaries between sovereign states and by territorial or common 'spaces'. Ironically, international law does not even comprehensively define and delimit all these 'spaces', e.g., there is no universally agreed vertical boundary between airspace and outer space (Ireland-Piper and Freeland 2020: 66). But even if it did, would it be effective for the purposes of environmental protection to treat space on Earth (such as the atmosphere, or oceans) as separate from outer space, since the same conduct may affect both?

The ecosystem-based approach (or principle) to protect the ocean commons is expressly incorporated in a new UN agreement adopted in 2023 (Agreement under the United Nations Convention on the Law of the Sea on the Conservation and Sustainable Use of Marine Biological Diversity of Areas Beyond National Jurisdiction, ABNJ). Although ABNJ do not modify principles of jurisdiction established in UNCLOS, it strengthens the environmental protection of oceans (if not outer space) from pollution by space debris (Pozdnakova 2023). Research in green criminology (Lawrence et al. 2022) and AGC (Lampkin and Wyatt 2023) shows the equal pertinence of the ecosystem-based approach for the protection of outer space environment. Nevertheless, adoption of such an approach will not, in itself, provide a basis for the exercise of criminal jurisdiction with regard to misconduct affecting ecosystems. It is necessary to identify (or establish) a proper legal basis in international law and to adopt suitable provisions in national criminal legal orders of states.

The Search for a Legal Basis for States to Exercise Jurisdiction in the Interest of Common Good

How to ensure that states act to promote and protect the common good? This long-standing issue may not be resolved through the international law rules of extraterritorial criminal jurisdiction alone. However, as we have seen above, states choosing to punish environmental space crime may be hindered by international law principles and conditions, such as a requirement for a nexus or connection between the state and the environment harm or its perpetrator.

Scholars have proposed several approaches to the application of international law principles of extraterritorial jurisdiction to protect the common interest. Firstly, Ryngaert (2020) argues that 'globally harmful character of particular conduct, while not in itself authorizing the exercise of extraterritorial jurisdiction, should guide the application of existing principles of jurisdiction' (Ryngaert 2020: 140–141). He suggests that where a state seeks to assert jurisdiction to protect global values or the common interest, a 'tenuous territorial link can suffice' (Ryngaert 2020: 140–141).

Secondly, since the presence of a connection between misconduct or an offender and a state may encourage the state to take action in the common interest (Ryngaert 2019: 211), we should re-visit and even re-invent the international law requirement of 'nexus', to enable (and possibly encourage) the exercise of criminal jurisdiction over environmental space crimes. Instead of looking for formal territorial or national (personal) link(s), one can focus on a broader range of *relationships* which may be connected to a state's territory, material environment (in which space activities are conducted), and on interconnected groups of people and interests across boundaries (McVeigh 2019: 193). Furthermore, looking at the jurisdiction over space crime as a *functional*, rather than territorially based, concept (Valverde 2019), one may reconsider the question of 'who' governs, and the related question of 'whose law' governs: e.g., not a state and its law but a local indigenous community whose land or waters were affected by space debris. This does not create any tension with international law requirement of 'nexus' between a state and the harm: a community invoking its 'law' or 'rights' before a domestic court or competent public body may simply be a way to realize this nexus.

This line of thinking is supported by an eco-justice perspective, addressing environmental space crime by human-, or even species-*rights-based* approaches (Lampkin and White 2023). Narrative criminology (Presser and Sandberg 2019) will help us tell a story of a spaceflight as not (only) a heroic endeavour and future of humankind – a narrative actively deployed by private actors and space nations alike! – but also as a story of environmental harm such as damaged fishing areas in the Arctic (Byers and Byers 2017). By improving understanding and raising awareness about the human and environmental impacts of space debris on outer space and Earth environment (Lampkin and Wyatt 2023: 13; Takemura 2019: 15), AGC research may also prove invaluable for the effectiveness of international law rules on jurisdiction. Crucially, green criminological and AGC research may contribute to the better application of those criteria for the exercise of extraterritorial criminal jurisdiction, which are already largely accepted in international law. Thus, a discourse on what constitutes environmental 'harm' and who are 'victims' of such harm, including mass harms and non-human victims (Hall 2015: 16–17; Lampkin and Wyatt 2023) may illuminate the notion of relevant 'effects' and thereby facilitate application of a broadly practiced 'effects-based' approach to jurisdiction (Ryngaert 2015).

Thirdly, it is pertinent to explore the potential of jurisdiction based on the principle of universality, which allows any state to exercise criminal jurisdiction against exceptionally grave international crimes, regardless of the nationality of the perpetrator or where the crime was committed (Ireland-Piper and Freeland 2020: 56–57). Ecocide has now been proposed as a 'new' international crime (Burns 2023). Outer space and the Earth environment are global commons and a 'common concern' of humankind (Tan 2000), and the need to protect these through universal jurisdiction ought to convince states to place environmental misconduct among such international crimes.

A Positive Duty to Protect Outer Space Environment by Criminal Law?

The *permissive* nature of state jurisdiction in international law (Ryngaert 2015) implies that states are generally free *not* to exercise criminal jurisdiction, unless required to do so by a treaty (Ireland-Piper and Freeland 2020: 54). It has been argued by scholars that, to protect the environment, states may be under an *obligation* to exercise jurisdiction on the basis of existing permissive principles (Ryngaert 2020). In outer space law, support for such an argument may be found in provisions of the OST which seek to ensure that the exploration and use of outer space are carried out for the benefit and in the interest of all countries and to prevent harmful contamination of outer space (Gorove 1972b). Non-binding UN documents emphasize the need for the development of norms to ensure peaceful, secure and *sustainable* use of outer space (COPUOS 2018). This supports the relevance and acceptance of the Sustainable Development goals – including environmental protection and prevention of space debris – in the international legal order of outer space.

Admittedly, the OST does not explicitly require states to criminalize space crimes or adopt any other sanctions for offences towards outer space or Earth environment. In international space law, only a 'soft law' provision on national space legislation recommends states to include in their domestic laws 'appropriate procedures' such as 'enforcement mechanisms [which] could include administrative measures, such as the suspension or revocation of the authorization, and/or *penalties*, as appropriate' (United Nation General Assembly 2013).

The ISS Agreement does not expressly require any state partner to *actually establish* any criminal jurisdiction over misconduct in orbit in its national law, as would be required by the principle of legality to enable national authorities of those states to punish such misconduct. However, the effectiveness of the Agreement would be compromised in the absence of an appropriate legal basis for criminal punishment in the domestic legal orders (Goles 2022).

Like the OST, the ISS Agreement also does not call upon states to endeavour to harmonize criminal liability provisions for offences committed in orbit. In the absence of a clear obligation in international law, states may not feel compelled to criminalize space crime in their domestic laws. Furthermore, states have significant discretion with regard to criminal proceedings in individual cases: indeed, jurisdictional consultations between states envisaged by the ISS Agreement following a criminal act on board the ISS illustrate the need for procedural rules on the exercise of criminal jurisdiction.

In any case, just like 'space', a 'crime' and 'environmental harm' are not universally accepted, harmonized concepts. The 1988 Convention for the Suppression of Unlawful Acts Against the Safety of Maritime Navigation (SUA Convention) may be a useful example of an internationally binding instrument which requires states to take necessary measures to make offences against safety of navigations criminally punishable, depending on their gravity, and lays down principles and criteria for such offences.

Conclusion

To protect the environment of outer space and Earth from harmful and unsustainable conduct, we need to strengthen collective orientation, to motivate states to act jointly to combat environmental crime in outer space (Viikari 2008). In the absence of clear provisions in international law, states are likely to avoid taking unilateral steps to criminalize and prosecute environmental space crime. Admittedly, factors such as regulatory competition among space nations may in practice preclude states from imposing strict rules on their own space activities. However, this chapter focused on other factors: issues pertaining to the legal basis in international law for the exercise of criminal jurisdiction to protect the global commons, including outer space, from environmental misconduct. It also explored the interaction and cross-fertilization possibilities between AGC and law approaches to environmental space crimes.

The discussion in this chapter shows that states may be concerned that, by exercising criminal jurisdiction over environmental space crime, they may trespass on other states' sovereign boundaries. Even more problematic, due to the permissive nature of jurisdiction in international law, competent states may be unwilling to punish environmental misconduct, e.g., because damage was caused to the environment of areas beyond national jurisdiction, rather than to their territory or interests. Rapid international harmonization of environmental criminal law in the space sector is, in this author's view, an unlikely development, although the work on ecocide in international criminal law and the proliferation of space sustainability rules may give some hope for strengthening space environmentalism (Lawrence et al. 2022). AGC research may provide invaluable help to the effectiveness of international law, inter alia, by exploring the concept of environmental harm caused by space activities and thereby facilitating the application of existing international law principles of criminal jurisdiction to the 'new' area of environmental space crime.

References

Agrawal K (2023) Is the 'Environment' Being Considered as Outer Space Industries Expand? From: https://papers.ssrn.com/sol3/papers.cfm?abstract_id=4470000 (Accessed: 21 March 2024).

Blount PJ (2007) Jurisdiction in Outer Space: Challenges of Private Individuals in Space. *Journal of Space Law*33: 299–340.

Burns R (2023) From Meaning to Ecocide: The Value of Phenomenology for Green Criminology. *Critical Criminology*:1–18.

Byers M and Byers C (2017) Toxic Splash: Russian Rocket Stages Dropped in Arctic Waters Raise Health, Environmental and Legal Concerns. *Polar Record* 53: 580–591.

COPUOS (Committee on the Peaceful Uses of Outer Space) (2018) *Guidelines for the Long-Term Sustainability of Outer Space Activities.* From: https://www.unoosa.org/res/oosadoc/data/documents/2018/aac_1052018crp/aac_1052018crp_20_0_html/AC105_2018_CRP20E.pdf (Accessed: 25 March 2025).

Csernus M (2023) Might Contain Traces of Lotus: The Limits of Exclusive Flag State Jurisdiction in the Norstar and the Enrica Lexie Cases. *Leiden Journal of International Law* 36: 947–970.

Draft Convention on Jurisdiction with Respect to Crime (1935). *American Journal of International Law* 29(S1, 1935): 439–442. Published online by Cambridge University Press, 12 April 2017. https://doi.org/10.2307/2213634

Eski Y (2023) Space, the Final Frontier to Exploit and Annihilate: Space Exploration, Inhabitation and Settlement. In *A Criminology of the Human Species: Setting an Unsettling Tone* (pp. 71–98). Cham: Springer Nature Switzerland.

Gerhard M (2009) Article VI. In Hobe S, Schmidt-Tedd B, Schrogl K-U and Goh GM *Cologne Commentary on Space Law.* Cologne: Carl Heymanns Verlag.

Goles K (2022) Canada Extends Criminal Law Jurisdiction to the Moon, Library of Congress Blogs. From: https://blogs.loc.gov/law/2022/08/canada-extends-criminal-law-jurisdiction-to-the-moon/ (Accessed: 25 March 2025).

Gorove S (1972a) Criminal Jurisdiction in Outer Space. *The International Lawyer* 6(2): 313–323.

Gorove S (1972b) Pollution and Outer Space: A Legal Analysis and Appraisal. *New York University Journal of International Law and Politics* 5(1): 53–66.

Hall M (2015) *Exploring Green Crime: Introducing the Legal, Social and Criminological Contexts of Environmental Harm.* London: Palgrave.

Hardenstein TS (2016) In Space, No One Can Hear You Contest Jurisdiction: Establishing Criminal Jurisdiction on the Outer Space Colonies of Tomorrow. *Journal of Air Law and Commerce* 81: 251–288.

Ireland-Piper D and Freeland S (2020) Star Laws: Criminal Jurisdiction in Outer Space. *Journal of Space Law* 44: 44–75.
Kerrest A and Smith LJ (2009) *Article* VII. In Hobe S, Schmidt-Tedd B, Schrogl K-U and Goh GM *Cologne Commentary on Space Law*. Cologne: Carl Heymanns Verlag.
Lampkin JA and White R (2023) *Space Criminology: Analysing Human Relationships with Outer Space*. London: Palgrave.
Lampkin JA and Wyatt T (2023) An Astro-Green Criminological Examination of Orbital Space Debris. *Criminology & Criminal Justice*, 17488958231 1691.
Lampkin JA (2020) *Uniting Green Criminology and Earth Jurisprudence*. Oxon: Routledge.
Lawrence A, Rawls ML, Jah M, Boley A, Di Vruno F, Garrington S, Kramer M, Lawler S, Lowenthal J, McDowell J and McCaughrean M (2022) The Case for Space Environmentalism. *Nature Astronomy* 6(4): 428–435.
Lyall F and Larsen PB (2018) *Space Law*. London: Routledge.
McLaughlin E and Muncie J (2013) *Criminological Perspectives: Essential Readings*. Los Angeles: SAGE.
McVeigh S (2019) Critical Approaches to Jurisdiction. In Allen S, Costelloe D, Fitzmaurice M, Gragl P and Guntrip E (eds) *The Oxford Handbook of Jurisdiction in International Law*. Oxford: OUP, pp. 192–205.
Mukherjee R (2020) Ship Nationality, Flag States and the Eradication of Substandard Ships: A Critical Analysis. *Maritime Law in Motion*: 581–606.
Pozdnakova A (2023) Pollution of the Marine Environment by Spaceflights. In Platjouw FM and Pozdnakova A (eds) *The Environmental Rule of Law for Oceans: Designing Legal Solutions*. Cambridge: Cambridge University Press, pp 76–92.
Presser L and Sandberg S (2019) Narrative Criminology as Critical Criminology. *Critical Criminology* 27: 131–143.
Ryngaert C (2015) *Jurisdiction in International Law*. Oxford: Oxford University Press.
Ryngaert C (2019) Cosmopolitan Jurisdiction and the National Interest. In: Allen S, Costelloe D, Fitzmaurice M, Gragl P and Guntrip E (eds) *The Oxford Handbook Of Jurisdiction in International Law*. Oxford: Oxford University Press.
Ryngaert C (2020) *The Selfless Exercise of Jurisdiction to Protect Human Rights and the Environment*. Oxford: Oxford University Press.
Sachdeva GS (2023) *Crimes in Outer Space*. Singapore: Springer.
Santos B de Sousa (2020) *Toward a New Legal Common Sense: Law, Globalization, and Emancipation*. Law in Context. Cambridge: Cambridge University Press.
Simester AP (2021) *Fundamentals of Criminal Law: Responsibility, Culpability, and Wrongdoing*. Oxford: Oxford University Press.
Takemura N (2019) Astro-Green Criminology: A New Perspective Against Space Capitalism. *Toin University of Yokohama Research Bulletin*, 40: 7–16.
Tan D (2000). Towards a New Regime for the Protection of Outer Space as the Province of All Mankind. *Yale Journal of International Law* 25: 145.
Tronchetti F (2008) The Non-Appropriation Principle as a Structural Norm of International Law: A New Way of Interpreting Article II of the Outer Space Treaty. *Air & Space Law*, 33(3): 277–305.
United Nations General Assembly (2013). Resolution 68/74 of 11 December 2013: Recommendations on National Legislation Relevant to the Peaceful Exploration and Use of Outer Space, Item 5. [Online]. From: https://www.unoosa.org/oosa/oosadoc/data/resolutions/2013/general_assembly_68th_session/ares6874.html (Accessed 22 March 2024).

Valverde M (2019) Deepening the Conversation between Socio-Legal Theory and Legal Scholarship about Jurisdiction. In Allen S, Costelloe D, Fitzmaurice M, Gragl P and Guntrip E (eds) *The Oxford Handbook of Jurisdiction in International Law.* Oxford: Oxford University Press, pp 161–181.

Viikari L (2008) *The Environmental Element in Space Law: Assessing the Present and Charting the Future.* Leiden: Brill.

White R (2021) *Theorising Green Criminology: Selected Essays.* London: Routledge.

17 Countering Sexual Harassment and Assault in Outer Space

Contributions from a Criminological Perspective[1]

Kerry Clamp

Introduction

Potential space crime and victimisation, its control, and responses to it lack significant criminological attention.[2] This chapter does not attempt to plug this gap; instead, it aims to highlight relevant criminological knowledge and learning that may be valuable in this underexplored field by responding to several exploratory questions. What are the challenges of securing justice and safety in the unique environment of outer space given the hostile environment and isolation from civilisation? What other frameworks could be considered in addressing crimes, considering the limitations of state jurisdiction and traditional legal mechanisms in outer space? What ethical considerations arise in our response to harmful behaviour when environmental factors shape that behaviour? What role could restorative practice play in shaping a moral order and respectful behaviour in outer space? The responses to these questions should be viewed as an exploratory mapping and discipline-conscious raising exercise in creating the basis for ethical space exploration and habitation. In exploring these questions, this chapter contributes to this emerging sub-field of study in two ways.

First, the chapter broadens the scope of the literature that has drawn similarities between outer space and Antarctica (see Williams 2021). It does so by examining the prevalence of sexual harassment and assault in hostile research outposts like Antarctica, and the profound impact of social isolation experienced during COVID-19 lockdowns. Drawing parallels between the unforgiving conditions of these extreme environments on Earth and the unique challenges faced in space exploration, this chapter helps us to understand why certain behaviours could be difficult to address in outer space and underscores the urgency of developing effective strategies for prevention and intervention. While other types of crime may occur, a focus on sexual harassment and assault serves as an exemplar for beginning to imagine a criminology of outer space.

Second, this chapter contributes to the discourse about the limits of state jurisdiction and regulatory gaps in addressing crimes committed in outer space (Hassan 2015; Sachdeva 2023). The unique environmental conditions of outer space pose significant spatial and temporal challenges to traditional

DOI: 10.4324/9781003437178-18

legal paradigms and mechanisms, such as the police and courts. Recognising these hurdles, alternative strategies to address harm and wrongdoing are explored, aligning with Clifford Shearing and his colleagues' call for a shift from a 'punishment mentality' to a 'risk mentality' (Bayley and Shearing 1996, 2001; Johnston and Shearing 2003). Such a shift includes proactive measures to develop safe workplaces on Earth rather than solely focusing on developing retroactive punitive measures in outer space. This approach does not negate the need for responses to address harm when it occurs, which may involve punitive actions, but underscores that reactive responses alone are insufficient in preventing harm. Restorative practice is presented as one effective approach to creating psychologically safe workplaces that promote a positive and healthy culture conducive to collaboration and organisational success. Even in workplace settings where sexual harassment and assault occur, restorative practice has been used to address individual incidents and then centralised those individual lessons to drive institutional reform.

Anticipating Problems: Why Terrestrial Crime and Justice Matter in Outer Space

Increasing human activity in space – spanning space tourism, space militarisation, interplanetary missions, and commercial activity – holds the potential for conflict between individuals, commercial entities, and nations. Currently, the nationality principle applies to crimes committed by state personnel, but challenges will arise in managing and adjudicating crimes committed by individuals against each other as space tourists, astronauts on deep space exploration missions, and those employed by private contractors providing operational and service support.

These developments in space are not flights of fancy. The first luxury hotel – Voyage Station – is expected to open in 2027 by Orbital Assembly Corporation, accommodating 400 guests and 112 crew members (Cottier 2021; Mafi and McLaughlin 2022).[3] NASA's Artemis project aims to land astronauts on the lunar South Pole by 2024, serving as a steppingstone for a human mission to Mars by 2040 – an interplanetary journey anticipated to take 500 days (Grande et al. 2023).[4] These imminent developments mean that theorising about crime and its management in outer space is not as far-fetched as it might initially seem.

Research in criminology, social psychology, and clinical psychology has established that past behaviour is a good predictor of future behaviour (Farrington 2020; Webb and Sheeran 2006; Wood et al. 2002). When considering the types of behaviour that may pose immediate concerns for crime and justice in outer space, it is prudent to explore any problematic behaviour within those organisations currently working towards increased presence in outer space. Recent history suggests SpaceX (owned by Elon Musk), Blue Origin (owned by Jeff Bezos), and to a lesser extent NASA (the US federal agency responsible for the civil space program), have exhibited a sexist work culture characterised by machismo, dominance, and superiority (Gohd 2022). This work-based culture has been implicated in enabling sexual harassment and assault to occur (Brodsky 2023; Lovell 2021). It is reasonable to expect this culture may persist in space, particularly if allegations of sexual harassment and assault not being taken seriously are true (see Bartels 2021a,b; Chang 2022).

What then are the issues likely to arise in outer space where this type of behaviour occurs? Somewhat novel challenges relate to: (1) the remoteness in which criminal acts occur and the difficulties associated with meting out a traditional criminal justice response in these settings, (2) the known impact of outer space on the physiology and psychology of people that may lead to uncharacteristic offending behaviour (otherwise unlikely on Earth by that individual), and (3) the involvement of multiple organisations in servicing research missions or commercial activities without a central coordinating body to respond to offensive or illegal behaviour.

In the following subsections, lessons from scientific missions to Antarctica and the impact of COVID-19 lockdowns are drawn upon to elucidate the challenges that could arise in outer space and the limitations of prevailing conceptions of 'justice as punishment' as an approach to dealing with sexual harassment and assault in outer space.

Lessons from Antarctica

Antarctic and Southern Ocean science is primarily pursued by researchers with a PhD in geoscience. Eighty-six per cent of these degrees have been awarded to white men since 1973, making it the least diverse field within the STEM disciplines (Nash 2021).[5] Consequently, the infrastructure supporting Antarctic fieldwork primarily caters to the needs of heterosexual males, creating substantial physical, social, and economic obstacles for underrepresented groups. For example, exploratory research conducted by Nash et al. (2019: 14) found persistent challenges for females such as ill-fitting clothing and hygiene concerns,[6] and gendered issues like unwanted attention and harassment 'from men due to the isolated, highly (hetero) sexualized environment'.

In 2022, the American National Science Foundation (NSF) published a detailed report confirming a pervasive culture of sexual harassment, sexual assault, and stalking at Antarctic research stations (NSF 2022). Almost three-quarters of female and half of male respondents reported sexual harassment was a 'fact of life', and just under half of female and a third of male respondents agreed that sexual assault was a problem. Additionally, 95% of female focus group participants knew someone who had experienced assault or harassment while stationed in Antarctica.

The factors contributing to sexual assault in Antarctica are multifaceted. The extreme conditions of living and working in Antarctica pose significant psychological stressors, including a 'lack of social variation, monotony of the physical environment, confinement and limited privacy, and emotional and physical deprivations' (Sandal et al. 2018: 2235). These stressors can exacerbate interpersonal conflicts, stress-related mental health issues, impaired decision-making, and risky behaviours (see, for example, Zhang et al. 2021), all of which can be compounded by alcohol and illicit drug use (Nash 2021; NSF 2022).

Furthermore, the isolated work setting in Antarctica combined with living in close quarters far from home, blurs distinctions between work and personal life. This blurring makes it difficult to establish clear and appropriate boundaries, intervene as a bystander, hold peers accountable, and report unwanted behaviours (Nash 2021). Moreover, the close-knit environment engenders a fear of social stigma, ostracism, or retaliation within the small community, which discourages victims from reporting incidents of sexual harassment and assault (NSF 2022). Even where victims want to report incidents, there is a lack of appropriate reporting channels in research stations, and limited access to essential services such as medical, legal, and counselling support (NSF 2022). These barriers mean female scientists adapt to the 'highly sexualised fieldwork environment' in which they work (Nash et al. 2019).

The failure to investigate and hold perpetrators to account fosters a culture of impunity. Certain work-based cultures (e.g., at McMurdo Station) have been identified as contributing to a permissive environment for sexual misconduct which influences how sexual harassment and assault are perceived and addressed (NSF 2022). For example, where victims have reported transgressions they have experienced apathy, as Nash explains:

> The onus is on the victim, the survivor, to say something, and then when they do say something, they're punished…There's a backlash—women have their rooms destroyed on station, or they get tormented for the entire season until they get so distressed that they're the ones that have to leave. The stakes are really high for women who speak up.
>
> (cited in Elliott 2023: no page)

In addressing sexual harassment and assault, organisations look to reinforce policy (Davis 2018), and increase the proportion of women in the workforce, particularly in high-level positions (Raj et al. 2020). Yet, Antarctica's unique context, setting, and organisational structure pose hurdles to these solutions. Antarctica is characterised by a complex multiorganisational operational environment with varied jurisdictions and authorities over participants, which poses a significant barrier to comprehensive policy enforcement (NSF 2022). Furthermore, actively recruiting women is difficult:

> in part because the extreme and isolated conditions result in a more limited candidate pool in general. In addition, many of the positions that need to be filled tend to be pulled from fields that are also male dominated (e.g., construction trades, skilled machine operators, etc.), thus further narrowing available female candidates.
>
> (NSF 2022: 14–15)

Addressing sexual harassment and assault in Antarctica therefore requires a comprehensive approach tailored to its unique context. The NSF (2022) report recommends providing comprehensive training on prevention, consent, and reporting procedures, along with promoting gender equity and respectful behaviour in research station communities. Yet, critics argue that these efforts do not go far enough. Nash et al. (2019) have long drawn attention to the need for a systemic response, one that addresses 'the culture and climate that maintain these abuses' (Raj et al. 2020: 292). Nash (2021: 567) explains that National Antarctic Programs (NAPs) often overlook sexual harassment in their policies and documentation, which not only shapes the field culture but also reflects organisational values:

> Programs should be seeing sexual harassment as a catastrophic risk to the organization treating it just like they would with any other legitimate safety issue that they cater for to manage risk around – likc falling in a crevasse. Like all the things people get trained to survive in an extreme environment, sexual harassment should be at the top of the list too, and it's just not.

The challenges mentioned in this section are likely to be amplified in outer space, where journeys are often long and require the successful collaboration of *all* spacefaring people. Understanding the impact of gender imbalances, the diversity of industries involved, and the difficulties the highly isolated environment poses for traditional responses demonstrates a need to rethink our approach. This involves not only thinking about how we might respond to transgressions, but also how we might prevent such behaviour from occurring in the first place.

The contextual factors explored in the following section, drawn from COVID-19 studies, are salient because they mirror the conditions expected during interplanetary space missions. Unlike the controlled Biosphere 2 experiments, which proactively simulated conditions to address foreseeable challenges (see Alling et al. 2002; Nelson et al. 2015),[7] lockdowns occurred in real-world settings that were not highly regulated. This offers valuable insights into potential issues that might arise with the advent of space tourism, and the commercialisation of outer space.

Lessons from COVID-19 Lockdowns

Research on human interactions and emotions during space missions (both short- and long-term) has shown environmental conditions such as increased radiation, micro-gravity, social isolation, and sleep deprivation can lead to a range of psychological effects, including anxiety, depression, and hostility during missions (Arone et al. 2021; Rohrer 1961). These conditions can also lead to post-mission personality changes and emotional problems (Aldrin 1973; Sengupta et al. 2022).

Similarly, COVID-19 lockdowns, characterised by extreme social isolation have also been associated with anxiety, depression, and hostility. Studies on the physiological and psychological impact of lockdowns reveal insights on aggressive behaviour. For instance, Lo and Li (2023) identified several contributing factors for aggression and violence including perceived crowdedness, a lack of personal space, reduced personal connectedness, blurred work–life boundaries, boredom, and restricted opportunities for physical withdrawal from crowded living environments.

Personal space, shaped by culture and environment, refers to the spatial boundaries that individuals require to be comfortable (Paulus 2018). Studies indicate a correlation between reduced personal space and increased violence and aggression in closed settings, such as prisons and psychiatric wards (Kumar and Ng 2001; Lawrence and Andrews 2004), public transport (Lewis et al. 2021), and online environments (Nassiri et al. 2010). Encroachment on personal space, even briefly, can cause stress and discomfort (Szpak et al. 2015), making it crucial to consider the impact of space confinement on spacefaring people.

Pfattheicher et al. (2020) also found boredom, characterised as an aversive motivational state induced by a lack of internal or external stimulation, can lead to violence directed towards oneself or others. While solitude has been demonstrated to enhance various psychological qualities (Leary et al. 2003), boredom can lead to a decline in psychological wellbeing. Lo and Li's (2023) study found that while men were much more prone to physical aggression, women scored higher in verbal aggression, anger, and hostility.

This raises the question of whether individuals can be held responsible in the same way as those who are deemed to be rational when we know environmental factors can play a significant role in the incidence of harmful behaviour? Of course, mitigations can be put in place to minimise factors that lead to negative psychological and interpersonal interactions as have been shown in the Biosphere experiments. But will commercial organisations, who place a high premium on profits, do the same? Or will they cut corners as we have seen on Earth? This raises concerns about the potential consequences for those spacefaring people who do not receive the same training and vetting as astronauts and are employed as workers in those organisations or use them as consumers (Sachdeva 2023).

Given the potential for sexual harassment and assault to occur in outer space and the constraints with traditional legal responses to that behaviour as outlined in this section, the next section underscores the need for a paradigm shift in our thinking. Given the near blank slate that outer space presents, criminology is well placed to offer alternative options to state-based justice pursuits. Specifically, the literature on adopting a risk mindset as opposed to a punishment mindset and how restorative practice can be used to develop a complementary skillset is presented. Restorative practice promises a dual benefit of helping to develop our thinking around how to develop proactive approaches to relationship management that will reduce the likelihood of harm occurring as well as a process through which to proactively respond to harm beyond a single incident.

Proposing Solutions: Some Lessons from Criminology about Non-State Responses to Reducing Risk

Legal studies dominate contemporary space scholarship, but criminologists argue that relying solely on government agencies and the law for security and order is flawed. They remind us that 'the government is merely a node within a security network and that punishment is no more than one of many instruments used to enforce order' (Schuilenburg 2015: 42). So, while the prevailing notion in public discourse suggests that the government and its agencies (such as the police) can effectively provide security, law and order, and crime control without the active involvement of the community, criminological research shows this is a myth (Garland 1996). In looking for effective models of community governance, Nash's idea of treating sexual harm and assault as a catastrophic risk within industries operating in extreme settings is taken seriously.

As the previous substantive section has shown, dealing with harmful conduct perpetuated by permissive cultural norms and specific environmental factors creates an ethical obligation to consider alternative approaches beyond punishment. Success hinges on the development of a respectful and inclusive work culture, alongside psychologically safe working conditions, and effective mechanisms for addressing harm. The next two sub-sections outline how restorative practice – which encompasses both techniques and processes for managing relations in a community – can be used to create the conditions for just workplaces and a means by which harm can be effectively addressed when it occurs (Moore and Vernon 2024).

From a 'Punishment Mentality' to a 'Risk Mentality': Developing Psychologically Safe Workplaces

The prevailing conception of justice is equated with punishment. According to Shearing and Johnston (2005), what makes punishment efficient, and therefore attractive, is its dual function in achieving both instrumental and symbolic outcomes. In respect of the former, punishment prevents disorder by deterring specific and general offending. For example, imprisonment not only incapacitates offenders but is also believed to deter them and others from future wrongdoing. In symbolic terms, punishment achieves justice by simultaneously rectifying an imbalance caused by the offence by imposing a corresponding disadvantage on the wrongdoer and condemning their wrongful actions. This efficient integration of symbolic and instrumental methods for maintaining order forms the basis of a 'punishment paradigm' and involves a backward-looking orientation because it involves action *after* an incident occurs (Shearing and Johnston 2005). Thus, two central characteristics of this paradigm are assigning blame and the inherent connection between order and punishment.

Yet, punishment relies on a perpetrator being identified and then found guilty. As this chapter has demonstrated, the experience of justice for sexual harassment and assault in remote and isolated settings within the punishment paradigm is limited. As such, Shearing and Johnston's (2005) call for a shift to a risk mentality seems to have some merit when considering the occurrence of sexual harassment and assault in outer space. Crucially, this requires a shift away from an exclusive focus on identifying, blaming, and punishing an offender to a complementary focus on pre-empting and averting misconduct to ensure there is no wrongdoer, hence, no blame, and consequently, no need for punishment. Underlying this approach is that action is not directed towards individual offenders, but towards the conduct of potential victims, to vulnerable situations, and to those routines that create criminal opportunity as an unintended by-product (Garland 1996).

Thus, treating sexual harm and assault as a catastrophic risk requires an intentional and concerted effort to develop psychologically safe workplaces and a just workplace culture. This is particularly important in those industries[8] currently working towards outer space presence on Earth to prevent harm from occurring in those workplaces in outer space. According to Moore:

> Establishing psychological safety in the workplace goes beyond changing rules and enforcing outcomes on individuals found responsible for workplace harm. It requires a stronger culture of dynamic safety, in which unacceptable behaviours and safety failures are certainly addressed, but workers are also actively involved in identifying and naming the behaviours that are to be encouraged and reinforced – and senior managers ensure that frontline managers have the skills to guide the requisite adaptive changes.
>
> (2023, personal communication)

In this respect, Moore and Vernon (2024) argue that having a different mindset is not enough. What is needed is a complementary skillset to facilitate an inclusive and respectful work-based culture that promotes wellbeing. If managers can better support all members of a work unit to establish, maintain, and deepen working relations more effectively, then each manager and their team will need to spend far less time – if any – on responding to damaged relations (AARJ 2024). This requires frontline managers to:

- 'Articulate a clear purpose or vision.
- Coach individual team members to align their work with that purpose or vision.
- Foster effective affiliation between team members through a system of mentoring.

- Enable appropriate participation by facilitating collective sense- and decision-making.'

(Moore and Vernon 2024: 196)

A system of relationship management represents a fundamentally different approach from behaviour management or order maintenance (AARJ 2024). The key requirement for effective relationship management – to support appropriate behaviours, minimise inappropriate behaviours, and provide learning and healing when inappropriate behaviour does occur – requires both skills and systems for working *with* people. In this regard, restorative practice should be viewed as 'a continuous exercise in reviewing, fine-tuning, and aligning workplace relations' (AARJ 2024: no page). Where conflicts do occur, this requires a slightly different approach, but one that continues to be practiced with those involved and in ways that can improve the workplace relations.

From a 'Punishment Mentality' to a 'Risk Mentality': Linking Recovery with Institutional Reform

The core features of the risk mentality – a forward-looking orientation and disregard for blame and wrongdoing – can also be applied to situations where a risk has materialised. An aligned response that encompasses such a mentality is restorative justice. Conventional restorative discourse:

> teaches us to put the problem, not the person, in the centre of the...dialogue...and heal the harm together...what the restorative practitioner seeks is that focus on the problem that will often lead to a wrongdoer in effect pointing the finger at themselves to say 'I/we need to change' if this problem is to be fixed, if this relationship is to be repaired.
>
> (Braithwaite 2024: 259)

The most well-known restorative justice process in criminal justice is conferencing – a meeting involving the victim, offender, and their loved ones – to respond to undisputed harm (i.e., where the offender has admitted guilt for the incident). Yet, conferences can also be used to respond to conflict associated with a pattern of poorly resolved incidents and issues where the institutional culture of an organisation is seen to contribute to harmful behaviour (Moore and Vernon 2024).

A striking feature of accounts given by victims of abuse occurring within the institutions where they reside, or work, is the profound harm inflicted when mishandling their reports. Often victims report that this mishandling is equally detrimental to their wellbeing as the initial abuse (Shepherd 2015; Vernon 2017). Hence, a meaningful solution for victims does not exclusively hinge on punitive actions but also on the establishment of protective measures to avoid the recurrence of similar incidents in the future (Shearing and

Johnston 2005). In this way, responding to realised risks employs the same mindset as responding to potential risks.

One example of where this has occurred is in response to Australian Defence Force (ADF) personnel who had experienced sexual abuse, physical abuse, sexual harassment, or workplace harassment and bullying by fellow ADF members (Vernon 2017). In 2013, a taskforce was established to provide an appropriate resolution to individual complainants and to drive cultural change within the ADF. The taskforce established a national redress scheme that offered the option of a restorative engagement conference to all personnel, in addition to reparation, criminal prosecution, or therapeutic options. Interestingly, only 1% of complainants wished to refer their case to the relevant policing agency, and while many victims were interested in a restorative process, they wanted to meet senior ADF representatives rather than those who had abused them. Complainants were reportedly:

> deeply troubled and traumatised by the failure of Defence to administer its duty of care to its members…Most wanted, above all, that others would not experience similar abuse. The consistent request from complainants was to have their personal account of abuse listened to by Defence; have their personal account accepted as true; be given acknowledgement that the abuse was wrong and should not have happened.
>
> (Vernon 2017: 146)

For an intervention to be effective, it must therefore comprehensively address patterned violence or abusive behaviour in addition to any specific incident/s. This includes recognising the physiological, psychological, relational, and cultural factors that may lead individuals of diverse gender identities to react differently to shame and exploring how scripts involving humiliated anger drive harmful actions (Moore and Vernon 2024). To respond to these concerns, this approach decentralises the learning process in each individual case and subsequently centralises these individual lessons to derive collective insights for systemic institutional reform (Moore and Vernon 2024). When the program officially concluded in April 2016, complainants reported immediate therapeutic benefits and satisfaction that their concerns had been thoroughly acknowledged and addressed (Vernon 2017).

Moore and Vernon (2024) argue that a program capable of seamlessly intertwining healing on individual, institutional, and systemic fronts has the potential to catalyse a virtuous cycle wherein individual recovery and institutional reform mutually reinforce each other. It is this chapter's contention that a risk mentality implemented with a restorative practice skill-set holds promise for addressing sexual assault and harassment prevalent in the space industry and reducing the likelihood of these incidents taking place in outer space.

Conclusion

As we venture into an era of increased human activity in outer space, the potential for conflicts, crimes, and harmful behaviours is not mere science fiction but a pressing concern. Current legal frameworks, such as the Outer Space Treaty, do not cover crimes committed by individuals against each other in outer space. Despite accumulating evidence in criminology identifying the limitations of punitive measures to respond to crime, there has been a sluggish response within the field to consider how we might deal with crime in outer space. While lawyers may primarily focus on devising legal instruments as solutions, criminology offers valuable insights and opportunities for the prevention and resolution of harm beyond legal frameworks. For this reason, the aim of the chapter was to broaden the scope of discourse and leverage some lessons and opportunities that criminology provides for effective preventative action and processes for dealing with sexual harassment and assault in outer space.

To chart a new path informed by criminology, thc chapter drew parallels between sexual harassment and assault in remote, socially isolated settings like Antarctica and COVID-19 lockdowns and the space industry. The factors contributing to sexual misconduct in these settings are connected to the harsh environmental conditions of social isolation, a workforce characterised by gender inequity, and poor organisational responses to harm by employees against other employees. Acknowledging the significant spatial and temporal challenges to a traditional criminal justice response in these settings, the chapter used Nash's call to treat sexual harassment and assault as a catastrophic risk to consider alternative ways that may help to reduce the likelihood of sexual harassment and assault from occurring in outer space.

Criminological research underscores the limitations of punitive measures in addressing harmful conduct, especially in environments characterised by permissive cultural norms and specific environmental factors. As such, a shift from a 'punishment mentality' to a 'risk mentality' was central to identifying alternatives. Restorative practice was presented as a promising approach because it emphasises proactive measures and collective responsibility in preventing and addressing harm. In extending these principles to the space industry, the chapter suggests that meaningful actions informed by a risk mentality can help reduce the likelihood of incidents such as sexual assault and harassment in outer space.

In a future where interplanetary missions and space tourism are on the horizon, the importance of such an approach cannot be overstated. The wellbeing and safety of those journeying into space, along with the success of these missions, depend on adopting innovative, forward-thinking solutions that prioritise risk assessment and prevention. By embracing a risk mentality and adopting restorative practice, this chapter argues we can pave the way for a more just and safer era of human activity in space. Together, these contributions pave the way for a more nuanced understanding of crime and justice in outer space, laying the groundwork for future research and policy development in this emerging sub-field.

Notes

1 The author would like to acknowledge the helpful feedback received on earlier drafts of this chapter from John Braithwaite, Stephen Farrall, David Moore, Sarah Dauncey, Paul Roberts, attendees at a European Society of Criminology Working Group on Restorative Justice on the topic, and the editors of this volume. The chapter is infinitely better because of their input. Nevertheless, any errors or omissions remain my own.
2 Although see Lampkin and White (2023) on environmental harms and Tiarks (2020) on restorative justice.
3 A second hotel, Pioneer Station, is also due to be opened as soon as 2025 for 28 guests at a time (see Mafi and McLaughlin 2022).
4 It is acknowledged that these plans may take much longer or not materialise at all, like Mars One (see Day 2015).
5 The unequal gender balance is also reflected at NASA, with females comprising 28% of senior executive and 16% of senior scientific employees (Davenport 2019).
6 Issues that have also been raised in other settings including the military and outer space (McCahey 2022). For example, in 2019 NASA had to cancel the first all-female spacewalk at the International Space Station because there was only one space suit that would fit a woman (Fortin and Zraick 2019).
7 In the Biosphere 2 project, there was an equal balance of men and women to reduce some of the issues found in Antarctica and on submarines with disproportionate gender representation (see Nelson et al. 2015).
8 Landon and Schneider (2017) state hospitality, mining, manufacturing, transportation, energy, and research industries are likely to be established in outer space first.

References

AARJ (2024) Workplaces. From: https://www.aarj.org.au/restorative-practices/workplaces/.

Aldrin E (1973) *Return to Earth*. New York: Random House.

Alling A, Nelson M, Silverstone S & Van Thillo M (2002) Human Factor Observations of the Biosphere2, 1991–1993, Closed Life Support Human Experiment and Its Application to a Long-Term Manned Mission to Mars. *Life Support & Biosphere Science* 8(2): 71–82.

Arone A, Ivaldi T, Loganovsky K, Palermo S, Parra E, Flamini W & Marazziti, D (2021) The Burden of Space Exploration on the Mental Health of Astronauts: A Narrative Review. *Clinical Neuropsychiatry* 18(5): 237–246.

Bartels M (2021a) Jeff Bezos' Blue Origin Faces Scathing Criticism of Safety and Culture. From: https://www.space.com/blue-origin-safety-culture-concerns.

Bartels M (2021b) SpaceX Faces Sexual Harassment Allegations from Five Former Employees. From: https://www.space.com/spacex-sexual-harassment-allegations-lawsuits.

Bayley D & Shearing C (1996) The Future of Policing. *Law & Society Review* 30, 585–606.

Bayley D & Shearing C (2001) *The New Structure of Policing: Description, Conceptualization and Research Agenda*. Washington, DC: US Department of Justice.

Braithwaite J (2024) *Simple Solutions to Complex Catastrophes: Dialects of Peace, Climate, Finance, and Health*. Switzerland: Palgrave Macmillan.

Brodsky T (2023) In a Galaxy…Not So Far Away: The Privatization of Space Flight and the Need for Labor Laws in Outer Space. *Hofstra Labor and Employment Law Journal* 40(2): 523–566.

Chang K (2022) Jeff Bezos' Rocket Company Accused of Toxic Culture and Safety Issues. From: https://www.nytimes.com/2021/09/30/science/jeff-bezos-blue-origin-safety.html.

Cottier C (2021) The First "Space Hotel" Plans to Open in 2027: How One Aerospace Company is Preparing for the Advent of Tourism in Outer Space. From: https://www.astronomy.com/space-exploration/the-first-space-hotel-plans-to-open-in-2027/.

Davenport C (2019) At NASA, 2019 Was the Year of the Woman, Yet Women Still are a Big Minority at the Space Agency. From: https://www.washingtonpost.com/technology/2019/11/26/nasa-was-year-woman-yet-women-still-are-big-minority-space-agency/.

Davis M (2018) The Commercial Advantage in Space's Grey Zone. From: https://www.aspistrategist.org.au/the-commercial-advantage-in-spaces-grey-zone/.

Day D (2015) Red Planet Rumble. From: https://www.thespacereview.com/article/2809/1.

Elliott C (2023) For Female Scientists, Antarctica Can Be a Sexual Harassment Hellscape. From: https://www.motherjones.com/environment/2023/02/female-scientists-antarctica-sexual-harassment-mcmurdo-nsf-report/.

Farrington D (2020) The Integrated Cognitive Antisocial Potential (ICAP) Theory: Past, Present, and Future. *Journal of Developmental and Life-Course Criminology* 6 (2): 172–187.

Fortin J & Zraick K (2019) First All-Female Spacewalk Cancelled Because NASA Doesn't Have Two Suits That Fit. From: https://www.nytimes.com/2019/03/25/science/female-spacewalk-canceled.html.

Garland D (1996) The Limits of The Sovereign State: Strategies of Crime Control in Contemporary Society. *The British Journal of Criminology* 36(4): 445–471.

Gohd C (2022) The Space Industry Has a Big, Ugly Sexual Harassment Problem: Will Things Finally Change? From: https://www.space.com/space-industry-sexual-harassment-problem.

Grande M, Guo L, Blanc M, Alve J, Makaya A, Asmar S, Atkinson D, Bourdon A, Chabert, P, Chien S, Day J, Fairén A, Freeman A, Genova A, Herique A, Kofman, W, Lazio J, Mousis O, Ori G, Parro V, Preston R, Rodriguez-Manfredi J, Sterken V, Stephenson K, Vander Hook J, Waite J & Zine S (2023) Enabling Technologies for Planetary Exploration. In M Blanc (Ed.) *Planetary Exploration Horizon 2061: A Long-Term Perspective for Planetary Exploration.* Oxford: Elsevier Science, pp. 249–329.

Hassan R (2015) From Earth to Heaven: States' Criminal Jurisdiction in the Space Station. *IIUM Law Journal* 23(3): 401–414.

Johnston L & Shearing C (2003) *Governing Security: Explorations in Policing and Justice.* London: Routledge.

Kumar S & Ng B (2001) Crowding and Violence on Psychiatric Wards: Explanatory Models. *The Canadian Journal of Psychiatry* 46(5): 433–437.

Lampkin JA & White R (2023) *Space Criminology: Analysing Human Relationships with Outer Space.* Switzerland: Palgrave Macmillan.

Landon J & Schneider E (2017) These 5 Industries Will be First to Do Business in Space. From: https://www.weforum.org/agenda/2017/11/industries-will-make-money-in-space/.

Lawrence C & Andrews K (2004) The Influence of Perceived Prison Crowding on Male Inmates' Perception of Aggressive Events. *Aggressive Behavior*, 30(4): 273–283.

Leary M, Kowalski R, Smith L & Phillips S (2003) Teasing, Rejection, and Violence: Case Studies of the School Shootings. *Aggressive Behavior* 29(3): 202–214.

Lewis S, Saukko P & Lumsden K (2021) Rhythms, Sociabilities and Transience of Sexual Harassment in Transport: Mobilities Perspectives of the London Underground. *Gender, Place & Culture* 28(2): 277–298.

Lo L & Li W (2023) Encaged and Enraged: A Study of How Level of Aggression Relates to Perceived Crowdedness, Risk, and Boredom. *International Journal of Conflict and Violence* 17: 1–11.

Lovell B (2021) Sex and the Stars: The Enduring Structure of Gender Discrimination in the Space Industry. *Journal of Feminist Scholarship* 18(Spring): 61–77.

Mafi N & McLaughlin K (2022) A Space Hotel Could Open as Soon as 2025. From: https://www.architecturaldigest.com/story/worlds-first-space-hotel-open-2027.

McCahey D (2022) Women in Antarctica Face Assault and Harassment – and a Legacy of Exclusion and Mistreatment. From: https://theconversation.com/women-in-antarctica-face-assault-and-harassment-and-a-legacy-of-exclusion-and-mistreatment-190620.

Moore D & Vernon A (2024) *Setting Relations Right in Restorative Practice: Broadening Mindsets and Skill Sets.* London: Routledge.

Nash M (2021) National Antarctic Program Responses to Fieldwork Sexual Harassment. Antarctic Science 33(5): 560–571.

Nash M, Nielsen H, Shaw J, King M, Lea M & Bax N (2019) "Antarctica Just Has This Hero Factor…": Gendered Barriers to Australian Antarctic Research and Remote Fieldwork. *PLOS ONE* 14(1): 17–22.

Nassiri N, Powell N & Moore D (2010) Human Interactions and Personal Space in Collaborative Virtual Environments. *Virtual Reality* 14: 229–240.

Nelson M, Gray K & Allen J (2015) Group Dynamics Challenges: Insights from Biosphere 2 Experiments. *Life Sciences in Space Research* 6: 79–86.

NSF (2022) *Sexual Assault/Harassment Prevention and Response (SAHPR): Final Report.* From: https://www.nsf.gov/geo/opp/documents/USAP%20SAHPR%20Report.pdf.

Paulus M (2018) Preschool Children's and Adults' Expectations about Interpersonal Space. *Frontiers in Psychology* 9: 2479–2487.

Pfattheicher S, Ljiljana L, Westgate E & Schindler S (2020) On the Relation of Boredom and Sadistic Aggression. *Journal of Personality and Social Psychology* 121(3): 573–600.

Raj A, Johns N & Jose R (2020) Gender Parity at Work and its Association with Workplace Sexual Harassment. *Workplace Health & Safety* 68(6): 279–292.

Rohrer J (1961) *Interpersonal Relationships in Isolated Small Groups: Psychophysiological Aspects of Manned Spaceflight.* New York: Columbia Press.

Sachdeva G (2023) *Crimes in Outer Space: Perspectives from Law and Justice.* Singapore: Springer Nature.

Sandal G, Van De Vijver F & Smith N (2018) Psychological Hibernation in Antarctica. *Frontiers in Psychology* 9: 2235–2243.

Schuilenburg M (2015) *The Securitization of Society: Crime, Risk and Social Order.* New York: New York University Press.

Sengupta S, Shroff D & Rasal P (2022) *Gender Diversity: Heading Towards an Inclusive Work Culture.* Mumbai: Anglo-Eastern Maritime Training Centre.

Shearing C & Johnston L (2005) Justice in the Risk Society. *The Australian and New Zealand Journal of Criminology* 38(1): 25–38.

Shepherd M (2015) Institutional Responses to Child Sexual Abuse: The Latest on the Proposed Redress Scheme. From: http://www.shepherdsfamilylaw.com.au/site2018/wp-content/uploads/2018/02/Shepherd-M_LSJ-December-2015-Final-copy.pdf.

Szpak A, Loetscher T, Churches O, Thomas N, Spence C & Nicholls M (2015) Keeping Your Distance: Attentional Withdrawal in Individuals Who Show Physiological Signs of Social Discomfort. *Neuropsychologia* 70: 462–467.

Tiarks E (2020) Interplanetary Spaceflight and Restorative Justice. From: https://room.eu.com/article/interplanetary-spaceflight-and-restorative-justice#:~:text=A%20common%20objection%20to%20restorative,could%20pursue%20excessively%20punitive%20responses.

Vernon A (2017) The Ethics of Appropriate Justice Approaches: Lessons from a Restorative Response to Institutional Abuse. *Law in Context: A Socio-Legal Journal* 35(1): 139–158.

Webb T & Sheeran P (2006) Does Changing Behavioral Intentions Engender Behavior Change? A Meta-Analysis of the Experimental Evidence. *Psychological Bulletin* 132 (2): 249–268.

Williams K (2021) Space Crime Continuum: Discussing Implications of the First Crime in Space. *Boston University International Law Journal* 39(2): 79–108.

Wood W, Quinn J & Kashy D (2002) Habits in Everyday Life: Thought, Emotion, and Action. *Journal of Personality and Social Psychology* 83(6): 1281–1297.

Zhang M, Zhao X, Liu Y & Yang J (2021) Perceived Stress and Possible Stressors in the General Public in China During the COVID-19 Pandemic. *Ethics, Medicine and Public Health* 18: 1–8.

18 Punishment for Offending in Space

The Ethics of Incarceration

Victoria Nagy

Introduction

In 2021 in Queensland, Australia, a 17-year-old Aboriginal teenage boy with an intellectual disability was locked in solitary confinement for over 500 days (Smee 2023). He was on remand, awaiting his trial on sexual assault charges against a 16-year-old girl. This period was during his 744 days spent in the youth prison. The solitary confinement consisted of 20 hours per day in his cell locked down. It was not his behaviour in prison that led to the decision to keep him in solitary; the chronic understaffing of the youth detention centre meant that about 70% of the time the boy found himself in solitary it was because there were not enough staff available to ensure the safety of him or other remandees.

Adults face similarly dehumanising conditions. In Tasmania's adult prison, prisoners sometimes get no more than 2 hours outside of their cells with the prison perpetually in lockdown. This is mirrored in other states around Australia and in Australia's offshore detention of asylum seekers and refugees on Papua New Guinea. Further afield in the UK, floating barges are being used to house asylum seekers in detention; locked away from communities without easy access to people off the barge, being at threat of bacterial legionella infection and abusive behaviour from staff has led to one person remarking that it is "exactly the feeling of being in prison" after details emerged of a death onboard (Al Jazeera 2023).

The widening net of incarceration to include large numbers of youth, adults, citizens and non-citizens, and those awaiting trials or being sentenced is a worldwide issue. With the number of people incarcerated growing, "tough on crime" responses being vote winners, and media sensationalism about criminal offending, there appears to be little appetite for radical change in how we deal with lawbreaking (i.e., criminal offending) or perceptions of deviancy (e.g., seeking asylum). Thus, if there is such an emphasis on incarceration for deviancy and criminality (actual or perceived) then it is important to ask how deviancy or criminality away from Earth would be treated; how would wrongdoing (especially if interpersonal and violent) be treated? I will consider how we conceive of and use incarceration, including solitary

DOI: 10.4324/9781003437178-19

confinement, the harms and ethics of imprisonment and what using similar methods away from Earth would mean for travel or habitation further afield.

Imprisonment

Prison as a place for punishment, through extended periods of deprivation of liberty inflicted by the state, is relatively modern as far as it being a wholesale solution to the problem of crime is concerned. Historical responses included corporal punishments in public or places where the punishment quickly took place. Punishment, under a utilitarian rationale, is necessary for the greater good of society, if the peace and wellbeing of the society has been threatened or impacted by criminal behaviour.

Considering how best to respond to crime and deviancy, both Cesare Beccaria and Jeremy Bentham argued that traditional forms of punishment (i.e., inflicting bodily harm) were unethical in an enlightened era. Beccaria stated that punishment could not be an act of violence but had to be public, swift (celerity), necessary, and proportionate. Taking a liberal view of the reason for incarceration over the use of corporeal punishment, proportionality is key. This guides most Western criminal justice institutions to this day; proportionality is built upon the principles of parity (equally serious crimes receive equally serious punishments), difference (unequally serious crimes receive unequally severe punishments), and positive slope (the more serious a crime is the more severe the punishment is) (Duus-Otterström 2019). Therefore, a more serious offence (e.g., a sexually violent murder) will receive a more serious punishment than a minor offence (e.g., a traffic violation).

Overall, there are usually five purposes to sentencing, namely,

- Just punishment
- Deterrence (specific/general)
- Rehabilitation
- Denunciation
- Community protection

Some of these focus on the individual (just punishment, rehabilitation, (specific) deterrence); denunciation's role is somewhat vaguer as it is about condemning or censuring the conduct of the individual. Deterrence is comprised of specific deterrence (deterring the individual from committing crime again) and general deterrence, that is, deterring the greater population from committing similar offences. The argument here is that seeing the punishment on the individual will lead to members of the community thinking twice about performing similar illegal behaviours. The last purpose (community protection) considers the offender's threat to the wellbeing of community members. Although oft cited that prison *is* punishment not *for* punishment (that is, deprivation of liberty is the punishment, rather than prison as a place where cruel and unusual punishments are inflicted), it is fair to say that retribution not restoration is the goal in public and politicians' minds alike when thinking about incarceration. Harsh prison sentences, dire conditions, lessened human rights are celebrated while simultaneously noting that this is done for the protection of, and in the name of, the community. Similar to deterrence, the harsh treatment of imprisoned men and women is also done in the name of education; for the public as much as the prisoner who need to know what happens to those who break the law.

As Bagaric (2020) argues, some of these purposes can be counterintuitive. Community protection is usually cited as an important factor for choosing incarceration over other punishment types, yet rehabilitation as a purpose of sentencing suggests that leniency may be required in order to ensure that rehabilitation can occur. However, rehabilitation is not reintegration in and of itself. Reintegration with the community is often difficult due to not only how individuals are treated during incarceration but also afterwards. Fears about community safety and a burgeoning interest in risk prevention have led to poor reintegration outcomes for formerly imprisoned people; growing recidivism rates in many Western nations where a retributive and tough on crime approach is championed is demonstrative of this issue (Leverentz, Chen and Christian 2020).

Community protection is the oft-cited reason for "tough on crime" policies in Western nations. The argument that this policy has come out of the war on crime/war on drugs waged by the United States (US) from the 1970s (Mauer 1999), has expanded to include declining belief that rehabilitation can be effective for community protection (Newburn 2007). While there has been a drive (hastened by the Covid-19 pandemic) to consider abolition of prisons, community pressures demand increased incarceration and punishment while simultaneously turning away from reintegrative or rehabilitative practices (Staff and Wenzelburger 2021).

There is also the ongoing debate about the use of solitary confinement. Initially built into the design of Pentonville Prison in England in 1842, solitary confinement (or as it was known initially, separate confinement) has continued to be utilised in prisons around the world. At the outset the idea behind solitary confinement was that inmates would be sectioned off from each other in order to prevent fraternisation, and to drive religious observance and self-reflection. Left in small cells, with only the most minimal light, no way to communicate with anyone but those whom the guards deemed appropriate (e.g., wardens, physicians, or religious officials), and possibly a Bible, prisoners would be required to spend hours on menial, repetitive but physically exhausting work. Very soon it became apparent that rather than having a reforming effect, solitary confinement produced and amplified mental breakdowns, delusions, hallucinations, panic, depression, and anxiety (Cox and Marland 2018). A 700-page report from 1922, *English Prisons Today*, detailed how little solitary confinement had changed in the English prisons in proceeding decades, highlighting how being locked into small cells with little or no human interaction would drive prisoners to insanity or suicide (Blom-Cooper 1987; Hobhouse and Brockway 1922). Scharff Smith

(2006) details how by the early 20th century the science of solitary confinement was settled in not only the UK but across the West; by the time of the Second World War solitary confinement was being phased out although it still existed in enough prisons that research on sensory deprivation and perceptual deprivation between the 1950s and 1970s demonstrated the ill effects on people. It was with booming prison numbers in the 1980s onwards that solitary confinement began to be used as a space-saving option in some prisons in the US, as a form of "safety" for prisoners in other European nations, and as an option for removing problematic prisoners from the general population across the board. In counterterrorism operations, it was also used to try and "break" prisoners for interrogation purposes post September 11. Thus, it has been clear from the beginning that solitary confinement's effects produce psychological effects that are recognised as cruel and unusual punishment by medical personnel and social justice activists (Coppola 2019).

Heading to Space

Current efforts in human space travel and habitation are highly controlled by private and public enterprise. Attention in public enterprise (e.g., nation-state space missions) has been heavy on ensuring that the right individuals are chosen for missions. For example, the planned 2024 NASA Artemis II 10-day mission, which will include a fly-by the Moon, includes 4 crew members: the commander a naval aviator and test pilot, the pilot had over 3,000 flight hours before joining the astronaut corps, the mission specialist is an engineer by trade, and the final member, while a rookie for space travel, was a F-18 pilot before joining the NASA program (Whitman Cobb 2023). Although there are now more scientists involved in space shuttle programs (via the role of mission specialist) to open the pool to more individuals, these missions are still predominantly led and crewed by military pilots. In private space travel (or space tourism as it is currently labelled) there is more wiggle room. In 2021 SpaceX launched four civilians into orbit, including a CEO, a physician assistant, an aerospace data engineer, and a pilot and geoscientist. Prior to this, Jeff Bezos headed to space on a Blue Origin mission that included the actor William Shatner as a passenger (Shivaram 2021). At the moment, space flight is prohibitively expensive for both public and private space travel and enterprise.

Not only is cost and previous experience a factor but also psychometrics. Extensive tests are undertaken of potential astronauts to ensure that their personalities are appropriate for space travel; a wide variety of non-pathological personality questionnaires are completed by applicants for general personality assessment. Such tests include the Personality Characteristic Inventory (PCI), NEO Personality Inventory Revised (NEO-PI-R), the NEO Five-Factor Inventory (NEO-FFI), or the Temperament Structure Scales (TSS). Factors such as aggression, neurosis, and goal orientation are part of these tests. NASA's "Right Stuff" definition of the ideal personality is a mix of these personality questionnaires and other non-testable characteristics such as technical aptitude, ability to tolerate isolation, or cultural sensitivity (Mittelstädt et al. 2016). Currently only those who have the "Right Stuff" in personality, as well as technical skills and experience with flight, can be considered for space travel. Testing indicates that there is little difference in personality between men and women who are part of astronaut programs (Musson, Sandal and Helmreich 2004). It could be assumed that those who choose to be astronauts or want to travel to space are somehow different in personality or characteristics than the rest of the population. Research has established that actors share similar personality traits to astronauts (e.g., high performance in stressful environments, tolerance for ambiguity, perseverance and taking sensible risks) (Chterev and Panero 2021; Nettle 2006). Similarly, those who work in Antarctica are often chosen for personality traits that are similar to those of astronauts; namely, low interpersonal issues and low neuroticism (Musson, Helmreich and Sandal 2004). However, amongst a general student population it was clear that neurosis is much higher amongst this group than either Antarctic station workers or astronauts (Musson, Helmreich and Sandal 2004).

Thus, space travel is highly restricted due to money, personnel, previous experience, personality, and technology. But there is a strong likelihood that this will change; more private enterprise entering the arena will likely have an impact on not only the technology but also the profile of those traveling to and residing in space, as well as the goals associated with this travel (e.g., mining, tourism, long-term accommodation). As such, considering not only if crime will happen but what needs to happen *when* it does is crucial. It is especially important that the discipline of criminology is part of these discussions as it has the research evidence to offer suggestions.

Space Crime and Punishment

As technical skills in conjunction with strict guidelines about personality are currently the focus of state-funded, and to a lesser extent private, space travel, this means that offending, and subsequent incarceration, is not a burning issue. Most space travel is still measured in days rather than months or years. The issue of criminal behaviour has, until now, not figured as a problem needing solving for this form of space adventure. The International Space Station (ISS) is at present the only location where missions are longer (lasting in general for 6 months, although a few individuals have stayed for almost a year). It is also the location where the first alleged space crime took place (until it was discovered that it had not).[1]

In 2019, news broke that the first case of space crime had been committed onboard the ISS. NASA astronaut, Anne McClain, was accused of identity theft and twice improperly accessing her wife's bank accounts while she (McClain) was onboard the ISS. McClain's wife, Summer Worden, alleged that McClain had accessed her bank records after login details had been changed in order to prevent McClain from accessing the account. McClain stated that they were accessing the account to make sure that there was money to pay bills. Media outlets quickly released articles about the "first crime in space". As it was a case about identity fraud and improper access of bank accounts, there was nothing salacious about the crime itself. Interest in McClain and the alleged crime quickly diverted to where it had occurred and the questions it raised for justice administration. The public wanted to know: Who prosecutes crime in space? Which jurisdiction does the crime take place in? What if the crime takes place on a private vessel or in a location that is run by a private enterprise? Responses from legal scholars and other experts highlighted that there were clearly defined responses for illegal behaviour onboard various space craft for commercial companies, private individuals as well as part of state-sponsored space programs. Each nation is responsible for any registered object launched into space as well as personnel from that state. Due to the nature of the treaties governing space, all crimes are raised to national (or international) level immediately, rather than how crime (interpersonal as well as property or digital crime) is traditionally handled (i.e. at a lower level). The Canadian Parliament recently passed legislation about crimes committed on the Moon via the Lunar Gateway program that aims to have humans habiting the Moon. The legislation outlines

> That if a Canadian crew member commits an act or omission – that if committed in Canada would constitute an indictable offense – on, or in relation to, a flight element of the Lunar Gateway, on any means of transportation to or from the Lunar Gateway, or on the surface of the Moon, that act or omission is deemed to have been committed in Canada. If that act or omission is committed by a crew member of a Partner State, it is deemed only to have been committed in Canada if it threatens the life or security of a Canadian crew member or is committed on or in relation to, or damages, a flight element provided by Canada.
> (Canadian Criminal Code, clause 296(4))

Thus, crime can very quickly have international diplomatic implications if not handled carefully. Defining offences as happening outside of the physical territories of a state but still within its jurisdiction may set a precedent for other countries to also codify that crime committed in space is a national rather than international criminal justice issue.

I will not spend valuable wordcount in this chapter describing the various treaties governing space exploration and responses to petty crime onboard the ISS or other spacecraft.[2] Suffice it to say that current laws governing it are not fit for purpose and will require change sooner rather than later (Grush 2019). Instead, I would argue that the enforcement of penalties is of more interest to us. As it stands, the assumption is that the crime would be tried and punished on return to Earth. However, if space travel will be to regions beyond the Moon, and habitation will also be not only at a distance but for an extended period, with no quick way to return the accused to Earth to face trial, ensuring swift justice will require trials and punishments to be handled away from Earth.

Space Incarceration

In this section I will focus on solitary incarceration. As there are no cases of space crime and punishment that we can turn to, this is highly theoretical, but I do hope that some of the thoughts offered in this section can help us start thinking about what to do when it is clear that not only has a crime been committed but there must be a response to it in order to ensure safety in the interim and that justice is served in the long term.

Case #1

A passenger aboard a flight to a Mars habitat begins first suggestively flirting then overtly sexually harassing a fellow passenger. As the quickest trip to Mars takes about 9 months to complete, when the orbits of the two planets align, this is a considerable amount of time that this person has to make their victim feel uncomfortable, and mentally and physically threatened. The passengers land on Mars and are ensconced in their habitat. The harassment continues and one night the victim is raped. The victim soon discloses to another person what is happening. They decide to escalate the issue to the commander of the habitat. While removing the alleged offender back to Earth is possible, it will be another 12 months before they can arrive back to Earth to face an investigation and trial. There is a three-month wait required for the two planets to align again for take-off, and another 9-month journey home. To ensure the safety of habitat dwellers and passengers on the returning flight the alleged offender needs to be kept isolated.

In this scenario, the individual who has allegedly been sexually harassing and then eventually raping another person would need to be kept in solitary confinement for at least 12 months. They have not been charged with a crime, but it is crucial that the alleged offender is not allowed to continue to have access to either the victim or others who may become victims. Learning from the isolated, confined and extreme (ICE) environment that is Antarctica demonstrates that sexual harassment can run rife amongst all inhabitants across a variety of levels of seniority, especially when left unchecked. Like sexual harassment elsewhere, it also disproportionately affects women. According to reports, almost 60% of women living and working in Antarctica had been victims of sexual harassment and assault (NSF 2022). Previous policies required victims in Antarctica to work with their assailants or require other women to work alongside the assailant even when there was a high likelihood of victimisation. Oftentimes when reporting their discomfort with this, supervisors would be unsupportive (Perry 2023). In only one case of an assault was the alleged offender fired and returned to the US to face trial (Perry 2023). As there is no law enforcement in Antarctica with no jails or prisons, responding to allegations of sexual harassment and assault are station dependent. The laws of the home country responsible for the station are followed within that station; however, there are caveats to this. With multiple nationalities working alongside each other there are jurisdictional issues that arise when the offender and victim are different nationalities but working in the same station. In order to prevent diplomatic tensions, this may result in overlooking harms and crimes committed by individuals not from the home nation of the station. Victims also report that with no jail to keep their assailant separated away from them, they do not feel safe disclosing or continuing to live and work at the station following a sexual assault (Perry 2023). For these reasons, allegations of sexual harassment and assault need to be taken seriously on Mars as isolation is even more extreme than back on Earth.

In this case, solitary confinement is the chosen option by the Mars habitat dwellers to help keep the victim safe from their attacker. However, physically confining a person to their quarters (which are small out of necessity for this Mars habitation) raises issues such as ensuring that an individual has adequate space and time for physical activity, ensuring that their mental health does not deteriorate, and ensuring that their human rights are not violated. Prolonged solitary confinement (defined as more than 15 days, with 22 hours or more in isolation (APT n.d)) is considered cruel, inhuman or degrading treatment and in some circumstances, torture. Pre-trial solitary confinement is used in some Earth jurisdictions due to either a lack of bail options, overcrowding in prisons and remand centres, or for the safety of other prisoners/remandees or the accused themselves. The overreliance of this in places like Iceland, where over 60% of those awaiting trial have been kept in solitary isolation, has been condemned by Amnesty International (2023) as also being a form of coercion from authorities seeking confessions. Therefore, the effect that the individual's solitary confinement on Mars may have not only on the individual but any investigation and trial must also be considered.

As the other crew members are neither criminal justice personnel, nor lawyers or judges but either military personnel, representatives of the state, or civilians there is also an element of extrajudicial punishment or vigilante justice that must be discussed. Solitary confinement in such a case where there are government representatives (military or non-military) would suggest that the pre-trial confinement is extrajudicial punishment as it is not being undertaken with due process; where it is civilians deciding the treatment of the alleged offender due to fears of further victimisation then vigilantism is a risk. Therefore, there is a risk to the individual's mental and physical health; questions of maintaining the alleged offender's human rights while separated from the rest of the inhabitants; and due process that if not followed could result in justice not being attained for the victim once the alleged offender is returned to Earth for trial.

Case #2

Every habitation on Mars requires at least one medical doctor to be available in case of emergencies or other health issues that require medical treatment. The current doctor is a seasoned space dweller, trusted for their expertise, professionalism and calm. Following an emergency situation, which has left several people injured and another person in a coma, the doctor decides that due to a lack of medical supplies, their own expertise, and possibility of removing the person from the surface of Mars and getting them back to Earth in time to receive life-saving intervention, that "active" euthanasia is the best option. The victim has never indicated that in the case of a severe emergency they wish to not be resuscitated, and as such this is (under current legislation of the home country) an act that would result in a charge for murder. In the case of an egregious crime such as this the individual should be automatically removed from their post and put into confinement to their quarters. However, the other injured personnel still require treatment, and the base cannot be without a medical doctor while they await the next possible departure window back to Earth which is currently over a year away.

A little bit sensational but also plausible.[3] Disregarding discussions of morality in times of medical emergencies and extremely injured people on habitats far away from Earth, at its core this is a case about a medical doctor committing an act that criminal justice authorities would classify as murder (at the most severe) or manslaughter (if mitigating circumstances were articulated and argued). However, unlike our previous case, where the individual is not vital to the health and wellbeing of others, in this scenario simply locking the doctor into solitary confinement is not an option. There are still injured crew who require medical care, as well as those who are well and not immediately in need of medical interventions. However, should any of the other injured personnel take a turn for the worse there is no guarantee that the doctor will not again decide to end a person's life. In a situation such as this the remaining personnel need to be able to work with and ultimately trust the doctor.

Assuming that the medical doctor's choice in this scenario is not due to a substance or mental health issue, geographic and professional isolation is a risk to poor performance in medical personnel (Willcock 2001). High levels of burnout are also associated with this field, with characteristics including emotional exhaustion and depersonalisation (Willcock 2001). Naturally, with more risk factors regarding burnout and underperformance, the more doctors are at a higher risk to themselves as well as their patients. This is separate to the other psychological and physiological responses that would cause stress in a space environment including lack of privacy; endured separation from friends and family; delayed communication; social monotony; environmental risks living in space; fatigue, and exhaustion; altered cycles of light and dark; and communication issues due to an inability to detect nonverbal cues (Palinkas and Suedfeld 2021). These have all been noted as effects felt by inhabitants in Antarctica and other ICE environments. Therefore, it is likely that extreme environments, such as living and working on Mars, may create not only situations but responses within individuals that lead them to offend in ways that they may be highly unlikely to if they remained on Earth.

But that does not answer the question of what to do in this situation. Is the doctor to be held separately? If they are the only one with the professional knowledge to help keep settlers/travellers healthy, can it be risked that they are out of action while incarcerated away from others? Can they be trusted and be supervised to ensure that they do not behave in such a manner again? And if they are allowed to continue working, how does this make the survivors (both of the incident and others living and working with the doctor) feel? Will there be shock, confusion, or anger that also needs to be addressed?

Possible Solutions

In both cases the immediate safety of others can be guaranteed through solitary confinement of the accused. In one scenario, awaiting return to Earth would mean approximately a year in solitary confinement that would have additional mental and physical health pressures beyond the stresses that would be associated with long distance space travel. In the other scenario, the skills of the accused mean that keeping them in solitary confinement may not be an option, but not doing so may result in further distress and harm. Following on from the utilitarian positions that have underpinned our purposes of sentencing in the developed world, and which have advocated for abolition of the death penalty, it is clear that crime in space cannot, with good conscience, be punished by means that are in large part abolished around the world. Although some developed nations still maintain the death penalty (namely the US, Japan, Taiwan and Singapore) and many developing nations that are part of the space race also have capital punishment on their statute books (for example, Russia and China), without trial this penalty cannot be applied by those in space with these alleged offenders. Due process is still a requirement to punishment, and return to Earth for trial is a condition of it. Without trial, imprisonment for punishment of other crimes is also not a possibility. Thus, in these scenarios, incarceration would have a role to play in pre-trial holding of accused offenders but not as a form of punishment in space.

The extended period of time that a person may be forced into solitary confinement would therefore be linked to the idea of community protection rather than explicit punishment for their actions. However, incarceration is not the only means by which community protection can be achieved. Restorative justice is one possible way to tackle the issue of safety (of all), justice, and reintegration of the individual back into the community to allow the (space) community to move forward in an ICE environment.

Restorative justice operates on three main principles: offender responsibility, allowing reintegration into the community, and reparation to the victim/s. There are various models of restorative justice available which consider things like cultural differences between the offender and victim, juvenile versus adult offending, or attempted restoration after serious violent interpersonal offending versus minor petty crimes. Ideally in all models of restorative justice, the victims, communities and offenders are all in dialogue about how to best deal with the aftermath of the offending, individual as well as collective responsibility for the offending, the impacts of offending upon victims, communities and offenders, making amends, and healing victims' hurt (Morris and Maxwell 2001). Instinctively, people may be wary or outright suspicious of restorative justice as it is not underpinned by retributive principles. However, consistently victims report feeling more involved with the process than traditional criminal justice responses, are more likely to receive reparation for the harm (e.g., an apology, explanation, or community work), and less likely to be angry or fearful (Morris and Maxwell 2001). These processes can also have greater impact upon the offender, who can get a clearer understanding of the harm they have caused (Johnstone 2003). It importantly also transfers decision-making authority from government to victims and offenders and the community around them both, thus allowing for more personalised and inclusive justice than traditionally. It makes punishment more active in that the offender needs to take responsibility for their actions and rectify the harm, rather than passively take punishment (Umbreit 1995; Zehr 1985).

There are possible issues with restorative justice being applied in the above scenarios, including:

- A lack of trained mediators on these missions who could run restorative justice models to a high standard and prevent secondary victimisation for the victim.
- Restorative justice models are applied as an aspect of punishment *after* conviction.
- The safety of other inhabitants may not be guaranteed.
- There are still enormous physical and psychological pressures facing individuals to allow survival on a space mission that are not there back on Earth.

The biggest issue facing the use of restorative practices in space is that they require the offender to admit to their criminal activity and take responsibility for their actions. It is for this reason that restorative justice is applied as a portion of a sentence and done with the consent of both victim and offender. Pre-emptively using restorative practices may circumvent due process, and the use of them may result in a case being dismissed against the accused when they return to Earth.

Conclusion

To rely on forms of responses to crime that we regularly do in the developed world, namely incarceration and solitary confinement as much for punishment as for pre-trial methods to hold accused individuals, runs the risk of exacerbating the problems we are currently contending with around the world. Exporting this problem into space and part of any communities humans create will lead to more harms than protections, especially in times and spaces when humans will be the most vulnerable.

Rather than punishment for offending, restorative practices such as mediation may need to be introduced to ensure that relative calm and safety is established within habitats in places like Mars as a way to prevent extended periods of solitary confinement for offenders, to ensure safety for victims, the community and the offender, and that essential services (such as medical care) can be maintained while awaiting departure back to Earth to face the justice system. Inadequate facilities in colonies months or years away means that restraining, trying, and punishing individuals away from Earth is problematic. Also problematic to the wellbeing of a victim or community is the acquittal of an individual while living away from Earth. Finding an accused not guilty via virtual trial may require the victim to live in close quarters to their attacker for an extended period of time; a not guilty verdict does not mean that the victim was lying or unharmed by the ordeal. Thus, mediation may be a way by which to ensure that the victim feels that their experiences are being validated, could help make a victim, offender or community feel safe, and return a colony or habitat to equilibrium to ensure the long-term health and safety of everyone before the accused returns to Earth.

Notes

1 McClain's wife, Worden, was indicted in 2020 for making false statements and as of March 2024 the case is still not resolved.
2 Suffice it to say that there are plenty of treaties that answer these questions to a certain extent including the *Outer Space Treaty*, the *Rescue Agreement*, the *Liability Convention*, the *Registration Convention* and the *Moon Agreement*.
3 The scenario is partially taken from a *Journal of the American Medical Association* account from an anonymous resident who admitted to administering a fatal dose of morphine while on his hospital rounds when attending to a young woman dying of cancer (Gaylin et al. 1988).

References

Al Jazeera (2023) Asylum seeker dies on UK barge housing migrants and refugees. Al Jazeera News. From: https://www.aljazeera.com/news/2023/12/12/asylum-seeker-dies-on-uk-barge-housing-migrants-and-refugees.

Amnesty International (2023) Iceland's abusive use of solitary confinement must end immediately. From:https://www.amnesty.org/en/latest/news/2023/01/iceland-abusive-use-of-solitary-confinement-must-end-immediately/.

APT (n.d.) Solitary confinement. Association for the Prevention of Torture. From: https://www.apt.ch/knowledge-hub/dfd/solitary-confinement#:~:text=Of%20particular%20concern%20to%20the,in%20excess%20of%2015%20days.

Bagaric M (2020) In search of a coherent approach to community protection in sentencing. *Monash University Law Review* 46(3): 79–115.

Blom-Cooper L (1987) The penalty of imprisonment. The Tanner Lectures on Human Values. From: https://tannerlectures.utah.edu/_resources/documents/a-to-z/b/Blom-Cooper88.pdf.

Chterev K and Panero ME (2021) Exploring similarities across the space and theater industries. *Frontiers in Psychology* 11(574878): 1–4.

Coppola F (2019) The brain in solitude: an (other) eighth amendment challenge to solitary confinement. *Journal of Law and the Biosciences* 6(1): 184–225.

Cox C and Marland H (2018) "He must die or go mad in this place": Prisoners, insanity, and the Pentonville Model Prison Experiment, 1842–1852. *Bulletin of the History of Medicine* 92(1): 78–109.

Duus-Otterström G. (2019) Weighing relative and absolute proportionality in punishment. In Tonry M (ed.) *Of One-Eyed and Toothless Miscreants: Making the Punishment Fit the Crime?*Oxford, Oxford University Press, pp.30–50.

Gaylin W, Kass LR, Pellegrino ED, and Siegler M, (1988) 'Doctors must not kill'. *Journal of the American Medical Association* 259(14): 2139–2140.

Grush L (2019) The first alleged crime committed in space raises questions about jurisdiction in orbit. From: https://www.theverge.com/2019/8/27/20833761/nasa-iss-international-space-station-alleged-crime-anne-mcclain-jurisdicition-framework.

Hobhouse S and Brockway AF (eds) (1922) English prisons to-day: Being the Report of the Prison System Enquiry Committee. From: https://archive.org/details/englishprisonsto00prisuoft/page/n5/mode/2up.

Johnstone G (2003) *A Restorative Justice Reader: Text, Sources, Context.* Cullompton: Willan.

Leverentz A, Chen EY and Christian J (eds) (2020) *Beyond Recidivism: New Approaches to Research on Prisoner Reentry and Reintegration.* New York: NYU Press.

Mauer M (1999) Why are tough on crime policies so popular? *Stanford Law and Policy Review* 11(1): 9–22.

Mittelstädt JM, Pecena Y, Oubaid V and Masschke P (2016) Psychometric personality differences between candidates in astronaut selection. *Aerospace Medicine and Human Performance* 87(11): 933–939.

Morris A and Maxwell G (eds) (2001) *Restorative Justice for Juveniles: Conferencing, Mediation and Circles.* Hart Publishing.

Musson DM, Sandal GM and Helmreich RL (2004) Personality characteristics and trait clusters in final stage astronaut selection. *Aviation, Space and Environmental Medicine* 75(4): 342–349.

Musson DM, Helmreich RL and Sandal GM (2004) *Baseline personality comparisons between astronauts and Antarctic personnel: Implications for generalization of psychological research findings.* 55th International Astronautical Congress. From: https://www.academia.edu/19267076/Baseline_personality_comparisons_between_astronauts_and_Antarctic_personnel_Implications_for_generalization_of_psychological_research_findings.

Nettle D (2006) Psychological profiles of professional actors. *Personality and Individual Differences* 40(2): 375–383.

Newburn T (2007) "Tough on crime": Penal policy in England and Wales. *Crime and Justice: A Review of Research* 36: 425–470.

NSF (2022) *Sexual Assault/Harassment Prevention and Response (SAHPR) Final Report.* National Science Foundation Office of Polar Programs, United States Antarctic Program From: https://www.nsf.gov/geo/opp/documents/USAP%20SAHPR%20Report.pdf.

Palinkas LA and Suedfeld P (2021) Psychosocial issues in isolated and confined extreme environments. *Neuroscience & Biobehavioral Reviews* 126: 413–429.

Perry N (2023) Women working in Antarctica say they were left to fend for themselves against sexual harassers. Associated Press. From: https://apnews.com/article/women-working-antarctica-sexual-harassment-assault-mcmurdo-ba0e550fddf1ab0afd031ff4d25143cb.

Scharff Smith P (2006) The effects of solitary confinement on prison inmates: A brief history and review of the literature. *Crime and Justice* 34(1): 441–528.

Shivaram D (2021) The 1st all-civilian crew has launched to orbit the Earth. NPR. From: https://www.npr.org/2021/09/15/1037015900/the-1st-all-civilian-crew-is-about-to-orbit-the-earth.

Smee B (2023) Five hundred days in solitary: Queensland teenager's case 'a major failure of our system'. *The Guardian*. From: https://www.theguardian.com/australia-news/2023/jun/26/500-days-of-solitary-aboriginal-teen-spent-extraordinary-period-in-isolation.

Staff H and Wenzelburger G (2021) The partisan politics of the penal–welfare nexus: A quantitative analysis of party influence on the relationship between penal and welfare policies. *Journal of Public Policy* 41(3):573–599.

Umbreit MS (1995) Holding juvenile offenders accountable: A restorative justice perspective. *Juvenile and Family Court Journal* 46(2): 3–42.

Whitman Cobb W (2023) Meet the next four people headed to the Moon – how the diverse crew of Artemis II shows NASA's plan for the future of space exploration. *The Conversation*. From: https://theconversation.com/meet-the-next-four-people-headed-to-the-moon-how-the-diverse-crew-of-artemis-ii-shows-nasas-plan-for-the-future-of-space-exploration-203214.

Willcock S (2001) The underperfoming GP. *Medicine Today*, 2(2): 128–133.

Zehr H (1985) Retributive Justice, Restorative Justice. New Perspectives on Crime and Justice. Occasional Papers of the MCC Canada Victim Offender Ministries Program and the MCC U.S. Office of Criminal Justice. Indiana: MCC U.S. Office of Criminal Justice.

19 The Problem with the Placement of Space Infrastructures

The Complex Politics of Paving the Way to Space

Karlijn Korpershoek

Introduction

The European Space Agency (ESA) has declared a "launcher crisis". Due to the termination of the Ariane 5 rocket and the delay of its successor, Ariane 6, the ESA has seen a drought in launches not experienced since the start of the century. While not being able to launch from Europe's spaceport, ESA and its member states are reliant on other launch providers. As a consequence, ESA is predicted to launch a satellite for its Galileo program for the first time with an outside partner: SpaceX. This decision has been a contentious one, as it will be the first time a satellite carrying classified equipment will be launched from anywhere else but Europe's spaceport (Reuters 2023). As much as collaboration is deemed to be a key characteristic of the space industry, the reality is that politics are always in play and having sovereign access to outer space is deemed preferable.

These politics are changing, with the commercialisation of the American market; the explosive growth of Indian and Chinese space sectors, and the war between Ukraine and Russia, there is a return to a discourse that re-emphasises national sovereignty in outer space (Schreiber 2022). This goes hand-in-hand with an increasingly crowded outer space landscape, with more commercial and national actors joining the outer space market which is ultimately projected to reach one trillion US dollars in the next two decades (Citi 2022; McKinsey 2023).

As the sphere of influence of outer space is growing, so is the engagement of social scientists, legal scholars, and humanities with the domain that historically has largely been dominated by natural scientists and engineers. Much of the literature focuses on what is happening "out there", where outer space is constructed as a place separate from Earth space (Beery 2016). This chapter aims to bring the discussion back to Earth, by taking the political and military implications of launch sites, and their entanglements with geopolitical power dynamics which create persisting inequalities.

All space missions begin on Earth, and this article addresses these physical locations where all rockets set off from: launch sites. It is through an exploration of Earth-based space infrastructures, with a particular focus on Europe's spaceport in French Guiana, that the intricate relationship between

DOI: 10.4324/9781003437178-20

politics and outer space comes to the fore. Through this lens, conversations about justice, ethics, and outer space can be informed through places on Earth where the industry already has had a specific impact.

In this chapter, I propose that looking at the politics of large space infrastructures provides a practical inroad to discussions on the ethics of outer space exploration that often remain largely hypothetical and elusive. I add a practical and ethnographic angle to the debate on justice and politics in and of outer space. The first section introduces the underlying theme of accessibility to outer space, and how an interrogation in existing infrastructures can dismantle the myth that the space enterprise is "a profoundly human aspiration that unites all people in all places" (Gorman 2007: 154). The second part focuses on Centre Spatial Guyanais (CSG), the spaceport of the European and French Space Agencies, and its historical ties to lingering colonial relationships. Through that deep dive, other places where similar dynamics are playing out will be addressed. Additionally, the military entanglement in launch sites will be discussed. The chapter ends by exploring whether and how power dynamics on Earth can be translated into a discussion about justice and ethics in outer space.

Who Has Access to Outer Space?

Outer space exploration in its current socio-technological configuration originated with the Space Race between the Soviet Union and USA as the Cold War raged on between these political power players. Many of the initial actions in outer space were about displaying technological and military superiority over the other (Rementeria 2022). It is commonly understood that the Apollo mission, putting American astronauts on the Moon, was the unofficial ending of the battle for "firsts". To avoid the increased militarisation of outer space, and more countries joining to claim outer space in one way or another, the United Nations (UN) drafted the Outer Space Treaty (OST) (Pachankis 2022). This 1967 document outlines the basic principles for the use and exploration of outer space. There are 114 countries currently party to the OST including the major spacefaring nations (United Nations Office for Outer Space Affairs 2023). The first article of the OST stipulates that "the exploration and use of outer space, including the Moon and other celestial bodies, shall be carried out for the benefit and in the interests of all countries" (United Nations Office for Outer Space Affairs (UNOOSA) 1967: 13). The exact meaning of this has been under contention, as it is not clearly defined what "benefits" exactly entail and who oversees defining that (Wiser and Aganaba 2023). However, the OST has fed into the narrative that space exploration generally is for the betterment of humanity and has resulted in it being framed as an "a-political" environment (MacDonald 2007) where collaboration takes precedence over political or military actions.

This discourse that "space is for the benefit of all" has already been extensively criticised by scholars as well as the public. The persistent colonial undertone of this narrative has been problematised for several of its underlying assumptions: little attention is given to counter-narratives and those who might believe that outer space should not be opened for socio-technological exploration at all (Bawaka Country 2020); that the timeline and mode in which developments take place should drastically be lowered (Smiles 2020); or those who believe that outer space should only be opened up for certain kinds of activities (Impey 2021).

Even though ethicists and social scientists are claiming a seat at the table, as for example NASA's ethics workshop shows (Pirtle et al. 2023), these efforts are still very early on. The question is whether they come in time, as developments in space sometimes go at speeds that are hard to keep up with for legislators and social scientists. For example, both China and the US have laid out detailed plans for missions to the Moon (Hickman 2019), and despite some initial legislations being created, there is far from a single agreed framework outlining the rules and legislations for a sustained human presence on the Moon (Boley and Byers 2020).

Yet, there are scholars turning their attention to these issues and paying notice to the political, financial, and ideological dynamics from which space exploration in its current configuration is born. Valentine, Olson, and Battaglia (2009) advocated for anthropologists to include outer space in the discipline's domain more than a decade ago, and many scholars have taken up this call. An important area of research is understanding the national ideologies that underlie the actions of different players within the space industry. For example, Olson (2018) explores how NASA constructs the solar system as a new domain for the exertion of technical and political control. Aima (2018) shows how the government of the United Arab Emirates uses outer space to draft a very particular perspective of the future, creating a new projection of nationalism in the present. Valentine (2012) focuses on the commercial aspect and illustrates that profit almost always supersedes the imaginations of a long-term future for stakeholders in commercial space companies.

What these studies show is how outer space exploration and Earth life have been feeding into each other from the onset and are intrinsically linked. Therefore, looking into that relationship should be at the cornerstone of talking about justice and ethics in outer space. For this, we go to the literal place of departure: launch sites. All space missions take off from Earth from launching pads, and as access to outer space is negotiated in the geographical locations where these are situated, they are where the entanglements of Earthly politics and outer space become immediately visible.

Launch sites often lie in the periphery of places, making them closely related to forms of "waste-lands" (Gorman 2007), and exhibit many similarities to other large-scale invasive infrastructures requiring particular geographical features. Often, there is even a clear overlap between the space sector and industries or projects to make use of land that has been constructed as "remote" and "distant" and thus fit for invasive projects. This is apparent in multiple launch and test site locations: Kiruna in Northern Sweden also has an ore mine; Woomera, Australia, was simultaneously a rocket test and military base; and Kourou, French Guiana used to be a penal colony before being turned into a launch state. These configurations of launch sites can inform us about the tension that exists between the narrative of a "space to benefit all", and the reality of the resources that are needed to get socio-technological access to outer space (Milligan 2023) and the impact these have on the people inhabiting the spaces close to those (Redfield 1996).

By turning to Earth-based infrastructures, the local entanglements into what is constructed as a "planetary" or "global" project become visible. Beery (2016) shows the various ways in which scale is politicised and used to create narratives of the future where the ones proposed by space agencies supersede those of other (local) actors. This chapter will now continue to explore one of those places, namely the launch site of the European and French Space Agency, which is in Kourou, French Guiana.

Europe's Spaceport…in South America

French Guiana has been an overseas department of France since 1946, when it was turned from a colonial possession into an integrated part of the European country. Nestled in between Suriname and Brazil on the Northern side of the South American continent, the territory is 7,000 kilometers and an ocean away from its national government in Paris. It is one of nine overseas regions of the European Union (European Commission 2022). This status means that the Euro is used, and that almost all French legislation, education, and social security are used. The nine locations for which this is the case, all aside from French Guiana islands, are called the "ultra-periphery" of Europe (European Commission 2011) and are often left out of dominant discourses describing or portraying the political entity (Boatcă 2020).

The geographical location and possibilities have given French Guiana a very specific place in France's colonial history. As the territory is mainly covered by dense Amazonian rainforest and does not have the right conditions for many of the activities that made France's Caribbean possessions successful, like sugar and coffee (Wood and Macleod 2018), the French government struggled with this part of their empire. After unsuccessful attempts to settle in the 1760s (Traver 2011), with many expeditioners dying of tropical conditions and illnesses, the French government decided to exploit the characteristics and the difficulty of adaptation for metropolitan France. In the 1850s, French Guiana was turned into a penal colony, as the territory, deemed to be a "green hell" (Renneville 2007), provided sufficient distance from the metropole while still relatively easily accessible by boat transportation.

For a century, prisoners of the French state were sent to the South American territory to sit out their sentence, which was prolonged by a rule called "doublage", which enforced an additional stay in French Guiana of at least half the prison sentence (Sanchez 2018). While imprisoned, many were forced to do hard physical labour and help set up infrastructures designed by the

French state. There were several camps, of which one of the most notorious was on the Ile de Salut, just off the coast of Kourou. Here, people were sent for social isolation. After several journalists had reported on the abysmal conditions at the penal colony, the French government decided to abolish the prisons and they were officially shut in the early 1950s.

The closure of the penal colony coincided with the start of the space race. The United States and the Soviet Union were the main actors in the socio-technological exploration of outer space, but while they were fighting out for the initial key milestones, other countries tried to make their own way to space. France was one of the key players from the European continent. In 1961, Charles de Gaulle set up the French Space Agency, making it the third oldest after NASA (US) and ROSCOSMOS (Russia, formerly Soviet Union). They were using a test site in Hammaguir, in the Algerian desert, to make their first steps into outer space, which would ultimately result in the launch of the first French satellite, Astérix, in 1965 (Dumoulin Kervran et al. 2021), making it the third country to successfully launch a satellite with their own vehicle.

However, as the space race went on, other significant geopolitical power shifts were happening simultaneously. Most important for this chapter, are the wars for independence of the European colonial empires. In Algeria, the war for independence resulted in the Evian Accords and as part of these agreements, France was told to remove their launching site from Algerian territory (Howard 2023). They were granted a transition period, in which several launches took place, but nevertheless CNES (Centre National d'Études Spatiales), the French Space Agency, had to look elsewhere for their space future.

So, in 1964, CNES created a shortlist of 14 places that could be suitable and feasible launch sites; among the majority were (former) French colonies or allies. These were ranked based on a criteria list that included logistical, scientific, and technological but also social and political criteria. Each location was given a score from 1 to 3, and a particular weight per category was added. Certain countries were immediately dismissed, like Somalia (for a lack of political stability), and the Seychelles and Desirade (for logistical reasons). In the end, French Guiana received the highest score out of all 14 (Chambaz et al. 2021; CSG Archives 1964). It scored very high on location, with a latitude of approximately 5 degrees North, which would mean that rockets launching from French Guiana can make use of the "slingshot" effect (ESA 2021) which reduces the fuel needed to reach Earth's orbit. Additionally, the weather conditions are very favourable; it is not on a fault line of tectonic plates nor in a tornado-ridden region, so usually, the conditions are stable enough to allow for launches to take place. Its positional characteristics were also scored highly: with the North and East facing the sea, a launch site in Kourou can allow launches in both directions.

But it was not just these technological and scientific reasons that made French Guiana stand out. The territory had a clear advantage over the runners-up, Brazil and Australia, namely that it is French territory which would make the development and management of the spaceport politically less complicated. The key "issues" identified were a lack of labour force as deemed necessary by CNES, and a potential issue with the local community not seeing the benefit of this launch site. The colonial mindset from metropolitan France regarding French Guiana becomes clear in how the latter would be solved: they would "understand what would be best for them" once CNES explained it to the local communities (CSG Archives 1964).

The sudden arrival of the French Space Agency and their requirement for large batches of land, meant that the communities already living in the region around the proposed launch site had to be relocated. This region of French Guiana was, and still is, inhabited by Creole and Amerindian communities, mainly living in large, wooden, and open housing. At the time, the main method of providing sustenance was a form of agriculture called "abattis", or slash and burn, where grounds are used to grow crops and then are burned to fertilise (Demaze and Manusset 2008; Tsayem Demaze 2008).

When the space industry was installed, they required the lands inhabited by these communities, so CNES forced the families to relocate into smaller, concrete apartments. As compensation, they were provided with plots of land, but these were on the other side of the Kourou River (Gorman 2007). This made them hard to access without a car, or a boat, and completely changed the communities' relationship to their agricultural practices which had previously taken place in their backyard. This history of expropriation related to the arrival of the launch site is a painful conversation up to this day. It is only recently that a book has been published on the experiences of one family (Chocho-Dufail 2023), and how the situation caused intergenerational trauma for the entire community. The author of the book tells me that it is still a painful topic, one often unknown to even the children of those expropriated.

The story of the expropriation shows that when the space agency turned up, it was not just with a particular vision of the future, but also with ideas on how to live in the present which were imposed on those already present (Izoard 2018). CNES took it upon themselves to create the urban infrastructure of Kourou, the town that would support the launch site and house its labour force. The hierarchies of the space center therefore echoed in the setup of the town, with the largest villas close to the seafront being built for the higher-ups in the space center. So, social stratification has been embedded in the setup of Kourou from the outset (Redfield 2000).

It was not just the housing that was designed by CNES, but also the social services in the form of the hospital, school, sports and entertainment facilities. These were all built with a specific target audience in mind, namely the workers coming over from metropolitan France (CSG Archives 1965). There are other examples where the arrival of a particular industry directly led to the creation of the urban infrastructure meant to support it. Karafantis and Leslie (2019: 3) demonstrate how the aerospace industry brought rise to "blue-collar and blue-sky communities of Southern California" in the 1960s. And then there is the famous example of Henry Ford's "Fordlândia", a town for rubber production in Northen Brazil where the social life and infrastructure were based on his corporate values (Barkemeyer and Figge 2012). The speed and ease with which these decisions were made, emphasise the power dynamics between French Guiana and metropolitan France. As the former is a formal part of the latter, no intervention from a foreign government was required to make such drastic changes to the inhabitants of the region.

Dual Use/Dual Presence

An additional point to consider is not just how past relationships shape the basis of launch infrastructure, but also the military and political power that is still exerted from and around space. Kourou does not only house the Centre Spatial Guyanais, but it is also home to a large base of the Foreign Legion of the French forces. Explicitly present due to its placement in the center of the town, the entanglements of outer space and the military in Kourou are evident. The section of the army arrived in the 1970s, a few years after the first launch from the base, and in addition to playing an important role in border protection and actions against illegal gold mining in the territory, the protection of the launch site is one of their key missions.

The military is part of a vital arrangement that ensures the safety of and around launches at the Centre Spatial Guyanais (Bene 2023; Cooper 2006). Operation TITAN, consisting of air, land, and naval divisions of the FAG (Force Armée de Guyane), oversees keeping the rocket site and risk area protected during the launch procedure. For each launch, there are about 250 soldiers mobilised (FAG 2023). Also, the transportation of rocket and satellite parts from the port of Cayenne to the launch site is overseen by the army. Additionally, the permanently stationed firefighters at CSG are part of the Brigade des Sapeurs-Pompiers de Paris; the only division in France which is military-trained (Boutinaud 2016).

Aside from military presence in Kourou, there is another military angle that has gained attention in the past few years. With the invasion of Ukraine by Russia, an end was made to the launching of the Soyuz rocket from French Guiana. This collaboration between ROSCOSMOS and CSG was signed in 2005 to boost diplomatic connections between the European Union and Russia in outer space (Ingold 2006). Additionally, it would allow the Russians to launch from a location closer to the equator than Baikonur, their launch site in a special zone in Kazakhstan. This was a successful collaboration with the first Soyuz rocket being launched in 2011; within a span of a decade 27 launches were made, of which 26 were fully successful. However, quickly after the start of the conflict, Russia decided to withdraw its presence from French Guiana, which consequently led to the cancellation of remaining flights (Vidal and Privalov 2023). Launch sites and their usage are intrinsically linked with the geopolitical developments and connections of its national government. As such, French Guiana and its launch complex got pulled into a conflict between Europe and Russia. In this case, the withdrawal was instigated by the Russians themselves. However, it raises important questions about what would have happened if it had been a forceful termination initiated by the French government.

The intersection between outer space and the military is heavily researched (Ayadi 2019; Byers 2020; Hussain and Ahmed 2019) but mainly focuses on the weaponisation of outer space (Mueller 2003; Pražák 2021) and the role of satellite-based technologies in surveillance (Steer 2017) and navigation (Ceruzzi 2021). The narrative that outer space is key in national security exists alongside the discourse that it is for "peaceful purposes". However, there is little academic attention to how this military power is exercised on the ground. Both in outer space and on Earth, outer space and the military are incredibly closely connected, with satellites surveilling the Earth as well as personnel patrolling the physical areas of launch sites.

Earthly Entanglements on Land Rights and Livelihoods

The dynamics displayed in Kourou are in no way exceptional to Kourou. Gorman (2005) and Klinger (2019; 2021) have explored how space infrastructures are often built in places that have historically been framed as a "terra nullius". A case that has repeatedly been covered in American as well as European media is that of the Thirty Meter Telescope to be built on Mauna Kea, Hawaii. The dispute around the usage of the mountain, which holds important spiritual meaning for local communities, has been ongoing for several years and has not been conclusively resolved (Maile 2009; Sammler and Lynch 2021). The SpaceX launch site in Boca Chica, Texas, has forced many people to sell their homes or be troubled by the sound of the regular rocket tests (Keates and Maremont 2021). In surrounding villages close to Baikonur, space debris left by rocket launches disturbs lives in large areas depicted as "fall zones" (Kopack 2019).

None of these narratives are simple; there will be local communities and individuals happy with the arrival of large-scale infrastructures in their region, as they are also a source of income and can be a source of pride (Szolucha 2023). The space industry does not hold a monopoly on complicated power dynamics where large-scale infrastructures impede on the livelihoods or land rights of others. Particularly in the US and Canada, the placement of pipelines has long caused issues for Indigenous communities (Grote and Johnson 2021; Hunt and Gruszczynski 2021). In the Amazon, including French Guiana, tensions exist between large-scale agriculture or energy projects and Indigenous methods of tending to land (Conceição et al. 2021; Grote and Johnson 2021).

These examples where the outer space industry is involved in disputes around land use on Earth are important as "land" rights in outer space are becoming a realistic prospect. Talking about the discussion of ethics of space infrastructures on Earth can help debunk some of the myths propagated by certain space actors that space exploration is "for the benefit of all mankind" (Smiles 2020).

The discussion on land rights on Earth and how these have created and perpetuated injustices can inform the conversations on how to treat exploration of the Moon and other celestial bodies. The recent controversy around the Peregrine Mission One launch sparked broad media attention in the US to the topic of the usage of outer space, and the question of who gets to go, and what for (Weibel 2024)? On board this mission, among others, were the payloads of two commercial companies, Elysium and Celestis. These two enterprises sell the option to send human ashes or DNA traces to outer space. One of Celestis' payloads, named Tranquility Flight, was set to go to the Moon with the materials of more than 65 people (and a dog) on board (Bartels 2024). There was active opposition by Indigenous groups in the United States, in particular by the Navajo Nation, to the latter as the Moon is seen as a sacred site. Harvey (2024) explains that: "the Moon is an ancient relative — Grandmother Moon is a term of reverence shared by many Indigenous Peoples — and we should be careful, diligent and respectful when visiting her" and that, thus, the US government and commercial companies should seek wider consultation on the usage of the "shared cultural space" that is the Moon. Even though the mission ultimately failed and did not land on the Moon, questions are raised about who gets to make decisions about outer space, and how those decisions are regulated. A concern, highlighted by this incident, over the legal framework of outer space, is that there is a chance that actors with current spacefaring capabilities will be able to act in ways that might impede the usage of other nations or peoples in the present, as well as the future. Since the creation of the Outer Space Treaty, not only more national actors have entered the field, but the industry has also seen exponential growth in commercial business involvement.

Conclusion

All in all, launch sites are integral in discussions about justice and ethics of the outer space industry. Conversations about anticipated space actions like asteroid and Moon mining often overshadow that there are places on Earth already implicated in debates on land rights and outer space. Rather, these studies should be considered, not as a legacy of the past, but one which persists to this day.

Rementeria (2022: 9) argues that the debate awaits on "how rights, environmental sustainability, space settlements, appropriation and free access, liability claims, insurance duties, and so many other concepts should be interpreted in view of the contemporary and expected uses of outer space". This chapter gives several starting points for this conversation.

By bringing outer space into the realm of Earth space, discussions can be opened up to a wider participatory framework where actors without access to outer space can be included in discussions (Yap and Truffer 2022). The politics of launch sites undo the myth that space exploration is inherently done "for the benefit of all" and rather points to the need to investigate who defines those terms in the first place. However, space exploration does also allow for a dismantling of power dynamics that perpetuate Earthly imbalances if the promise of the Outer Space Treaty is drastically upheld. Launch site and other space infrastructures should be included on any debate on justice and ethics of space exploration, as without them, space activities as we know them now would be impossible to undertake.

The chapter informs debates around the politics and ethics of outer space generally and launch sites specifically. By addressing the relationship between Earthly power dynamics and their potential translation into outer space, the connection between land rights and potential legislation for asteroid mining comes to the fore. I draw on anthropological and historical insights that can help inform debates around legislative models without proposing normative or definitive options for the governance of outer space. Scholars from different disciplinary backgrounds have demonstrated that a broader approach to outer space governance and ethics is needed to address the complex power dynamics that feed and sustain space exploration (Yap et al. 2023).

References

Aima R (2018) Life on MARS: Dubai projects a new vision of nationalism. *World Policy Journal* 35(1): 10–15.

Ayadi J (2019) Military implications of the use of outer space: a European perspective. From: https://iihl.org/wp-content/uploads/2019/10/Ayadi.pdf.

Barkemeyer R and Figge F (2012) Fordlândia: corporate citizenship or corporate colonialism. *Corporate Social Responsibility and Environmental Management* 19(2): 69–78.

Bartels M (2024) Human remains are headed to the moon despite objections. *Scientific American*. From: https://www.scientificamerican.com/article/human-remains-are-headed-to-the-moon-despite-objections/.

Bawaka Country (2020) Dukarr lakarama: listening to Guwak, talking back to space colonization. *Political Geography* 81:102218.

Beery J (2016) Unearthing global natures: outer space and scalar politics. *Political Geography* 55: 92–101. doi:10.1016/j.polgeo.2016.04.003.

Bene K (2023) The role of the French foreign legion in Latin America in the defence of the European space program. *Academic and Applied Research in Military and Public Management Science* 22(2): 55–66.

Boatcă M (2020) Thinking Europe otherwise: lessons from the Caribbean. *Current Sociology* 69(3): 389–414.

Boley A and Byers M (2020) U.S. Policy puts the safe development of space at risk. *Science* 370(6513): 174–175.

Boutinaud P (2016) La BSPP dans la défense du territoire national. *Revue Défense Nationale* 786(1): 34–40.

Byers M (2020) Arctic security and outer space. *Scandinavian Journal of Military Studies* 3(1): 183–196.

Ceruzzi P (2021) Satellite navigation and the military-civilian dilemma: the geopolitics of GPS and its rivals. In Geppert ACT, Brandau D and Siebeneichner T (eds) *Militarizing Outer Space*. London: Palgrave Macmillan, pp. 343–367.

Chambaz B, Dezoteux B, Redon D and Azoulay G (2021) *Traversée culturelle dans les archives de l'espace*. Vol. 2: Le centre spatial guyanais. Paris: CNES.

Chocho-Dufail J (2023) *Mémoires des expropriés de Kourou et Sinnamary*. Cayenne: Symbioz/CTG.

Citi (2022) Space: The dawn of a new age. From: https://icg.citi.com/icghome/what-we-think/citigps/insights/space_20220509.

Conceição KV, Chaves MED, Picoli MCA, Sánchez, AH, Soares, AR, Mataveli GAV, Silva DE, Costa JS and Camara G (2021) Government policies endanger the Indigenous peoples of the Brazilian Amazon. *Land Use Policy* 108:1–15.

Cooper NJ (2006) The French foreign legion: forging transnational identities and meanings. *French Cultural Studies* 17(3): 269–284.

CSG Archives (1964) (25W614) Recherche de sites de lancement. From: Archives at Centre Spatial Guyanais.

CSG Archives (1965) (25W602) Note sur les problemes urbanisme posés par l'implementation du CSG au Kourou. From: Archives at Centre Spatial Guyanais.

Demaze MT and Manusset S (2008) L'agriculture itinérante sur brûlis en Guyane française: la fin des durabilités écologique et socioculturelle? *Cahiers d'Outre-Mer* 61(241–242): 31–48.

Dumoulin Kervran D, Lamy J and Verlin J (2021) Outposts of science: placing science bubbles at the margins of (post)colonial French territories. *HAL Open Science*. From: https://halshs.archives-ouvertes.fr/halshs-03389148.

ESA (2021) Earth from Space: Kourou, French Guiana. From https://www.esa.int/Applications/Observing_the_Earth/Copernicus/Earth_from_Space_Kourou_French_Guiana.

European Commission (2011) Eight ultra-peripheral regions: eight jewels in the EU's Crown. From: https://ec.europa.eu/regional_policy/en/newsroom/news/2011/11/eight-ultra-peripheral-regions-eight-jewels-in-the-eus-crown.

European Commission (2022) EU policy for the EU Outermost Regions. From: https://ec.europa.eu/regional_policy/en/policy/themes/outermost-regions/.

FAG (2023) TITAN, dispositif de protection du dernier lancement d'Ariane 5. From: https://www.defense.gouv.fr/operations/actualites/fag-titan-dispositif-protection-du-dernier-lancement-dariane-5-0.

Gorman A (2005) The cultural landscape of interplanetary space. *Journal of Social Archaeology* 5(1): 85–107.

Gorman A (2007) La terre et l'espace: rockets, prisons, protests and heritage in Australia and French Guiana. *Archaeologies* 3(2): 153–168.

Grote KM and Johnson JT (2021) Pipelines, protectors, and settler colonialism: media representations of the Dakota Access Pipeline protest. *Settler Colonial Studies* 11(4): 487–511.

Harvey A (2024) Stop sending human remains to the Moon. *Nature*. From: https://www.nature.com/articles/d41586-024-00106-8.

Hickman J (2019) Research viewpoint: international relations and the second space race between the United States and China. *Astropolitics* 17(3): 178–190.

Howard C (2023) In the air: technology, colonialism, and the space race in postwar surrealist journals. *Dada/Surrealism* 24(1): 1–29.

Hunt K and Gruszczynski M (2021) The influence of new and traditional media coverage on public attention to social movements: the case of the Dakota Access Pipeline protests. *Information Communication and Society* 24(7): 1024–1040.

Hussain MZ and Ahmed RQ (2019) Space programs of India and Pakistan: military and strategic installations in outer space and precarious regional strategic stability. *Space Policy* 47: 63–75.

Impey C (2021) Unbound: ethics, law, sustainability, and the new space race. *Studia Humana* 10(4): 1–17.

Ingold O (2006) Soyuz in French Guiana: a strategic perspective. *Space Policy* 22(2): 140–148.

Izoard C (2018) Ma famille vivait là où se trouve maintenant la base Soyouz. *Z: Revue itinérante d'enquête et de critique sociale* 12(1): 186–187. doi:10.3917/rz.012.0186.

Karafantis L and Leslie SW (2019) "Suburban warriors": the blue-collar and blue-sky communities of Southern California's aerospace industry. *Journal of Planning History* 18(1): 3–78.

Keates N and Maremont M (2021) Elon Musk's SpaceX is buying up a Texas village. Homeowners cry foul. From: https://www.wsj.com/articles/elon-musk-spacex-rocket-boca-chica-texas-starbase-11620353687.

Klinger JM (2019) Environmental geopolitics and outer space. *Geopolitics* 26(3): 666–703.

Klinger JM (2021) Critical geopolitics of outer space. *Geopolitics* 26(3): 661–665.

Kopack RA (2019) Rocket wastelands in Kazakhstan: scientific authoritarianism and the Baikonur cosmodrome. *Annals of the American Association of Geographers* 109 (2): 556–567.

MacDonald F (2007) Anti-Astropolitik — outer space and the orbit of geography. *Progress in Human Geography* 31(5): 592–615.

Maile DU (2009) On the violence of the Thirty Meter Telescope and the Dakota Access Pipeline. *Cultural Critique* 73: 88–124.

McKinsey (2023) A giant leap for the space industry. From: https://www.mckinsey.com/featured-insights/sustainable-inclusive-growth/chart-of-the-day/a-giant-leap-for-the-space-industry.

Milligan T (2023) From the sky to the ground: Indigenous Peoples in an age of space expansion. *Space Policy* 63: 1–10.

Mueller K (2003) Totem and taboo: depolarizing the space weaponization debate. *Astropolitics* 1(1): 4–28.

Olson V (2018) *Into the Extreme: U.S. Environmental Systems and Politics Beyond Earth*. Minneapolis: University of Minnesota Press.

Pachankis Y (2022) Reading the Cold War through outer space: the past and future of outer space. *International Journal of Scientific & Engineering Research* 13(6): 826–829.

Pirtle Z, McBrayer K and Beauchemin A (2023) Artemis, ethics and society: synthesis from a Workshop. From: https://ntrs.nasa.gov/citations/20230012799.

Pražák J (2021) Dual-use conundrum: towards the weaponization of outer space? *Acta Astronautica* 187:397–405.

Redfield P (1996) Beneath a modern sky: space technology and its place on the ground. *Science, Technology & Human Values* 21(3): 251–274.

Redfield (2000) *Space in the Tropics: From Convicts to Rockets in French Guiana*. Berkeley: University of California Press.

Rementeria S (2022) Power dynamics in the age of space commercialisation. *Space Policy* 60: 1–13.

Renneville M (2007) Les bagnes coloniaux: de l'utopie au risque du non-lieu. *Criminocorpus, revue hypermedia*. From: https://journals.openedition.org/criminocorpus/173.ff10.4000/criminocorpus.173.

Reuters (2023) EU fine-tunes plan to launch Galileo satellites on SpaceX. From. https://www.reuters.com/technology/space/eu-fine-tunes-plans-launch-four-galileo-satellites-spacex-2023-11-07/.

Sammler KG and Lynch CR (2021) Apparatuses of observation and occupation: Settler colonialism and space science in Hawai'i. *Environment and Planning D: Society and Space* 39(5): 945–965.

Sanchez JL (2018) Le bagne commence à la libération. *HAL Open Science* 3: 58–63.

Schreiber NH (2022) Man, state, and war in space: neorealism and Russia's counterbalancing strategy against the United States in outer space security politics. *Astropolitics* 20(2–3): 151–174.

Smiles D (2020) The settler logics of (outer) space. From: https://www.societyandspace.org/articles/the-settler-logics-of-outer-space.

Steer C (2017) Global commons, cosmic commons: implications of military and security uses of outer space. *Georgetown Journal of International Affairs* 18(1): 9–16.

Szolucha A (2023) SpaceX's biggest rocket flies for the first time: but do we understand what this actually Means? From: https://thegeekanthropologist.com/2023/04/20/spacexs-biggest-rocket-flies-for-the-first-time-but-do-we-understand-what-this-actually-means/.

Traver BJ (2011) *After Kourou: settlement schemes in French Guiana in the Age of Enlightenment*. Doctoral thesis, Pullman: Washington State University.

Tsayem Demaze M (2008) Croissance démographique, pression foncière et insertion territoriale par les abattis en Guyane française. *Norois* (206): 111–127.

United Nations Office for Outer Space Affairs (2023) *Outer Space Treaty*. From: https://www.unoosa.org/res/oosadoc/data/documents/2023/aac_105c_22023crp/aac_105c_22023crp_3_0_html/AC105_C2_2023_CRP03E.pdf (accessed: 10 July, 2024).

Valentine D (2012) Exit strategy: profit, cosmology, and the future of humans in space. *Anthropology Quarterly* 85(4): 1045–1067.

Valentine D, Olson V and Battaglia D (2009) Encountering the future: anthropology and OUTER Space. *Anthropology News* 50(9): 11–15.

Vidal F and Privalov R (2023) Russia in Outer Space: a shrinking space power in the era of global change. *Space Policy*: 1–9.

Weibel D (2024) The sacred Moon: navigating diverse cultural beliefs in lunar missions. *The Space Review*. From: https://www.thespacereview.com/article/4733/1.

Wiser L and Aganaba T (2023) An evolving space governance system: balancing interests in five policy debates. *Acta Astronautica* 203: 537–543.

Wood S and Macleod C (2018) *Locating Guyane*. Liverpool: Liverpool University Press.

Yap XS and Truffer B (2022) Contouring 'earth-space sustainability'. *Environmental Innovation and Societal Transitions* 44: 185–193.

Yap XS, Janssen MA, Aganaba T, Tutton R, Korpershoek K, Profitiliotis G, Rabitz F, Subbiah MS and Wagenknecht L (2023) Four alternative scenarios of commons in space: prospects and challenges. *International Journal of the Commons* 17(1): 390–410.

20 The Borderlands between Law and Ethics

Jurisdiction, Enforcement, and Teleology

Brian Patrick Green

Introduction

Location-based, border-delineated, jurisdictional legal systems work well enough on Earth, and are the international standard. But in space, these systems immediately fail for two primary reasons: it is difficult to delineate borders in space and it is difficult to enforce laws when power trails off from weak (near Earth) to non-existent (anywhere sufficiently far).

While these legal questions of jurisdiction and enforcement are difficult enough, ultimately ethics itself becomes a puzzle as we are faced with questions of the ultimate meaning and purpose of human life in the universe. What are law and ethics before the question of the meaning of human existence itself? This chapter will briefly investigate these three phenomena – jurisdiction, enforcement, and the role of ethics before the universe itself – in an effort to determine what is new about space, how ethics and law might best integrate in this new domain, and what this means for human action and existence in space (Green 2023A).

Borders and Jurisdictions

"Space", "Ethics", and "Law" are vast and essentially contested categories (Gallie 1955). Not only is the terrain conceptually difficult, but space itself has few or no natural borders to draw jurisdictions upon. Planetary and other solid surfaces are the rare exception – humans have those figured out – but otherwise, the void is a difficult place to draw a line. Particular orbits might be reserved for the safe passage of objects, but it probably does not make sense to stake a property claim or "own" them like terrestrial territory, much less demand sovereignty over them.

Planets and moons without solid surfaces (ocean-planets or gas-giants) present the same problem as space itself. While perhaps someday some entity might claim a complete object in space, perhaps an asteroid, there seems to be something wrong with the idea of owning a planet, whether solid, liquid, or gas, particularly the latter two cases. Speaking from the "gut", pre-rationally, it would seem wrong for one person or even a small group of people to own

DOI: 10.4324/9781003437178-21

something so much bigger than they are. A narrower "claim" might be possible, e.g., perhaps to mine hydrogen, but staking claim to things comparable in size to or larger than the Earth itself, our source and giver of life, violates strong ethical intuitions. Even the thought of it smells of hubris and transgression, making ourselves as though we were equal to ancient gods. The territory of ocean and gas planets likewise is difficult to stake claim to, due to their fluid nature. Objects with under-surface oceans, such as Europa, with layers of surface and non-surface, would be even more complex to claim or own.

On Earth, legal jurisdictions end with territorial waters and with altitude: territorial airspace. As terrestrial creatures we find it easy to demarcate terrestrial territory, but not liquids, gasses, or voids. When vessels and structures are in these domains, we give jurisdiction to the nations who sent or created them. This makes jurisdiction in space *different* from that of terrestrial land, but similar in some ways to ocean and air, analogically speaking. The Outer Space Treaty follows this analogy (UNOOSA 1967: VIII).

It is, however, at least conceptually possible to think about borders in orbits. We already make terminological distinctions between orbits, for example, low Earth orbit (LEO), geosynchronous Earth orbit (GEO), planetary, lunar, solar, and galactic orbits. Conceptual and verbal distinction are a necessary precursor for ethical and legal distinction. Additional relevant distinctions might include descriptions of orbits: their altitude, angles, eccentricity, and so on. Lastly, Lagrange points, as islands of relative gravitational stability, are special cases. While it might not make sense to be able to "own" or claim sovereignty over a Lagrange point, it might make sense to be able to reserve space in that area, with due consideration to safety, etc.

What do these above thoughts have in common? They are conceptually, technically, ethically, and politically complex enough to classify, monitor, and maintain in an orderly manner that innovative systems are needed to do so, especially at the legal level. Territorial borders on Earth took a long time to become internationally recognized and "standard", and even today some still do not seem to understand the concept (e.g., those who invade others). Space will begin as difficult conceptual terrain, but become easier as norms become clearer.

Borders can be physical, ethical, and/or legal. Law should be ethical and what is ethical should be physically (and socially) possible, because "ought implies can" (Green 2022: 16, 32; Kant 1996: 92; Kant 1933: 637; Stern 2004). Possibility is always grounded in reality and in our understandings and delineations of reality. *Unclear physical boundaries lead to unclear moral and legal boundaries.* If we can clarify and agree upon the physical boundaries, then ethics and law can start to become clearer.

"Borders" in space are conceptually possible. Current regulatory agencies cover orbits, communications, etc. (e.g., the International Telecommunications Union, ITU), but as space becomes more and more crowded, competition will increase in scale and type. And as competition increases, *clarity* of borders and *believability* of enforcement will become critical.

In the American Wild West the lack of fences made the open range a way of life, but also left the land open to smugglers, squatters, and outlaws. "Interplanetary distances" might "make good neighbours", but when the delineations are so unclear; it's hard to know "when to start shootin'". And the Outer Space Treaty is clear: space is for peaceful purposes. No shootin' allowed. But in the absence of clear boundaries there is another absence: clear deterrence by a legal authority. In other words, when it comes to following the law we have to ask: who's gonna make you?

Deterrence and Enforcement

If a crime is committed on, e.g., the International Space Station, actual enforcement depends on (among other things) timeliness of detection, type of crime, severity, jurisdiction (whose module?), national origin of the accused perpetrator, community judgment, and possibly, ultimately, the ability to apply force.

In the absence of clear national jurisdictional control, jurisdiction devolves to the Outer Space Treaty, and other relevant space treaties, which serve as the body of international law on space. The laws and jurisdictions of the sending countries are enshrined in the treaty but some things are bigger than mere national law (UNOOSA 1967: VIII).

If the OST is broken, enforcement is by the nation in question. In other words, enforcement of the OST and other space treaties is voluntary. Easy to sign on to, not so easy to manage compliance. What if the subject nation is uninterested in enforcing law? And the OST does not cover many new and more relevant topics, for example, "space littering": the creation of space debris.

On Earth, formerly, littering was the norm since the beginning of time. Unwanted items were literally "thrown away", and the term appears even in distantly related languages like Marshallese (where the word "jolok" literally means to "throw away" trash), thus perhaps indicating something of the temporally distant origin of the concept (Abo et al. 1976: 114, 479). With small populations and only biodegradable materials, this approach was not a problem. With large populations and materials like glass, metal, and plastic, littering becomes a huge problem. In the United States, there was a huge societal effort to reduce littering during the 1960s, 70s, and 80s, continuing even today.

Unfortunately, these anti-littering norms were just a bit too late to become norms for early space exploration and use. Generation of space debris (now all-too-common in LEO & GEO) was formerly accepted, but has gradually become less acceptable as spacefaring nations have come to recognize that space debris is extremely dangerous. More than mere littering on Earth, littering in orbit leaves objects moving at kilometre-per-second velocities right where other valuable objects are likely to be. Perhaps now the generation of space debris – "littering" – should be called a crime, akin to dumping dangerous objects on a freeway.

In November of 2021, the Russian government conducted a direct-ascent anti-satellite missile test, which generated a rapidly expanding cloud of debris (US Space Command 2021). This was a form of profligate littering for political signalling; but signalling what exactly? Deterrence. Russia's message was clear: we can destroy our satellites, we can destroy your satellites; do not think that you are safe from us. From the perspective of February of 2022, when Russia invaded Ukraine, the signalled deterrence from November towards other spacefaring nations became clearer: do not interfere, your space assets are at risk (Schmitt 2022).

Adding a moral quality to the issue at hand, from good actors, deterrence convinces bad actors not to act badly in the first place; from bad actors, deterrence convinces good actors not to oppose them. Interestingly, deterrence works both within legal jurisdictions and beyond them, hence, Russia's ASAT test, attempting to provoke fear among those who might oppose Russia, despite Russia having no legitimate claim of jurisdiction over other nations or space. Both within legal regimes and without them, power, and the fear that comes along with it, can be motivating.

Related to enforcement, deterrence serves to psychologically prevent outcomes that the deterring party does not want, without resorting to more physical means. Others see the deterrence and decide not to act in a way that brings enforcement. Under situations of strong deterrence, enforcement can almost become superfluous: the power differential makes clear that fighting is not worth it. If bad actors know they will be caught and severely punished then they might be deterred from acting badly. And if good actors fear negative repercussions they may be deterred from doing good things.

However, both forms of deterrence assume very clear boundaries of good and bad action: exactly what space does not yet have. Unclear borders mean unclear deterrence. Some nations' expertise at liminal competition and warfare benefits them well in space. This is a lesson for others to learn and adapt to.

Beyond deterrence preventing the need for enforcement there is also actual enforcement when laws and norms are broken. This might require force, violence, even war. Space defence and war – admittedly horrible futures that should be avoided if at all possible – are unfortunately still genuine areas for investigation. Lack of clear boundaries makes Just War Theory difficult to apply, particularly when it comes to just cause, last resort, non-combatant discrimination, proportionality, as well as all three *jus post bellum* criteria: restoring political, social, and environmental peace (Green 2022: 88–90). This is partially due to dual-use problems in space and the high-energy space environment: anything in space involves such high speeds and energies that they are almost intrinsically disproportionate and risk indiscriminacy and injustice.

In the next section, some ethical tools for thinking about space will be examined more closely, to see what bigger picture questions might enlighten or obscure this already complex arena.

Can Ethics Help Clear Up Ambiguities?

Having a strong set of conceptual tools can help to illuminate where justifiable limits might be on behaviour. But they still need to be connected to the specifics of space itself. Once normative and legal expectations are clear with respect to space, then deterrence and enforcement can gain more legitimacy and efficacy. In the book *Space Ethics* I lay out 65 tools for assisting ethical decision-making in space, but here I want to look at just three of them, with the hope that they might start to indicate some ethical boundaries for law in space.

Dual-use technologies are ones which have potential for both peaceful and violent use. This is a special threat in space due to the energies involved with fuel and speed. Not to mention the space environment itself, which is typically extremely unfriendly to human life. There is an old Latin dictum: "*abusus non tollit usum*", which means "abuse does not remove use". It is an argument in favour of allowing dual-use technologies to remain in society, since bad uses do not make good uses impossible. Bricks can be thrown to break windows, but their legitimate use as building materials should not be banned. Knives can be used for murder, but they are also needed to prepare food (Green 2020).

Abusus non tollit usum made sense in the past, and continue to makes sense now, in most cases. However, one of the reasons it made sense was because in the past technologies were not so powerful. Those not-so-powerful technologies remain with us today, and remain permissible in many cases. However, as humankind has grown in power, the risks of abuse have increased. Certainly guns, explosives, and other "useful" things are heavily restricted in many places in our world. And so, expanding the scope again, perhaps with some contemporary technologies abuse should take away use, particularly for techs with catastrophic misuses, such as nuclear weapons, chemical and biological weapons, lethal autonomous weapon systems, militarized AI, self-replicating nanotechnology, and so on (Green 2016: 4; Green 2022: 90).

The risk equation states that risk = harm x probability, and as previous philosophers of technology have noted, if a particular use of technology presents a socially unacceptable risk, then that use should not be done (Davis 2012: 8). The gambled loss isn't worth the possible pay-out. Davis appeals to technologies such as nuclear power, which, when gone awry, risk contaminating large areas of land – a potentially unacceptable outcome.

And what about for space? For example, the exact same technologies can be used to divert an asteroid away from hitting the Earth (a good use), as can be used to intentionally divert an asteroid towards hitting the Earth (a bad use). Should humankind, then, never explore this kind of technology at all, given its risk of abuse? Or should the technologies be explored but never assembled unless needed? Or always be kept under state control? Or United Nations control? What kinds of boundaries should we put on this kind of dual-use technology? As a first approximation it might make sense to research the technology to at least understand its feasibility, and perhaps even conduct tests, as NASA's 2022 DART (Double Asteroid Redirection Test) impact mission successfully piloted (Surowiec 2023). However, when it comes to actual use of such technology – even for beneficial uses such as planetary defence or asteroid redirection for mining purposes – it would make sense for an international organization such as the United Nations to be involved because the use of such technology potentially puts at stake the welfare of everyone on planet Earth. But for now, this is not the case.

Enforcement beyond law is necessary if law refuses to place itself where it needs to be. Just because boundaries – physical, ethical, or legal – are unclear does not mean that they should not or do not exist. But it does mean that an enforcer needs to step up, hopefully with some boundary clarifications. And preferably the enforcer and clarifier are good, both technically and morally, and legitimate, which in the international order means gathering something like a consensus on behaviour and leadership. Until clarity may exist, here are two tools-of-thought for trying to make ethical sense within unclear and risky circumstances.

In the early modern period, moral uncertainty was a source of genuine puzzlement and thinkers developed a structure containing five approaches to moral uncertainty (which today we might also think of as related to risk). When judging on the permissibilties of action among a number of experts, one can follow opinions ranging from the weakest with respect to risk to the strongest. The *minus probabilissimus* standard allows for permissibility if one opinion affirms, but no others. *Probabilism* permits an action if only a few opinions affirm (more than one, less than half). *Aequiprobabilism* permits if there are equal opinions on each side. *Probabiliorism* permits an action only if most opinions affirm its permissibility. And *tutiorism* follows the most morally conservative option: only permitting when all opinions are in unanimous agreement (Green 2022: 41; Mahoney 1987: 138–143).

While we may not use these words these days to convey such ideas, the concepts are absolutely in play in society all the time. For example, with respect to climate change humanity is following *probabiliorism*, that is, most scientists (our moral experts here) agree that climate change is happening, that human activities are causing it, and that we ought to stop it. Most opinions affirm, and so the action to stop climate change ought to be carried out. As a counterpoint, technological research and development follows a *minus probabilissimus* or *probabilism* standard. That is, if one inventor wants to create a new technology, generally she or he can do so without outside interference – they are the one expert opinion that decides the case. Similarly, if a small group want to do the same thing, they can do so – they are the few opinions that affirm the choice. There are exceptions to the above, for example with nuclear technologies, but with respect to space the example is clear: if one man wants to start a company that builds reuseable rockets, he can do that, and indeed he has. Regulatory agencies are involved, but no one seriously tried to shut the whole operation down *ab initio*. That is simply not the standard that we apply to tech innovation, which still has a presumption of public benefit to it and therefore gets the lowest moral uncertainty restrictions. As technologies become more powerful, these uncertainty standards might be revised.

Another approach when dealing with uncertainty is more explicitly risk focused instead of expert focused. There are six main approaches to risk (Green 2019; Green 2022: 40–44). The first is the polluter pays principle: an action can proceed unhindered, and if damages occur then the guilty party can pay for them (COMEST 2005: 7). Second, the prevention principle: if risks are clear and unacceptable, then an action should not proceed at all; harms should be prevented (COMEST 2005: 7). Third, the precautionary principle: if risks are unclear and unacceptable then actions should not proceed until clarity and acceptability can be established (EUR-Lex 2016). Fourth, the gambler's principle (alluded to above), which says: "don't bet more than you can afford to lose", that is, be averse to the highest risks, lower risks are permissible (Davis 2012: 8; Jonas 1984: 38). Fifth, prudent vigilance: proceed with the desired action, but go forward watchfully and be prepared to stop if necessary (Gutman 2011; Presidential Commission 2010: 27, 123). Sixth, the proactionary principle: rush ahead with action because it is an emergency (More 2004).

Examples of these risk standards include normal tech development (which typically follows a polluter pays principle: one is liable for damages that stem from one's products or actions) and wartime research and development, for example the Manhattan Project, which followed the proactionary principle and sped ahead, risking atmospheric ignition (and global Armageddon) because the risk of Axis nuclear superiority was simply unacceptable.

Related to space, a few further clarifications might help. First, lower risk standards are suitable if actions can be reversed through compensation or remediation. For example, if pollution can be cleaned up easily, then polluter pays can make sense. On the other hand, if cleanup is extraordinarily difficult, then prevention or precaution would be in order. The space debris question would clearly call for a prevention standard in our contemporary world, and should have called for a precautionary approach at least since Kessler and Cour-Palais wrote their landmark article on orbital debris (Kessler and Cour-Palais 1978). But because too many chose the wrong risk standard, we now have to deal with the repercussions of those choices. And recalling moral uncertainty, once the possibility of the dangers of space debris became clear, the most stringent standard should have been invoked: tutiorism, i.e., production of space debris should have become strenuously avoided unless all experts agreed that it was permissible. But space debris is not the biggest problem in space ethics and law. The even bigger question is: what *should* we be doing here at all?

What Is the Big Ethical Picture for Humankind in Space?

The ethics and philosophy of space raises deep questions like:

> Why is the universe the way it is?

What is the purpose of humankind in the universe?
Is the universe "about something"? (Lupisella 2016)
Is the universe "doing something"?
Are life and intelligence "about something"?
Should humans be doing something?

If the universe is "doing something" then should we help it or oppose it?

These questions get at the idea of teleology – the study of purposes, goals, ends…the causes for the sake of which things occur (Green 2022: 209–211). Teleological questions might seem too obscure, but they are important because, through our choices, ethics charts the path from now to the future. If we do not have a conception of the future and what we are supposed to be doing, we will wander off, get distracted, and not accomplish anything good. While humans are typically at peace with the notion of other humans having purposes (our language and tool-use depend on it, after all) when the subject goes beyond humans and into biology or physics opposition quickly arises because the mechanistic worldview holds no place for teleology. Teleology, however, is about the future, and if we cannot know what the future should look like then we cannot take steps to make the future that way. Ethics breaks down, and with it the rationale for law. Without reason to appeal to, only will exists, and voluntarism – rule by the will of the strongest – dominates. (Ideally laws should be rational and exist for the sake of promoting the common good, but this more rationalistic (as opposed to voluntaristic: defined by will and force) conception of law is sometimes secondary, only appearing when circumstances put the law itself in question (Aquinas 2009: 7).)

We might wonder if the universe itself has a say in this rationalism vs. voluntarism debate, perhaps through what we call logic, science, or math. Is a rationalistic foundation possible for ethics and law? Even if so, then humanity still gets to affirm or reject the offer. But before we can decide to affirm or reject, we need to know what the offer is – if anything. We need to figure out what we are supposed to be doing in the universe.

In the past, religions traditionally fulfilled this purpose-giving function, to greater or lesser degrees, but their efficacy has weakened, and not necessarily been replaced by helpful concepts and behaviours. How can we find purpose in the absence of traditional answers? Two approaches come to mind: top-down and bottom-up.

The top-down approach: the universe seems to be growing in complexity over time, while entropy also increases. For example, hydrogen gas, left around for 13.8 billion years, turns into people…which is really very suspicious (Green 2023B; Swimme and Tucker 2011: 2, 13, 36). Does this indicate a teleology? Or merely directionless evolution driven by mechanism? Determining teleology from this direction seems intriguing but inconclusive. The top-down approach seems too abstract and contentious, being the subject of innumerable conflicting religions and theories, but the bottom-up approach might be the concrete ground upon which we can stand, not-too distant from that which is demonstrable in human life itself. Might the bottom-up approach be more promising?

In this case, the "bottom" is humanity: our own perspective. What can we learn from looking at ourselves? Rationalist ethical values at the human social level are possible, indeed, are logically demonstrable (Hou and Green 2023). Human societies (not all individuals) generally seek to, 1) survive: maintain existence in the short term, 2) reproduce: maintain existence through the long term, 3) live in society: maintain order and necessary division of labour in the short term, 4) educate: convey culture and necessary information to each other for the long term, and 5) seek truth: find facts, logic, and reason wherever possible.

Why these five moral values? These five values are rationally demonstrable by *reductio ad absurdum*, e.g., consider their opposites. Human groups generally do not seek to: collectively commit suicide, collectively never reproduce, completely isolate all members of society from each other, avoid learning anything (indeed, as imitative creatures this is nearly impossible), or purposefully seek to believe falsehoods (Hou and Green 2023: 13). If groups of humans choose to do this, we might think they were troubled, misguided, weird, crazy, and/or evil. And if one group forced these on another, we would know it is evil: oppression and genocide.

The *reductio* is clear: accepting the opposite of the moral value destroys the possibility of humanity or reasoning at all. Any reasoning in favour of extinction undermines the precondition for human reasoning at all: humans to perform the act of reasoning. Therefore, any argument in favour of extinction is self-refuting, knocking out its own preconditions, an absurdity. Likewise with any reasoning favouring seeking falsehood rather than truth: such seeking undermines rationality at all and therefore is self-refuting. And for society and education, humans need each other. Any argument in favour of social atomization or human feralness (raising children with no culture or education) destroys the precondition for human survival (we need each other, even if just to reproduce) and reasoning (reasoning requires language, which requires more than one person capable of speech).

As might be expected for values that are logically necessary, the five values of this rationalist ethic can be found in many cultures and philosophies, though often more implicitly than explicitly. They appear, for example, in Confucius as an emphasis on family (Confucius 2018: 15), social order (Confucius 1979: 63, 143), and education (Confucius 1979: 94), and in Aristotle in emphases on family (Aristotle 1941A: 1127), social order (Aristotle 1941A: 1129), and truth (Aristotle 1941B: 689) (and this list is not at all exhaustive). These ideas appear in more contemporary philosophy as well, for example, Hans Jonas argues in his *Imperative of Responsibility* that, contra Kant (2017), the only truly categorical imperative is that humans must exist, because without ethical agents, ethics is not possible (Jonas 1984). Every other ethical rule is secondary. However, in the 13th century Aquinas laid down these five values in one quick paragraph of his *Treatise on Law*, question 94, article 2, and then moved on (Aquinas 2009: 40–41).

Seeking teleology is, however, a fraught business. To Aristotle, the greatest truths were the hardest to find (Aristotle 1941B: 690–692), and the human mind seems to be particularly desirous of teleology, even finding it where it does not exist (Keleman 1999: 464, 467; Keleman 2004: 295–296; Keleman and Rosset 2009: 138). We ought to be careful, then, when looking for these cognitively attractive concepts.

We should not stop ourselves from seeking these things, however, just because we are not sure whether they exist. Seeking teleology seems to be intrinsic to our nature, and it might indicate that the universe is "interested" in the subject too, since we are particularly unusual expressions of the universe. Therefore, actions which facilitate this effort could be construed as being good and actions which impede this effort could be construed as being bad. We might not know if the universe has a teleology, but apparently *we* do (our *telos* is to seek *telei*) and we are not happy without it, so we should probably seek it, whether or not we know it is to be found, just in case we find something interesting.

Returning to more mundane subjects as borders and enforcement, if our goal is something like "determine if the universe has a teleology, and if so, what?" then, we might consider doing the following things regarding space: exploration to seek facts about the universe, i.e., truth. This requires: living in society (because sophisticated division of labour is necessary for spacefaring), education (to maintain a truth-seeking society capable of spacefaring), reproduction (to maintain society over time), and survival (to maintain any society or individuals at all).

Furthermore, we should set up our global legal regimes with these core values in mind. Not everyone has to follow every one of these values as a goal, but enough people need to do so or the mission will fail, and those working in ways to oppose these values (for example promoting misinformation, crime, or other things destructive to these values) should be restricted in their ability to cause disruption.

Exploration is a bold, yet proper goal for humanity. Scientific discovery provides facts about nature which allow for new technologies, and new technologies, in turn, allow for new types of sensors and experiments to further the scientific endeavour. Good philosophy and religion interpret facts for the sake of wisdom, to seek how best to use our new knowledge and power for moral ends. Wisdom provides guidance, which returns us to borders and enforcement. With wisdom in mind, hopefully we can determine how to reasonably delineate space for the sake of benefit to all. Some general rules might include:

1. Explore space in an orderly fashion, seeking truth. This includes enforcement of a legal order, not just for the sake of suppressing evildoers, but also to facilitate the positive goal of gaining knowledge about the universe.
2. Work collaboratively, since this is more likely to be effective at achieving the goal. While individual geniuses did much in the past, we are now

more in the realm of collaboration, even globally and with AI, for the sake of achieving our ends. We need to facilitate global collaboration on these challenges before us.

3 "Don't litter" – that is, do not destroy the preconditions for one's own existence, or the existence of your own goals, including the five moral values. While avoiding littering is particularly important in space, there are many other rules to derive from protecting the preconditions for our own existence, from protecting the environment and helping families and children to limiting certain weapons and dual-use technologies.
4 Conduct not only scientific and technological research, but also teleological research, making careful effort not to fool ourselves, especially since we are so easy to fool (Feynman 1974). But this philosophical question of meaning – just because it is difficult – does not mean we should avoid it. Indeed, meaning is absolutely necessary for human life and flourishing. We need reasons to live, meaningful work, relationships, mysteries to understand. And it is precisely because this task is difficult that we should pursue it, as John F. Kennedy noted in his 1963 speech at Rice University: some things should be done "not because they are easy, but because they are hard" (Kennedy 1963).

Principles require specification, and the above principles remain too general and must be carefully specified. There are resources for pursuing this operationalization of abstract principles into concrete decisions, but I can only refer to them; it is too much for here, yet, this is absolutely vital work for creating an ethico-legal regime in space (Flahaux et al. 2023; Velasquez et al. 2021). Without clarity of ethics and law the space frontier will remain too much the "Wild West" of the 1800s American frontier. Borders are possible and enforcement is possible, but we need a rational foundation for these regimes, and not the mere assertion of positivist-voluntarist law. The world is split into a multitude of states, each with incompatible laws expressing the wills of their lawmakers, whether they be dictators or democracies. But will unmoored from reason can go astray; laws need to be built on rational foundations that support long term human survival and flourishing. The values presented here can help to ground ethics and law in reason, and direct us towards a future aligned with each other, our human nature, and the environment.

Conclusion

In conclusion, space, ethics, and law have much to converse on, but the dialogue is longer than we might think. We cannot simply begin with law, nor with ethics, but we ultimately must get into the fundamentals of human nature and the universe itself. This is not a place for the faint-hearted, but then again, neither is space, nor exploration, nor any rational thought. If we want to try to get space right, then we must actually do the work to get it right.

The good news is that a lot of work has already been done. Just look at the UN Sustainable Development Goals: they align well with the five suggested values above (Hou and Green 2023: 5; UNDESA 2024), not to mention various human rights treaties, codes of ethics of technology companies, and other organizations (Google 2024; Hou and Green 2023: 13–14; IBM 2024; Microsoft 2024; OpenAI 2024; Salesforce 2024). The bad news is that there is a long way to go, and getting it wrong harms people. Construed more positively: there are a lot of opportunities here for improvement. So let's get to work.

References

Abo T, Bender BW, Capelle A, and DeBrum T (1976) *Marshallese-English Dictionary*. Honolulu: University of Hawaii Press.

Aquinas T (2009) *Treatise on Law*. Freddoso AJ (trans.). South Bend, Indiana: St. Augustine's Press.

Aristotle (1941A) *The Basic Works of Aristotle (Politics)*. New York: Random House.

Aristotle (1941B) *The Basic Works of Aristotle (Metaphysics)*. New York: Random House.

COMEST (World Commission on the Ethics of Scientific Knowledge and Technology) (2005) *The Precautionary Principle*. Paris: United Nations Educational, Scientific and Cultural Organization (UNESCO). From: https://unesdoc.unesco.org/ark:/48223/pf0000139578.

Confucius (1979) *The Analects*. London: Penguin.

Confucius (2018) *The Doctrine of the Mean*. Muller AC (trans.). From: http://www.acmuller.net/con-dao/docofmean.html.

Davis M (2012) Three Nuclear Disasters and a Hurricane. *Journal of Applied Ethics and Philosophy*4: 1–10. From: https://eprints.lib.hokudai.ac.jp/dspace/bitstream/2115/50468/1/jaep4-1_micael%20davis.pdf.

EUR-Lex (2016) Precautionary Principle. From: http://eur-lex.europa.eu/summary/glossary/precautionary_principle.html.

Feynman R (1974) Cargo Cult Science. From: https://calteches.library.caltech.edu/51/2/CargoCult.htm.

Flahaux JR, Green BP, and Skeet AG (2023) *Ethics in the Age of Disruptive Technologies: An Operational Roadmap: The ITEC Handbook*. From: https://www.scu.edu/institute-for-technology-ethics-and-culture/itec-book-pdf/.

Gallie WB (1955) Essentially Contested Concepts. *Proceedings of the Aristotelian Society* 56: 167–198. From: https://www.jstor.org/stable/4544562.

Google AI (2024) Artificial Intelligence at Google: Our Principles. From: https://ai.google/responsibility/principles/.

Green BP (2016) Emerging Technologies, Catastrophic Risks, and Ethics: Three Strategies for Reducing Risk. *IEEE ETHICS 2016 Symposium Record/IEEE Xplore*. From: http://ieeexplore.ieee.org/abstract/document/7560046/.

Green BP (2019) Six Approaches to Making Ethical Decisions in Cases of Uncertainty and Risk: The Principles of Prevention, Precaution, Prudent Vigilance, Polluter Pays, Gambler's, and Proaction. From: https://www.scu.edu/ethics/focus-areas/technology-ethics/resources/six-approaches-to-making-ethical-decisions-in-cases-of-uncertainty-and-risk/.

Green BP (2020) Ethics Is More Important than Technology. From: https://www.scu.edu/ethics/all-about-ethics/ethics-is-more-important-than-technology/.

Green BP (2022) *Space Ethics.* Lanham, Maryland: Rowman & Littlefield.
Green BP (2023A) Defining Law and Ethics at the Space Frontier (Law and Emerging Tech – Paper Series Episode 1). From: https://www.youtube.com/watch?v=6FA-gRoeiSs.
Green BP (2023B) Why Space is No Problem for Religion … But Time is. *Theology and Science.*
Gutman A (2011) *The Ethics of Synthetic Biology: Guiding Principles for Emerging Technologies.* The Hastings Center Report (July-August): 17–22. From: https://onlinelibrary.wiley.com/doi/pdf/10.1002/j.1552-146X.2011.tb00118.x.
Hou BL and Green BP (2023) Foundational Moral Values for AI Alignment. From: https://arxiv.org/abs/2311.17017.
IBM (2024) AI Ethics. From: https://www.ibm.com/artificial-intelligence/ethics.
Jonas H (1984) *The Imperative of Responsibility.* Chicago: University of Chicago Press.
Kant I (1933) *Critique of Pure Reason.* Smith NK (trans.). London.
Kant I (2017) *Groundwork for the Metaphysics of Morals.* Bennett J (ed.). From: https://www.earlymoderntexts.com/assets/pdfs/kant1785.pdf.
Kant I (1996) Religion Within the Boundaries of Mere Reason. di Giovanni G (trans.). In Wood AW and di Giovanni G (trans. & ed.) *Religion and Rational Theology.* Cambridge: Cambridge University Press.
Keleman D (1999) Function, Goals and Intention: Children's Teleological Reasoning about Objects. *Trends in Cognitive Sciences* 3 (December): 461–468.
Keleman D (2004) Are Children 'Intuitive Theists'? *Psychological Science* 15(5): 295–301.
Keleman D and Rosset E (2009) The Human Function Compunction: Teleological Explanation in Adults. *Cognition* 111(April): 138–143.
Kennedy JF (1963) We Choose to Go to the Moon. From: https://www.rice.edu/jfk-speech#.
Kessler DJ and Cour-Palais BG (1978) Collision Frequency of Artificial Satellites: The Creation of a Debris Belt. *Journal of Geophysical Research: Space Physics* 83(A6): 2637–2646.
Lupisella M (2016) Conversation at the Social and Conceptual Issues in Astrobiology Cconference.
Mahoney J (1987) *The Making of Moral Theology.* Oxford: Clarendon.
Microsoft (2024) Our Approach. From: https://www.microsoft.com/en-us/ai/our-approach.
More M (2004) The Proactionary Principle, Version 1.0. From: http://www.extropy.org/proactionaryprinciple.htm.
OpenAI (2024) OpenAI Charter. From: https://openai.com/charter.
Presidential Commission for the Study of Bioethical Issues (2010) *New Directions: Ethics of Synthetic Biology and Emerging Technologies.* Washington, D.C. From: https://bioethicsarchive.georgetown.edu/pcsbi/sites/default/files/PCSBI-Synthetic-Biology-Report-12.16.10_0.pdf.
Salesforce (2024) Ethical Use Policy. From: https://www.salesforce.com/company/intentional-innovation/ethical-use-policy/.
Schmitt B (2022) The Sky's Not the Limit: Space Aid to Ukraine. From: https://cepa.org/article/the-skys-not-the-limit-space-aid-to-ukraine/.
Stern R (2004) Does 'Ought' Imply 'Can'? And Did Kant Think It Does? *Utilitas* 16 (1): 42–61.
Surowiec J (2023) NASA's DART Data Validates Kinetic Impact as Planetary Defense Method. From: https://dart.jhuapl.edu/News-and-Resources/article.php?id=20230301.

Swimme BT and Tucker ME (2011) *Journey of the Universe.* New Haven: Yale University Press.

UNDESA (United Nations Department of Economic and Social Affairs) (2024) Sustainable Development: The 17 Goals. From: https://sdgs.un.org/goals.

UNOOSA (United Nations Office for Outer Space Affairs) (1967) Treaty on Principles Governing the Activities of States in the Exploration and Use of Outer Space, including the Moon and Other Celestial Bodies (The Outer Space Treaty), United Nations General Assembly Resolution 2222 (XXI). From: https://www.unoosa.org/oosa/en/ourwork/spacelaw/treaties/outerspacetreaty.html.

US Space Command Public Affairs Office (2021) Russian Direct-ascent Anti-satellite Missile Test Creates Significant, Long-lasting Space Debris. From: https://www.spacecom.mil/Newsroom/News/Article-Display/Article/2842957/russian-direct-ascent-anti-satellite-missile-test-creates-significant-long-last/.

Velasquez M, Moberg D, Meyer MJ, Shanks T, McLean MR, DeCosse D, André C, Hanson KO, Raicu I, and Kwan J (2021) *A Framework for Ethical Decision Making.* From: https://www.scu.edu/ethics/ethics-resources/a-framework-for-ethical-decision-making/.

21 The Ethics of Outer Space Intelligence Operations

Benjamin Segobaetso

Introduction

Certainly, the question of moral limits in the activities of intelligence officers during the industrialization of celestial bodies is a complex and significant issue. It raises ethical concerns that should be carefully considered. I devote this chapter to ethical questions related to intelligence operations in outer space not only because (on a practical level) risk factors may be higher in the interplanetary context but also because (on a theoretical level) this set of concerns aptly validates or illustrates the necessity for qualifications to particular ethical frameworks, notably consequentialist ones, which will be central in the subsequent sections. The increased risk is due to the high reconnaissance capability from satellites employed by a growing number of nations and private companies. These satellites are orbiting the planet, collecting information about what other countries are doing down on Earth and in outer space. Space reconnaissance involves an unusual kind of intelligence gathering and it is becoming more relevant as Cislunar and Lunar critical infrastructures get underway and top military officials are now describing these spheres as new domains of operation (Bender 2022). These new space economy and security developments require an examination of whether the ethics of intelligence gathering would or should change in some practical or theoretical way to apply to the new frontier, from terrestrial contexts to the extraterrestrial contexts.

It also remains to be seen whether the exponential growth of human activities in outer space will add grounds for making consequentialism an ethical theory central to the moral approaches considered necessary for celestial development (Arneson 2005). Before exploring consequentialism, a number of issues around the potential application of two common forms of intelligence operations (*discriminate reactive intelligence* gathering and "PRISM-style" *indiscriminate pre-emptive intelligence* practices) will be introduced in attempts to explain the speculative scenario of the envisioned landscape (Tavaglione 2016). Subsequently, several moral frameworks will be probed as well to see how they will apply to this new frontier, and a special

DOI: 10.4324/9781003437178-22

consideration will be given to the Just Intelligence Theory, which is an adaptation of the Just War Theory to intelligence (Bellaby 2012).

By and large, moral consideration of intelligence gathering in outer space activities is a tall order since average civilians are not privy to the day-to-day operations of our intelligence agencies' actions in outer space. This is because these operations by their very nature must be kept clandestine. Spacefaring nations in advanced democracies may accept this as they expect their outer space governments' policies to engage oversight mechanisms to safeguard the effectiveness and diligence of intelligence operations (Bellaby 2012). In this case, the citizens are probably resigned to the notion that they do not need to know what these agencies are doing daily in outer space (Chomeau and Rudolph 2006). By the same token, this also means that citizens are not able to focus ethical scrutiny on the agencies' principles and actions (Jensen 2016). Although our understanding of functions assumed by intelligence agencies in outer space is at best unclear, the increasing human activities in outer space may soon challenge our perspective on intelligence gathering in outer space. This is especially evident as the NASA Lunar Gateway program which is governed by the Artemis Accords gets under way. This program, if successful, may be the starting point of the potential industrialization of some outer space spheres (Casini et al. 2020).

The initiation of the NASA Lunar Gateway project is geared towards steering the US and its allies into the next era of Moon exploration as well as carrying out space launch operations from celestial bodies into deep space. It prompts several countries from around the world to gear themselves up to sending humans further into our solar system, beyond the International Space Station (ISS) (Casini et al. 2020). This project, which positions the Moon as representing a crucial steppingstone in humanity's pursuit to travel into deep space, may pioneer the potential industrialization of some celestial bodies and certain spheres of the interplanetary system. However, the initiative is bound to be accompanied by several ethical challenges and sovereignty issues which prompt the need for research on the ethical implications of these unfolding outer space developments. For example, the rush for outer space domination by spacefaring countries and private companies, creates competition that could potentially trigger spacefaring countries and commercial entities to engage in space based *discriminate reactive intelligence* gathering and "PRISM-style" *indiscriminate pre-emptive intelligence* practices against one another.

Discriminate reactive spying may take place as the number of mega constellation satellites projects in the low Earth orbit (LEO) and the geosynchronous orbit (GO) increases. This form of intelligence operation in outer space has the potential to attract substantial academic attention as has been the case here on Earth due to its tendency to focus on targeted individuals or groups (Tavaglione 2016). It may have to be recognized, though, that in some cases, there may be nothing inherently ethically wrong with discriminate reactive targeting practices in outer space, especially with increasing human activities in the interplanetary system. For example, as several spacefaring countries continue to amass more critical infrastructure establishments in deep space, intelligence collectors may selectively infiltrate the satellite information of certain countries that have consciously made themselves a threat to other spacefaring nations or some private entities' critical infrastructure. Therefore, there may be nothing wrong with infiltrating those countries or private entities that have made themselves a threat to others by invading their privacy or even worse. The moral dilemma that may subtly arise when more projects like the Lunar Gateway start to proliferate, include the question of how we can identify a certain behaviour as threatening in outer space. This may only be possible if the selected spacefaring countries' satellites are proactively infiltrated and spied on. It also remains ethically concerning whether there would be a certain amount of harm that will be allowed to enable a discriminate reactive spying to be pre-emptively enacted (Cantarella 2016). Other challenges point to how the blurred boundaries in outer space would play out in terms of intelligence operations when spy satellites float in geosynchronous orbit (GO) and mingle with commercial telecommunications satellites. It will be difficult to determine with precision which individual satellites or satellite groups can be considered threats to other countries' national security.

There is need for moral justification for spying activities in outer space not merely because anything deceiving seems apparent to necessitate moral scrutiny but due to our increasing economic and military dependence on space. This dependence may create potential vulnerability that could be an attractive target for adversaries to employ offensive or counter offensive space intelligence operations. The outer space operations may be intended to deceive, disrupt, deny, degrade, or destroy other systems; therefore, the phenomenon may warrant the activation of discriminate reactive spying measures to thwart the aggressors' efforts. However, these measures need to be ethically justifiable and conducted through an ethical framework for intelligence collection. The framework could perhaps reflect a version of *primum non nocere*, whether it be the Hippocratic injunction to "first, do no harm" or a responsibility to minimize harm.

The increase of critical infrastructure in outer space also presents an opportunity for spying on indiscriminate targets *en masse* without any basis for suspicion of wrongdoing in attempts to detect suspect individuals. This is PRISM-style indiscriminate pre-emptive spying (IPS) and is a concept distinct from *indiscriminate intimidative* spying (Tavaglione 2016). The latter is usually criticized for infringing moral sanctions against attacking privacy and for subsequently wronging its targets (which in an outer space context could comprise the entire international community), even though they may not know that their privacy is being intruded upon. However, it should not be assumed that indiscriminate pre-emptive spying will always contravene moral and legal prohibitions during the unfolding outer space developments, because that assumption leans on absolutism. However, we should consider the possibility that being wronged without being harmed might be another moral risk during the industrialization of the Moon and certain planets (Hardy 2016).

As the outer space development unfolds, it can generally be assumed that the targets of IPS will not know they are being spied upon; therefore, the espionage may have no direct physical or psychological effect on them. As it will have no such effect, it will cause no direct harm to them. Consequently, if there is something morally problematic with IPS, as human activities continue to increase in the interplanetary system it may be a wrong without direct harm.

If we are to examine the ethical implications of outer space intelligence operations from the assumptions that are favourable to IPS during the interplanetary development era, we will have to admit that moral principles are open with respect to: revision, valid objection, forfeiture, or annulment on some occasions. There may be instances in outer space where prohibitions against infringing privacy are outweighed by stronger counteracting reasons. However, to limit the harm done, principles of just intelligence as an ethical framework for intelligence collection would have to be met (Skerker 2018). The just intelligence conditions in this context regarding the practice of secret intelligence in outer space may be identified by referencing the Just War framework. This approach is adapted from Just War theory's accustomed principles that determine ethical constraints on armed conflict (Bellaby 2012). From this framework, similarities can be illustrated to derive *jus ad intelligentiam* and *jus in intelligentia* ethical concepts to govern the acquisition and use of secret intelligence (Bellaby 2012). For example, when considering the application of the Just Intelligence framework during the industrialization of the Moon and certain planets (IOMP), it would potentially be prudent to combine the criteria for *jus ad bellum* and *jus in bello*. This could potentially establish the ability for outer space governing bodies to weigh the justification of particular intelligence operations (*jus ad intelligentiam* and *jus in intelligentia*). Such operations may need to consider the following:

> A *just cause* (in the service of national security); *right intention* (acting for national security reasons rather than partisan ones); *proportionality* (the ethical risks of the operation should not exceed the harms the operation is meant to prevent); *right authority* (authorized by a legitimate, transparent, accountable body); and a *reasonable prospect of success*. They also need to be *discriminate* (minimizing harm or privacy intrusions to nontargets) and *necessary* (ensuring there is no other way to achieve the objective at a lesser ethical risk).
>
> (Skerker 2018: 509)

Thus, *ad intelligentiam* and *in intelligentia* criteria will be equally relevant to ascertain; for example, whether an agency may collect signals intelligence from, say, the entire LEO or from a particular satellite in GO. A just cause, right intention and legitimate authority are arguably important for both operations. It may be morally permissible during the interplanetary industrialization era to endorse a variation of some standard on which the discrimination and proportionality condition are mutually reliant (Omand and

Phythian 2018). For example, a collection procedure that perpetrates little harm (such as the gathering of communication metadata) can be applied in a more indiscriminate fashion – plausibly across a whole orbit area. Directing spacefaring intelligence analysts to listen in on a particular satellite's communications would require a highly scrutinized type of discrimination. Analysts may have to be required to substantiate that a target is highly likely to be of "intelligence value" rather than "possibly of intelligence value" depending on the ethical standard applied.

There may be some instances where a certain level of discrimination would potentially be unavoidable during the industrialization of the Moon and other celestial bodies when it comes to privacy encroachments. There may also need to be a differentiation made on outer space intelligence collection by liberal states from illiberal states. At the very least, liberal states engage collection procedures, sanctioning broad filtering so that human analysts are only likely to see communications from satellites or infrastructures that have a high likelihood of being "just" intelligence targets. Some scholars such as Omand and Phythian (2018) are less concerned about the particularity between storing and accessing data, as they maintain that storage does not significantly intrude upon any person's rights. Although it is a valid point that *accessing* an individual's private information is more insidious than merely storing it, this distinction may not be practical during unfolding outer space developments because of the sensitivity of critical information that may be stored for intelligence purposes. That on its own poses a danger since the storage may render access more feasible than previously. I will expand on the principles of Just Intelligence as an ethical framework for intelligence collection in the next section when I touch on Just War Theory.

Can Deontological and Consequentialist Theories Be Used to Justify Cases of Intelligence Collection during IOMP?

We can apply divergent moral/ethical theories to human activities in outer space and the outcomes could be so nuanced as to fall outside the spectrum of good and bad. In the remaining sections, I will focus my analysis on the examination of how consequentialist and deontological theories would fare in relation to intelligence operations during the IOMP. Deontological theory, more specifically, Kantianism, can be characterised by strict adherence to duty or obligations (Korsgaard 1986). Such theories are absolute in their evaluation of actions because they instil a very stiff standard for determining whether actions are morally acceptable (Korsgaard 1986). Where context, circumstances, and states of affairs are up for consideration in a theory like consequentialism, Kantianism downplays the importance of the consequences. The theory holds that some types of actions are wrong no matter what the consequences. Furthermore, for Kant, all actions should be taken in accordance with some fundamental principle or maxim. This fundamental principle or maxim is what determines whether an action is moral or immoral. According to such a theory, a morally good person will always do his or her moral duty in accordance with said fundamental principle or maxim, regardless of the consequences. Doing one's duty might not produce the most good for the-most-number of people and it might not even produce good for the individual performing the action, but it will make the acting agent a morally virtuous person.

The unfolding ethical implications of outer space developments can be assessed through Kant's famous categorical imperative, whose key concept is the universal law formulation (Myskja 2008). This formulation invites us to envisage the world in which the maxim is embraced by all the people. However, this absolute maxim gets problematic if, for example, a belligerent party wants to destroy another party's satellite and asks if it is protected against anti-satellite technology. According to the categorical imperative, the morally right action would be to tell the belligerent party the truth (Myskja 2008). Lying to the party and deceiving them would be morally wrong even though the satellite may bear the destruction.

It is hard to see how Kant's absolute theory of morality would assist spacefaring nations to effectively carry out intelligence collection operations during the IOMP. This is because intelligence by designation, is absolutely reliant on its proficiency to deceive and manipulate to be effective. Intelligence disciplines will not go far in outer space if all the officers adhere to Kant's categorical imperative. This theory of morality will render intelligence gathering in outer space impractical or even dangerous (Myskja 2008). "A collector's ability to ethically deceive, manipulate, empathize, and build trust with others for the purpose of getting them to spy on their native country leaves no room for such deontological theories" (Doty 2018: 5).

On this note, the discipline of intelligence operations in outer space will customarily violate the universal law and humanity formulation because "[factors] such as lying, deception, and manipulation cannot be universally applied to the whole of humanity because it treats prospective targets of operations as means to an end rather than an end in and of themselves" (Doty 2018: 5).

To refine our conception of possible justifications of ethical intelligence operations in outer space, and to properly direct intelligence collectors in the discipline, we may have to hold some ambitions to shift away from moral absolutist theories, taking account of others that are similar to Kant's. The ethical applications of signals intelligence and geospatial intelligence in outer space during the IOMP may necessitate more adaptable theories of moral justification.

A Consequentialist Perspective on the Ethics of Intelligence Operations during the IOMP

One of the more adaptable approaches that may theoretically shed light on matters of intelligence collection in outer space is consequentialism. The consequentialist approach may be more applicable to signals intelligence and geospatial intelligence interplanetary operations because it is not as absolute or as rigid, in regard to moral rules that can guide action (Vrist Ronn 2016).

As a class of normative, teleological ethical theories, consequentialism nonetheless presumes that "the only thing that is of absolute value in evaluating the rightness or wrongness of an action is the consequences of that action, not the action itself" (Doty 2018: 5). In this regard, this normative theory grants that a deed is morally fitting or ethically legitimate if it produces the best consequences. So, if the consequences of an action in outer space produce the greatest amount of good or wellbeing for spacefaring groups, then that action will have to be deemed as having met the requirements to be considered right, or morally justified. This could be permissible in terms of the ethical theories of consequentialism and utilitarianism.

The structure of consequentialism will render it applicable for assessing the conducts of collectors and their sources during the interstellar industrialization era. Espionage carried out in this era could be an unclear profession, where the boundaries of moral and immoral justifications will be blurry. Since intelligence operators are trained to lie, deceive, and manipulate others and their sources for the greater good of their country, it can be assumed that they will do the same during outer space development. These actions would potentially be considered wrong by some individuals and civil groups. However, under a consequentialist theory, and where the consequences of such actions are optimal, they can be justified.

During the era of the IOMP, space security and foreign policy will be central in terms of fostering the most good. Thus, if to better advise policy makers our outer space intelligence collectors are encouraged, and even compelled, to lie, deceive, and manipulate foreign nationals, a consequentialist theory will endorse the assertion that the ends justify the means. To further this argument, we must reflect on various forms of consequentialism and their implications for outer space developments.

Before delving into at least three forms (Act Utilitarianism, Rule Utilitarianism and Rule Consequentialism), I would like to acknowledge that Rule Consequentialism, if considered for addressing the impending ethical concerns raised above, could potentially apply to those who formulate policy or institutional codes of conduct. Act Utilitarianism, on the other hand, has some elements that could allow it to be applicable to agents who are on the ground collecting the intelligence; however, this applicability can also be considered under Rule-Consequentialism.

The Act Utilitarianism Perspective in Outer Space Development

In a broader scope, Act Utilitarianism is the category of Utilitarianism that showcases that the rightness or wrongness of each action individually is conditional on the utility of that action. The utility is defined as the level of good or happiness supported by the consequences (Eggleston 2014). Act Utilitarianism is juxtaposed with Rule Utilitarianism, which establishes moral/ethical demands in terms of creating a precedent for a particular deed that ought to be emulated in the future.

During the IOMP, Act Utilitarianism has the potential to embrace the justification of cases of intelligence collection by promising overall good and happiness. Nonetheless, prior to embracing Act Utilitarianism, we must reflect on the implications that it can have across the full spectrum of intelligence collection cases. For instance, a spacefaring country could undoubtedly exploit cases similar to the "ticking bomb thought experiment", through which morally shocking or appalling acts could be justified if they produced the most happiness or wellbeing. It is debateable whether such acts may be justified under certain circumstances, for example, in catastrophic scenarios in outer space. The difficulty is that Act Utilitarianism encourages an unsafe degree of keenness to perform such things. This is the key motivation behind why societies turn to Rule Utilitarianism.

The Rule Utilitarianism Perspective in Outer Space Development

During the IOMP, Rule Utilitarianism could establish a precedent for a given action in the form of a rule that will be adhered to by all in the future (setting a standard or code of conduct) for actions to be considered right and just. As Doty (2018: 5) puts it:

> The colour yellow at a stop light means slow down. For a Rule Utilitarian, always following the rule of slowing down when the light is yellow will generally lead to the best consequences and most good/wellbeing, than say, making a "game time decision" whether to speed up or slow down every time the light turns yellow – something an Act Utilitarian would do. Rule Utilitarians feel that following rules will generally lead to better overall consequences most of the time rather than making decisions about actions on a case-by-case basis, even though making decisions on a case-by-case basis might lead to more overall good and wellbeing in specific instances. Rule utilitarianism attempts to resolve the issue raised by the ticking time bomb thought experiment which plagues Act Utilitarianism.

As the development of outer space continues and with the establishment of the NASA Lunar Gateway, the rigidity of Rule Utilitarianism may be viewed by some analysts as disqualifying it from being a beneficial moral/ethical theory for assessing issues of intelligence collection. This is mainly because the domain of spying is a challenging sphere for absolutes. It is multifaceted and nuanced because it involves all different kinds of individuals who have different values and moral perspectives. Spacefaring parties will have to be flexible, adaptable, and socially aware to carry out intelligence operations within such complex situations. In this regard, some analysts could view absolute ethical rules as having the potential to render it challenging to operate in such a manner of being flexible, adaptable, and socially aware.

However, an important aspect to be cognizant of is that Rule Consequentialism could be of great value in adopting the implementation of rules and standards. Moreover, Rule Consequentialism can come in at the policy-making level (and maybe to some extent trickling down to certain aspects of how agents are expected to conduct their activity by applying the rules on a case-by-case basis). The case-by-case application could make more sense for the intelligence agents, but it could be dangerous if Rule Consequentialism is not applied to set up codes of conduct for the agents (Doty 2018).

With all its faults, Act Utilitarianism at this juncture may still be viewed by some analysts as having the capability to offer a much better justification of intelligence collection in outer space (Vrist Ronn 2016). The theory could be viewed by some academics as adequately adaptable in regard to intelligence gathering for outer space developments because it recognizes context; therefore, it could consider the circumstances when ascertaining whether or not actions are justified. The adaptability inherent in Act Utilitarianism enhances its potential applicability in outer space operations, particularly during the actual gathering of intelligence. With that being said, Act Utilitarianism in outer space may allow for ambiguity and excessive flexibility in some concerning behaviours it could potentially justify in pursuit of the greater good. For example, there could be ambiguities on whether the ends to the forms of behaviour carried out by a space farer/s in outer space would really validate their means.

Some of the ethical considerations in relation to matters of outer space intelligence collection that will potentially take place during the IOMP involve questions of the self. How does a space-faring intelligence collector reconcile a career of deceit and manipulation against an honest normal life? The notion of intermingling, taking solace, and empathizing with morally questionable personalities under the context of outer space security is daunting. It also prompts the question of how will this impact the intelligence officer's sense of self, integrity and dignity? Would there be a particular psychological profile required for such work and what impact would the work have on the officers' mental wellbeing? During the IOMP, a primary concern against consequentialism and utilitarianism may be that these approaches necessitate individuals to compromise their personal integrity.

If the paramount consideration in an ethical dilemma is the outcomes resulting from a specific action, and if the optimal outcome is the one that maximizes good and happiness, then the action leading to such favourable consequences must be embraced and standardized. This could be done despite what the dignity and personal values of the intelligence collector require.

Consider for example the analogy of a collector who discovers an outer space signals intelligence that could provide extremely beneficial and desperately needed information for the defence of the collector's country. However, the source who has the rest of the needed information agrees to give the information in exchange for a grossly despicable act. However, as an individual, the collector is unreservedly against the demanded act because they see it as unconscionable or morally reprehensible. In accordance with the tenets of Rule Utilitarianism, should the collector transgress their individual dignity and engage in the unthinkable act for the greater good of their country? Alternatively, does the collector decline the source's demands, thereby causing a detriment to national security due to the country's failure to obtain the information? Moreover, the collector would bear responsibility for their country's suffering due to their reluctance to act.

This analogy prompts significant reflections on the intelligence officer's duty versus personal integrity amid the evolving developments in outer space. It also prompts us to contemplate the ethics of unlimited liability which calls for intelligence collectors to put their service first before self to the point of death (Coleman 2013). Indeed, utilitarianism, especially in its most stringent interpretation, which requires unwavering compliance with rules, diminishes the significance of personal integrity. The principles of utilitarianism will necessitate outer space intelligence officers to act in a manner that maximizes overall good or wellbeing. However, in pursuing this goal, they would be compelled to relinquish their personal values or commitments that contribute to their own individual good or happiness; which seems unreasonably demanding. However, the situation may worsen if the action required by the source is so abhorrent that it would compromise the integrity of every collector. As an illustration, there are established procedures allowing a collector to obtain authorization to transfer their source to another collector if the demands of their assignment become excessively burdensome. Nonetheless, if the action required is so abhorrent that it would compromise the integrity of every collector, the situation becomes more intricate as the collector receiving the assignment must be thoroughly briefed on the issues.

Additionally, they must not have any personal integrity concerns regarding carrying out the conditionally required action. Operations of this calibre subsequently emerge as matters of strategic national policy, requiring resolution by decision-makers whose ethical stance may favour absolute Rule Utilitarianism. Despite the troubling implications for collectors' personal dignity and self-perception, their directives must be respected.

When contemplating the dignity of outer space intelligence collectors during IOMP, additional concerns arise regarding their loyalty to their sources. Indeed, the concept of loyalty toward sources prompts consideration of the extent to which they merit such allegiance, given their diverse range from sensible individuals to perverse, morally bankrupt ones.

The allegiance of collectors to their sources may originate from a commitment to advancing the welfare of others for altruistic reasons, grounded in the conviction that the wellbeing of others is important in and of itself. Collectors may also find themselves obligated to maintain loyalty to the source due to a sense of duty, driven by the aim of adhering to an internalized moral code, which includes fulfilling the obligation to prioritize the wellbeing of another out of a sense of moral duty. The presence of two distinct types of guilt feelings, altruistic guilt (AG) and deontological guilt (DG), can give rise to feelings of reluctance, remorse, and moral conflict. "AG arises from having harmed, through one's own action or omission, an innocent victim, while DG arises from the transgression of an internalized norm" (Mancini and Gangemi 2021: 12).

The concerns relating to officers' personal integrity and possible estrangement from their sense of self could be genuine and may certainly be relevant to cases of signals intelligence and geospatial intelligence collection in outer space. In deliberations of this kind, it's plausible that individuals' conception of a fulfilling life is often envisioned as one where advocating for the greatest good is seen as a virtuous pursuit (Doty 2018). This becomes especially significant in the context of emerging outer space security issues, as evidence suggests that it would be unjust to overlook considerations of utilitarianism and personal integrity, particularly concerning the integrity of individual value systems. This aspect warrants careful consideration during the IOMP. As we advance the industrialization of outer space, it behoves us to question whether it would be optimal to adopt a utilitarian perspective that also respects the individual integrity, sense of self, and identifiable values of collectors in intelligence collection scenarios. In this scenario, one could live a subjectively consequentialist life by abstaining from attempts to derive a modified version of Rule Utilitarianism from a foundational commitment to some form of contractualism, or from viewing morality as a fixed set of rules (Arneson 2005).

Replacing Act and Rule Consequentialism with a Sophisticated Rule Consequentialism

Given there are some slippery slopes in attempting to engage both Act and Rule Consequentialism in outer space developments, I examine how replacing both approaches with a different version of consequentialism would work. I attempt to achieve this by applying a Sophisticated Rule Consequentialism (SRC), which is related to a version of Rule Consequentialism but is not inherently susceptible to three standard objections. These are (1) Rule Consequentialism is blameworthy or at fault for rule worship, (2) the principle is idealistic in an adverse sense, and (3) Rule Consequentialism tends to overlap into Act Consequentialism or (if construed to circumvent the overlap), is clearly full of loopholes (Arneson 2005). As Railton (1984: 153) articulates:

> A Sophisticated Consequentialist is someone who has a standing commitment to leading an objectively consequentialist life, but who need not set special stock in any particular form of decision making and therefore does not seek necessarily to lead a subjectively consequentialist life.

In this regard, applying a SRC approach when engaging in outer space developments would authorize collectors to be largely concerned with the main set of circumstances; nonetheless, considering also personal dignity and sense of self as equally important. A Sophisticated Consequentialist approach will also assist collectors in outer space to avoid concerns of estrangement from their sense of self and personal values while also promoting the best conditions or set of circumstances. SRC is portrayed as a moral framework, particularly as a form of indirect consequentialism, wherein the moral rightness of actions is evaluated not by their consequences but by alternative factors (Arneson 2005). SRC's supplementary element to Rule Consequentialism is psychological, as it deems an action to be considered morally right if it arises from a motivational structure or a set of deliberative tendencies that usually result in the most beneficial overall outcome achievable by the agent.

Further Objection to Act Utilitarianism in Relation to Outer Space Intelligence Gathering

The IOMP should capture the attention of consequentialists because the seemingly boundless expanses of outer space hold the potential for a multitude of significant consequences, both positive and negative, surpassing those that solely impact Earth (Baum 2016).

Ensuring positive outcomes during the development of celestial bodies is crucial for the advancement of human civilization and may entail the establishment of Moon human settlements. Indeed, certain consequentialists, such as Baum, may contend that in present-day decision-making, there ought not to be an overemphasis on directing investments towards space developments. Rather, according to Baum and similar consequentialist perspectives, significant social investments should prioritize preventing civilization-ending catastrophes (Baum 2016).

Baum's context proposes that prioritizing the mitigation of devastations capable of annihilating the human civilization is paramount. This strategy guarantees that forthcoming generations can responsibly and ethically advance lunar industrialization and explore various avenues of outer space development while addressing intergenerational justice concerns and other relevant issues. It may be objected that if we utilize consequentialism or utilitarianism approaches during the IOMP, there might be a moral slippery slope. However, an argument can be made that utilitarianism, by its very nature, would mitigate such dangers unless we deviate from its core tenets, as exemplified by the fallacy Peter Singer encountered in his utilitarianism. For instance, in *Animal Liberation*, Singer (1973) allocates a substantial portion of the book to advocating for the theory. Act or Rule Utilitarian may struggle to endorse Singer's perspectives on outer space developments due to the potential risks outweighing the benefits, primarily because his approach lacks a policy orientation. An argument can be made that Singer's (1973) approach lacks the capacity to offer practical recommendations.

There is no straightforward way to determine whether outer space intelligence operations would be an influence for peace and security, or a basis for disaster and space warfare. The status of intelligence operations as a risky endeavour remains uncertain, with some academics cautioning that distinguishing espionage from preparations for war poses a significant challenge (Rovner 2022). For example, those subjected to intelligence operations cannot be certain that their adversaries' motives are solely benign, limited to the collection of information (Rovner 2022). This leads us to understand why the ethical reflections can be aided or advanced by turning from intelligence issues to potential outer space pre-emptive warfare issues. Conventionally, foreign intelligence agents have been perceived as below contempt because the presence of intelligence agents tends to instil fear of attacks, while spy worries have a recollective history (Rovner 2022). As Rovner (2022: 1) puts it:

> Fears of German espionage roiled Great Britain before the First World War. Fear of British espionage led to a panic in the Soviet Union in the interwar period, culminating in a brief but intense war scare in 1927. Fear of communist infiltration animated McCarthyism in the United States in the early Cold War causing some to worry that a hot war was coming. In these and other cases, spies were viewed as both sinister and repugnant.

Conclusion

The observations made in this section lead us to the following conclusions regarding the ethics of outer space intelligence collection during the IOMP. Deontological theories such as Kant's may not be adequate to guide intelligence ethics in outer space because their absolutist approaches have the potential to disqualify standard intelligence collection activities that can potentially produce the most-good. On the other hand, Act Consequentialism can potentially provide a much more progressive outlook than the absolutist deontology, and it may help outer space intelligence collectors to engage in operation activities that bring about greater good and are morally justified. However, there is recognition that Act Consequentialism could, on some occasions, justify excessive acts that are inherently morally dreadful and that violate the dignity of those who perform them.

Rule Consequentialism could present a more appealing alternative to conventional Act Consequentialism when applied to the development of outer space. Of significant note is Rule Consequentialism's comprehensive capacity to address particular practical challenges faced by Act Consequentialism, such as perceived excessive demands or challenges in maintaining consistent adherence. However, Rule Consequentialism does encounter numerous challenges, which, when applied in the context of outer space, may result in adherence to rules that lead to suboptimal outcomes. This approach may necessitate adherence and sacrifices based on codes even when others are not following them. As a result of these and various other concerns, Rule Consequentialism tends to lose favour, given the significant arguments against its viability (Arneson 2005).

To prevent such justifications, a Sophisticated Rule Consequentialist approach to the ethics of outer space intelligence during the IOPM could be highly considered. Since, it has been established that SRC demonstrates superiority over its primary competitors in moral theory, it is my conclusion that it deserves further examination and respectful consideration for outer space developments. An illustration of this has been showcased as entailing how SRC surpasses Act Consequentialism, since the latter often produces judgments regarding our moral obligations that contradict our carefully considered moral convictions. It is also worth reiterating that Sophisticated Rule Consequentialism has been proven to establish its superiority over absolutist deontologies, which assert that there are certain actions we must absolutely refrain from, regardless of the consequences. The above leads me to conclude that although SRC seems to have favourable qualities that puts it in the forefront for consideration for outer space intelligence collection during the IOMP; further research on this conclusion still needs to be conducted.

References

Arneson R (2005) Sophisticated Rule Consequentialism: Some Simple Objections. *Philosophical Issues* 15(1): 235–251.

Baum SD (2016) The Ethics of Outer Space a Consequentialist Perspective. In Schwartz J and Milligan T (eds) *The Ethics of Space Exploration*. Switzerland: Springer Cham, pp. 109–123.

Bellaby R (2012) What's the Harm: The Ethics of Intelligence Collection. *Intelligence and National Security* 27(1): 93–117.

Bender B (2022) Moon Battle New Space Force Plans Raise Fears over Militarizing the Lunar Surface. From: https://www.politico.com news 2022/03/12space.

Cantarella M (2016) Intelligence Ethics in the Digital Age. From: https://scholar.google.ca/scholar?hl=en&as_sdt=0%2C5&q=Cantarella%2C+M.+%282016%29.+Intelligence+ethics+in+the+digital+age.+Pg.+85-86.+&btnG=.

Casini et al (2020) Lunar Analogue Facilities Development at EAC: The Luna Project. *Journal of Space Safety Engineering* 7(4): 510–518.

Chomeau JB and Rudolf AC (2006) Intelligence Collection and Analysis, Dilemmas and Decision. In Goldman J (ed.). Ethics of spying: A reader for the intelligence professional. Lanham, MD: Scarecrow Press, pp. 114–125.

Coleman N (2013) Unlimited Liability: Soldiers and Unlimited Liability. From: https://www.militaryethics.org/Unlimited-Liability/15/.

Doty R (2018) *How Can Human Intelligence Collection Be Morally Justified?* PhD Thesis: University of Kansas.

Eggleston B (2014) Act Utilitarianism. In Miller D and Eggleston B (eds) *The Cambridge Companion to Utilitarianism*. Cambridge: Cambridge University Press, pp. 125–145.

Gendron B (2005) The French Vocational Baccalaureat Diploma: Space of a Plural Transition for the Youth. *European Journal: Vocational Training*, 36: 33–46.

Hardy J (2016) Targeting Thresholds. The Impact of Intelligence Capability on Ethical Requirements for High-value Targeting Operations. In Galliott J and Reed W (eds) *The Ethics and the Future of Spying*. Abingdon Oxon: Routledge, pp. 177–190.

Jensen M (2016) The Virtues of Bond and Vices of Bauer: An Aristotelian Defence of Espionage. In Galliott J and Reed W (eds) *Ethics and the Future of Spying*. Abingdon Oxon: Routledge, pp. 31–42.

Korsgaard C (1986) The Right to Lie: Kant on Dealing with Evil. *Philosophy and Public Affairs* 15(4): 325–349.

Mancini F and Gangemi A (2021) Deontological and Altruistic Guilt Feelings: A Dualistic Thesis. *Frontiers in Psychology* 12(651937): 1–16.

Myskja K (2008) The Categorical Imperative and the Ethics of Trust. *Ethics and Information Technology* 10(4): 213–220.

Omand D and Phythian M (2018) The Technoethics of Contemporary Intelligence Practice: A Framework for Analysis. In Miller S, Regan M, and Walsh P (eds) *National Security Intelligence and Ethics.* Abingdon Oxon: Routledge, pp. 39–60.

Railton P (1984) Alienation, Consequentialism, and the Demands of Morality. *Philosophy & Public Affairs* 13(2): 134–171.

Rovner J (2022) Spies as Agents of Peace, Themes: Innovation and Espionage. From: https://engelsbergideas.com/essays/spies-as-agents-of-peace/.

Singer P (1973) Animal Liberation. In Garner R (eds) *Animal Rights.* London: Palgrave Macmillan, pp. 7–18.

Skerker M (2018) Principled Spying: The Ethics of Secret Intelligence. *Ethics & International Affairs* 32(4): 509–511.

Tavaglione N (2016) A Dilemma for Indiscriminate Pre-emptive Spying. In Galliott J and Reed W (eds) *The Ethics and the Future of Spying*. Abingdon Oxon: Routledge, pp. 120–131.

Vrist Ronn K (2016) Intelligence Ethics: A Critical Review and Future Perspectives. *International Journal of Intelligence and Counter Intelligence* 29(4): 760–784.

22 Respect for the Non-Living in Early-Stage Space Expansion

Chelsea Haramia

Introduction

From the monocellular to the complex, we know that, at least in the Solar System, our planet appears to be unique in its ability to produce and support an abundance of life. Outside of our planet, we observe sterile environments and empty space. But to look outside of our atmosphere and see absence, lack and lifelessness is to observe the cosmos through the lens of unchecked assumptions. For instance, there is often the assumption that, without life, something is missing. While it is true that no human can observe the world from a completely neutral perspective, we are at the same time capable of examining our unchecked assumptions and recognizing potential biases that can lead to distorted understandings of value in the world. In this chapter, I discuss the life-bias that is found in common attitudes toward space exploration. I first explore the foundations of value attributions that ground claims to moral respect. I then show that sterile space environments may possess intrinsic value and deserve respect even though they do not possess living organisms. I also demonstrate that, even if the life-bias is legitimate, we might still have strong responsibilities to extend considerations of respect to the non-living and the artificial. I conclude that the burden of proof is on those who would deny respect to sterile and artificial elements of outer space to show why this must be so. Otherwise, we may default to an attitude of disrespect for the integrity of the extraterrestrial world as we expand into space, even on cosmic sites and systems devoid of life.

Respect

Justice often requires that we do not disproportionately or otherwise unduly harm that which is deserving of respect. Respect is owed to that which is valuable in and of itself, all by itself. For example, humans are typically considered to be valuable in this way and are thereby owed a baseline level of respect. What else is deserving of this kind of moral consideration? As we consider plans and forecasts for humans' expansion into space, we must confront this question head-on. Debates about what deserves respect have

DOI: 10.4324/9781003437178-23

typically not included a sustained focus on our obligations out in the cosmos. But we know that we as humans are capable of carefully evaluating this question. We are capable of acting justly, and we are capable of acting unjustly. Moving forward, we should hope to forestall moral failures in the realm of outer space, to pursue justice, and to seek to treat that which is intrinsically valuable with respect, to the best of our abilities.

It is worth noting that there are prudential as well as ethical reasons to consider how we ought to be treating that which occupies outer space environments. Rushing into under-considered projects and building ill-conceived social and physical systems risks not only injustice and moral wrongdoing, but also undermining that which is or could be highly useful to us humans regardless of whether it is valuable in its own right. There is already much discussion of the possibility of *instrumentally* valuable aspects of outer space—objects or environments that can be used for others' benefit, and that we might use to benefit ourselves and our futures. Perhaps there is even too much focus on the use-value of outer space, insofar as it supersedes the debate about what in the cosmic realm is valuable in and of itself, aka *intrinsically* valuable.[1] This is an important question because there are arguably moral limits on how we are permitted to treat that which is intrinsically valuable and deserving of moral respect, even if using it might benefit us. There will be times when morality demands that humans with the power to expand their projects into outer space must sacrifice some of their goals or interests to avoid mistreating that which they ought to be respecting. Which goals or interests must be sacrificed is a separate question. The question at hand is: Who or what could deserve moral consideration in outer space?

Of course, historically, many humans have failed to afford adequate consideration and respect even to other humans, and this failure has reliably been a product of unjust bias. Clearly, the issue of bias is highly relevant to discussions of moral consideration. Bias can cloud our analyses and lead directly to injustice. For example, while present-day humans are certainly not immune to bias, there is today a widespread ostensible rejection of the attitudes that position entire groups as automatically inferior and thereby less deserving of respect. This rejection is typically taken to be an example of moral progress. Such progress has led to the extension of respect to far more than just humans. If nothing else, ethicists and others have demonstrably been working to defend an increasingly widening scope of moral consideration.

Accounts of Value

Consider a brief account of the history of moral consideration within Western ethical approaches to determining that which is deserving of respect, starting with the approach that arguably has the narrowest scope. *Ratiocentrism* is the view that possessing rationality is what makes something valuable and worthy of respect.[2] According to this view, if you can evaluate and act on reasons, then, just by virtue of you existing as a rationally autonomous creature, you matter morally. You deserve respect. However, this view brings with it some highly counterintuitive counterexamples. There are many people who are not rational in this way, from babies to the elderly to the severely cognitively injured or disabled. Some people never have the potential to meet this bar, and some lose that ability permanently. However, there is a common intuition that these individuals are deserving *not only* of basic respect but also of *extra* consideration, care, and protection. Yet the ratiocentrist must conclude that such individuals have no value outside of their use. We would be permitted to exploit or oppress them as long as doing so did not disrespect rational humans. This is clearly an unpalatable implication. Furthermore, ratiocentrism also seems to entail that a vastly more rational alien society could be justified in mistreating humans in certain circumstances, simply because we are not as smart as them. Again, this seems counterintuitive. Finally, ratiocentrism entails that many or most non-human animals have no moral worth and such a view, if true, could be used to justify a wide array of animal mistreatment.

Animal ethicists in particular are known for rejecting the ratiocentrist view, and many argue for a much broader moral community bound not by presumed intelligence but by the capacity for sentience (Bentham 1789/1998; Regan 1985; Singer 1975).[3] If you are sentient, then you are capable of *experiencing* being better or worse off. According to *sentientism*, then, virtually all humans and entire swaths of the animal kingdom are deserving of respect because we are the kinds of things that can experience suffering or improvement. On this view, there is a baseline moral imperative not to cause undue suffering to anything that can suffer, and sometimes to positively improve the welfare of those who are capable of both pleasure and suffering. In an intuitively plausible way, then, this view rejects the claim that there is nothing in itself wrong with the mistreatment of non-human animals.

At the same time, there is an increasing focus on environmental ethics per se, and the environment does not comprise merely sentient animals. Plants and other arguably non-sentient organisms are contained in the natural environment that many humans believe to be worthy of respect and protection. In order to account for this, some have adopted a biocentrist view of moral consideration.[4] According to *biocentrism*, that which is alive is intrinsically valuable and deserving of respect. In astrobiology ethics, Charles Cockell, a prominent biocentrist, argues for the intrinsic value of microbes, and specifically communities of microbes. According to Cockell (2016), microbes are alive and thereby have interests, rendering them intrinsically valuable. If this is correct, it means we have strong duties to respect any extraterrestrial microbes we might come across—such as any potential Martian microbes.

One might challenge this view by arguing that it implausibly implies that stepping on a community of Martian microbes is just as bad as, say, killing some humans. However, one viable response to this charge is to maintain that intrinsic value comes in degrees and that duties of respect can depend heavily on details and context. However, philosopher Kelly Smith (2009, 270) anticipates this kind of response when considering the intrinsic value of Martian microbes, stating that:

> [t]he problem here is that such a move robs the intrinsic value account of its original motivation. The whole point of distinguishing intrinsic from instrumental value in the first place was to prevent any kind of haggling over who is more important than whom in different circumstances.

But there is an important challenge to Smith's account of the motivation behind views of intrinsic value. Instead of side-stepping moral haggling, perhaps the whole point of distinguishing that which is intrinsically valuable from that which is instrumentally valuable is to figure out what is *true* of the world. If it is true that microbes possess intrinsic value, then that is important for us to figure out. It is also important to note that this value by itself would not rule out the possibility of differing degrees of intrinsic value. Nor does anything about the world require that ethical analysis be straightforward and haggle-free, as much as we might wish it were so. The point is that understanding what has intrinsic value is only one part of the moral equation—understanding it does not guarantee that we then have automatic answers to the heavy question of what to do when values compete.

Indeed, we should take seriously the possibility that ethics is, in many cases, messy and complex. As Johnson-Schwartz (2020: 136) writes:

> intrinsic value is not a magical wand that immediately resolves heretofore ambiguous cases or that fabricates overriding obligations of protection or preservation. Instead, [one] can claim that an entity being intrinsically valuable merely provides this entity with a seat at the table of moral deliberation. All this establishes is that this entity must be taken into consideration for its own sake, even if it cannot speak for itself, and even if its value is vanishingly small.

So, while biocentrism need not require a commitment to the claim that all living things must be equally valuable under all circumstances, it does fundamentally reject the claim that there are no morally relevant considerations involved in the act of, say, destroying a community of microbes. Microbes, if they are intrinsically valuable, deserve a seat at the table of moral deliberation.

But even biocentrists cannot account for our duties to the non-living, if they obtain. One way to think about obligations to non-living entities is to look at ecosystems, because they contain non-living entities. Certain proponents of *ecosystem ethics* argue that ecosystems are legitimate targets of moral analysis; they are deserving of moral consideration and respect (e.g., Fox 2007; Rolston 2012). Leopold's (1949) famous *Land Ethic* is notable for its focus on valuing ecosystems, though the worth of these ecological spaces lies primarily in their capacity to house biotic communities. His approach does not afford special, non-instrumental status to non-living things. Yet, ecosystems are ecological spaces that involve processes of both biotic *and abiotic* materials, working together in certain integrated ways, which indicates that they are not reducible merely to the value of their living parts. Of course, they are likewise not reducible to rational or sentient parts, and a given ecosystem may not even have such parts to begin with. Thus, neither ratiocentrism, sentientism nor biocentrism can account for the possibility that we have duties to respect ecosystems in and of themselves. There must be something beyond these other attributes that can ground duties of respect insofar as it is true that we ought to respect ecosystems.

Life-Bias

I will return to issues in ecosystem ethics later in this chapter, but for now I wish to note that ecosystems have living material as a crucial component, even if ecosystemic moral considerations do not bottom out at their living parts. All of the approaches to moral consideration in Western analytic ethics presented above are therefore biological in an important sense. Intuitively, life matters morally.[5] While I do not deny this, I maintain that we should think carefully about whether it is the case that *only* life or life systems matter morally. There is an often unacknowledged and unchecked assumption that non-living components of our world do not matter morally in and of themselves. They have use value but not intrinsic value and are not owed respect simply by virtue of existing as they are.

Evidence of this life-bias in outer space discussions has been developed quite clearly by ethicist Johnson-Schwartz (2019: 354), who summarizes life-bias as "the belief that *a space environment is worthy of protection if and only if it is home to life (or traces of past life)*". She focuses on debates in planetary protection ethics. Planetary protection takes seriously the fact that extreme precautions must be taken to protect both Earth and other planetary bodies from potential harmful disruptions caused by living organisms traveling between these sites. Every time an extraterrestrial sample is retrieved on Earth, great care is taken not to allow the potential backwards contamination of Earth by whatever is contained in the sample. If, for example, the sample contained extraterrestrial microbes, we would not know how those organisms might interact with or affect environments here on Earth and must thereby exercise caution by keeping the sample well-contained. Planetary protection is a two-way street, and, arguably, great care must also be taken to prevent Earthly microbes and organisms from disrupting outer space environments that are visited by human or robotic travellers. This is known as forward contamination. The United Nations' Outer Space Treaty of 1967 addresses the issue of planetary protection and cautions against both backward and forward contamination in space exploration.

Johnson-Schwartz (2020: 126) exposes that several major players in planetary protection debates use language indicative of a life-bias, such as Astronomer and SETI founder Carl Sagan, who appears to lament that the Viking lander on Mars showed "no footprints, not even a cactus or a kangaroo rat" (Sagan 1980: 121), highlighting that "for as far as we could see, there was not a sign of life". Cockell (2005) goes as far as to entertain arguments for the claim that life is so important that humans ought to seed and populate lifeless planets with organisms, thereby importing value to those environments. Similarly, McKay (2013: 159) contends that a life-filled planet is more valuable than a dead one, when he states that "[l]ife is not the only source of value, but it is perhaps the most important source", following that with the claim that "life is unique in that not only can its value be preserved, it can be spread". Similarly, Lupisella (1997: 93) notes that, even if Mars turns out to be "dead" that does not mean that policy and planning efforts were "wasted" because they could still be repurposed for efforts that are able to recognize "the intrinsic value of fundamentally different life forms", implying that efforts spent on lifeless discoveries are only worthwhile if they help to lay the foundation for *life-bearing* discoveries.

Thus, there is evidence of the unexamined presumption that a lack of life entails a lack of value, and, as we expand into space, we should ask whether such bias is justified—specifically, a bias against sterile objects and environments in outer space. Arguably, humans have a long history of exhibiting bias against that which is deemed sufficiently unlike whatever it is those in power wish to exploit, whether it be other humans, other animals or elements of nature and the environment. These attitudes have often involved the idea of a standard against which other things are measured and, often, found wanting. But this attitude itself may be the problem. Even if human beings are truly intrinsically valuable, this does not mean that we are *at the same time* the standard against which all else must be measured to determine worth. And even if planet Earth contains much that has intrinsic value, that does not by itself mean that only Earth-like settings contain value.

When faced with hierarchical dualities, the category deemed inferior is often described in terms of what it *lacks*—and whatever it lacks usually happens to be that which the dominant party has. De Beauvoir's (1949/2010: 6) classic work in feminist philosophy bears this out with respect to sexism: "humanity is male and man defines woman not in herself but as relative to him". Women have historically been cast as, e.g., *less* intelligent, *more* emotional, weak*er*, etc. Less or more than what, one might ask? The answer is the standard, i.e., men, and the comparison to men is so entrenched that it can easily go unstated.[6]

We see this same kind of framing in the above progression of accounts of intrinsic value, where purportedly inferior groups can be described through their failures. Claims to intrinsic value have historically been theoretically unavailable for the *less* rational, the *non*-sentient, or the life*less*—even though these things all possess many features and attributes that, say, rational sentient life lacks. A non-rational ant is a sophisticated olfactory navigator. A non-sentient plant can photosynthesize. A non-living mountain range can mark the powerful collision of continental plates. When we categorize something first and foremost by what it is not, we are defaulting to a duality by indicating the simultaneous existence of that which seemingly exists all by itself, against which that in the initial category is being compared. And such dualities form the foundation of many claims of injustice. Sexism, ableism, racism, speciesism, and the like are founded on unjust hierarchies that cast the inferior group as missing something that the dominant group possesses.

Thus, we should be on the look-out for potential unjust dualities as we branch out into space, and there is reason to think that the attitudes that form the life-bias follow these lines of thinking. There is a tendency to conceive of that which lies beyond our planet as *non*-living and life*less*. We saw above Sagan highlighting a *lack* of signs of life on Mars, and Lupisella referring to Mars as *dead*. Environmental ethicist Holmes Rolston III (1986: 172) takes issue with this framing in his notable work defending the intrinsic value of non-Earth planets; he writes:

> [t]hese other worlds are not places that failed. Nature never fails. Nature only succeeds more or less with its projective integrity…We ought not condemn Mars because it failed to be Earth, although we may value it less than Earth…Learning to appreciate these alien places for what they are in themselves, not deprecating them for what they failed to be, will provide an ultimate test in nature appreciation.

More recently, philosopher Tony Milligan (2015: 107) echoes this as he writes:

> it is a familiar human trait, but also a failing, to encounter *what is* in terms of *what it is not*…It does not seem unfair to suggest that it is this sense of lack, this sense of what isn't there, which shapes and to some extent distorts perceptions and feeds a dismissal of calls for planetary protection…And yet what *is* there is also remarkable in its own right.

And Johnson-Schwartz (2019: 356) helpfully synthesizes both of these authors' points to conclude that:

> [w]hat Rolston and Milligan suggest is that the life-bias, which prompts us to compartmentalize environments primarily as either 'lifeless' or 'life-bearing', diverts our focus away from the positive, 'remarkable' features of these environments and thus negatively affects our ethical evaluation of these environments.

Therefore, the mere fact that there is a tendency to believe that living entities and environments are morally superior to non-living entities and environments does not excuse us from failing to examine this belief and to question the role that bias may be playing.

One worry is that lifeless entities and environments seem to lack any of the features heretofore used to ground claims of moral worth and consideration. But perhaps that is precisely the problem. A life-bias would cause us to see lifeless space environments first and foremost as what they are not, rather than what they are. If this bias occludes our ability to recognize intrinsic value in outer space, then we ought to reject it and acknowledge in our moral deliberations the possibility that sterile space environments can still be worthy of respect and ought to be considered in and of themselves. This worth would not entail that we could *never* harm or disrupt sterile space environments, instead, it would entail that we would need good reason to do so—just as we would need good reason to justify why a given act of harming a human could be permissible.

Ecosystemic Value

The challenge to the claim that non-living entities and environments deserve respect is that the life-bias is not *mere* bias. Perhaps common attitudes toward living versus non-living things are tracking real moral properties—namely, intrinsic value that is present in living organisms and systems and not present in non-living organisms and systems. For the sake of argument, let us grant truth to this challenge and see where we nonetheless end up.

As noted above, ecosystem ethics posits the intrinsic value of ecosystems, which entails that we have duties of respect to abiotic material *not* in and of itself, but rather insofar as it is part of ecosystemic integrity. Because ecosystems comprise both biota and abiota, life is necessarily part of ecosystems. As such, we have expansive ethical duties here on Earth, but one might contend that such duties do not extend into outer space insofar as outer space is sterile and not part of ecosystemic processes. There is the possibility that the intrinsic value of ecosystems lies in the systemic *holism* present in their ecosystem processes (Rolston 2012). Ecosystemic value, then, would not be reducible to the sum of value of each individually valuable entity in the ecosystem. Rather, ecosystemic value would manifest in higher-level systemic processes—the integrated functioning of the variable materials in ecosystems—resulting in systemic integrity and a whole that is greater than the sum of its parts.

We are used to thinking of ecosystems as their own small parts of planet Earth—a wetland, a rainforest, an alpine tundra, etc. Without denying such ecosystems their own identities and particular value, we can at the same time recognize that their boundaries are not discrete, and that they are integrated in and affected by much that ostensibly lies outside the narrower conception of "ecosystem".[7] From a more distanced perspective, then, we might also recognize our *planetary* ecosystem. Systemic processes across Earth are not contained to individual locations and are integrated in ways that lend to the study of planetary processes that generate global effects. Thus, it may make sense to ask how we ought to treat not only individual, smaller ecosystems within planet Earth but also how we ought to treat the planet itself and to respect its larger, planetary-level ecosystemic integrity.

But why assume that the bounds of ecosystemic functioning stop at Earth's atmosphere? Perhaps we should not assume this if we are taking seriously the ways in which planet Earth influences and is influenced by the larger Solar System. For example, planetary bodies as near as the Moon and as far as Jupiter and beyond have prevented objects from colliding with Earth. These interceptions have allowed for the development and continuation of various evolutionary trajectories here on Earth that have given rise to organismic flourishing and to integrated biotic and abiotic functioning. This is but one of many such causal examples of Earth's dependence on the Solar System for its current ecosystemic functioning. And of course, humans and their effects are not bound by Earth's atmosphere either. As we saw before with the issue of planetary protection, microbial life might hitch a ride on our outer space excursions, and the possibility for unintentional organismic expansion is present in any mission. In our own biological form, we have already travelled into near-Earth orbit and beyond our planet to our nearest neighbour, the Moon.

If we continue on this trajectory, it is easy to see how not only the Solar System's past but also its future can ground claims of extra-planetary solar-ecosystemic functioning. Earth may be our home, but the Solar System is our home, too. Milligan (2013: 4) considers humans' expansion to the Moon and, without denying that we may use it to a certain degree, writes that "if we *merely* use it, or regard it *merely* as a resource, then we risk losing sight of the ethical task of *being at home* in a larger region of space". Astronomer Woodruff Sullivan (2013: 172) also acknowledges this idea of outer space as our home when he writes that we should: "extend the concept of eco-logy (home-study, etymologically) such that home refers to more than just our neighbourhood, or our watershed, or even our biosphere, but indeed refers to the entire Solar System". He adds that, according to what he calls *planetocentric ethics*, "each planetary body has intrinsic value as part of an ecosystem writ large, and can be modified only with strong justification" (Sullivan 2013: 172). If this is correct, then many outer space environments in our solar system are owed our respect regardless of whether they individually contain life. Even sterile planetary bodies are nonetheless part of our solar ecosystem and ought therefore to be afforded moral consideration. Thus, even if we grant the truth of the life-bias, this does not rule out duties of respect to the abiotic insofar as solar-ecosystemic integrity can be intrinsically valuable.

Furthermore, our physical imprint includes not only footprints (and detritus) on the Moon but also imprints of the robotic extensions of ourselves that we have sent out into the cosmos—from rovers that drill holes on Mars to the Voyager spacecraft traveling into interstellar space. It is becoming increasingly clear that ecosystems do not comprise merely natural phenomena (whether biotic or abiotic) but also artificial phenomena. We are poised to proliferate our outer space impacts, given this technological trajectory—with non-human objects preceding and extending humans' bodily reach.

We already see strong integration between ourselves, our technology, and our natural environments here on Earth. As ecofeminist Vandana Shiva (1992) has argued, there are better and worse ways of manifesting these integrations, and many modern technological developments can disrupt rather than promote balanced, respectful, and sustainable ecosystemic functioning. As we extend ourselves further into space, we will need to consider how our technological artefacts—including robotic devices, artificial intelligence programs, and other technologies—affect and are affected by this larger ecosystem of which they are a part. This would mean that artificial entities within an ecosystem must be factored into the moral equation insofar as they are part of its systemic integrity. But perhaps this is not all that justifies their moral consideration.

The Value of the Abiotic

If the above is correct, then ecosystems are not of necessity strictly natural. Of course, this was the case well before digital technologies existed. Artificial human artefacts—e.g., housing, agricultural tools, weaponry, etc.—have since their inception been part of the smaller ecosystems of which humans have been part. But newer technologies, such as robotics and artificial intelligence, threaten to be more than just a new wave of artificial development. Among other things, these technologies are able to expand well beyond humans' own capabilities for expansion and to disrupt certain presumptions undergirding the previous discussions of moral considerability. That is, some new technologies have attributes that are strikingly similar to attributes valued by the ratiocentrist and assumed by many to be held only by humans—namely, the attributes of rationality and autonomy—enough so that there is a debate as to whether certain new technological entities, such as those with robotic artificial intelligence, have rights and are thereby owed a kind of moral respect (e.g., Gunkel 2018; Turner 2019).[8]

For example, Coeckelbergh (2010, 2018) has argued that we ought to treat robots as though they have rights insofar as we relate to them in the same way we relate to things that are typically afforded rights. Human beings are the most typical candidates for bearers of rights who are thereby owed respect. When robots occupy outer space as human proxies, they transmit information and remote sensory experiences back to our planet. Such robots are in a position to do the fieldwork for us, as it were, and to analyse and explore outer space environments in our stead. They may be extensions of our eyes, ears, minds, and general abilities to observe and interact with outer space. They will also act according to certain rational demands and exercise a kind of autonomy in their decision-making. Such robots would also have capabilities that far outmatch those of humans. Not only would their machine learning systems allow them to perform tasks that humans cannot, their physical characteristics would also be far more resilient in the harsh conditions of outer space than our fragile and demanding bodies ever could be. Under these circumstances, it is not difficult to see how we could end up relating to outer space robots in a way that opens up the question of whether they deserve respect in their own right.

Furthermore, as abiotic technology extends further into space, there are arguments to consider regarding the possibility that sterile space environments in fact deserve respect in their own right, regardless of their ecosystemic status. Rolston (1986) set the stage for this work by arguing for the projective integrity of the natural world and of planetary bodies in particular.[9] Outer space environments are the sites of truly remarkable natural projects. For example, Jupiter is not a failure for not developing the conditions for life just as a blue whale is not a failure for not developing the capacity for flight. Instead, Jupiter has integrity in and of itself and has succeeded as a site of one of nature's notable accomplishments. Nature has many projects, and, according to Rolston, these projects often demand respect. This respect, if it obtains, is owed due to the intrinsic value of nature's projective integrity. The cosmos is full of nature's projects and hence a place where we might discover projects with integrity and value. If this is correct, we ought to be factoring outer space environments into the moral equation, regardless of whether they house life or the potential for life. They, too, might deserve a seat at the table.

Conclusion

Let us consider, then, a likely early stage of future space expansion—one where humans and other organisms are largely or entirely absent due to our fragile forms, forms that are heavily dependent on life-supporting conditions that are so rare in our solar system. The places in question would involve a mixture of sterile planetary bodies and abiotic robots. These will be sites of largely or entirely non-living phenomena, but this does not mean we should automatically default to the assumption that these sites lack intrinsic value, as we saw above. They might have value in their own right, and they might also have value as key elements of ecosystemic integrity. Notably, the two are not mutually exclusive, and the reality of one would not preclude the reality of the other. Thus, we should consider and address the possibility that certain of nature's extraterrestrial projects and certain of humans' extraterrestrial artefacts command respect even when they have no living components. This means that humans on Earth must act with due consideration when proposing actions or devising systems that might exploit, destroy, devalue or harmfully disrupt these cosmic spaces.

I would like to be clear about the ethical conclusions I am arguing for in this chapter. I have not argued definitively that ecosystems or sterile environments or robots with machine learning technology are intrinsically valuable and deserving of moral respect. Those debates are far more involved than what I have presented here. Rather, I wish to claim that these debates deserve our attention. Even if we are unsure exactly what deserves moral consideration in and of itself, we can be sure that we have an obligation to *think carefully* about what truly deserves moral consideration in and of itself. We have a responsibility to deliberate well on the question of how we ought to act. We are the moral agents at the table, capable of making ethically informed decisions. And we can know that there are certain arguments we ought to be thinking about even if we remain unsure here and now about the correct conclusions.

What I contend, then, is that it is wrong to plan for space expansion without directly engaging with these debates. In particular, early-stage space expansion that precedes larger-scale human occupation is especially susceptible to the elision of these crucial questions, given our demonstrable life-bias in discussions to date. I have shown that there are viable candidate justifications for extending our assumptions about the possession of intrinsic value to the non-living, both natural and artificial. I have also shown that there is a legitimate risk that bias will continue to cloud our reasoning by allowing for the unchecked and unargued for assumption that sterile space environments lack anything of intrinsic value. Since so many of the debates about ethics and justice in space are focused on what to do when humans and other organisms occupy outer space environments, I wish to highlight the need to focus also on how to act appropriately in early-stage space expansion when outer space environments house the sterile and abiotic—and when we might easily default unthinkingly to a life-bias.

I conclude that the burden of proof falls on anyone who engages in early-stage space expansion to provide justification for disruptive actions in the face of legitimate moral debates about the potential intrinsic value and considerability of sterile natural and artificial elements of outer space environments. We should be sceptical of anyone with high confidence that they know, in these early stages, what exactly values and rights look like in outer space. As Johnson-Schwartz (2016: 106) notes:

> not only does our lack of knowledge about the space environment impede us from speaking with confidence about intrinsic values in space, it also hinders us from speaking with confidence about virtue and duty in space. This gives rise to two inveterate features of the methodology of space ethics: that it must develop slowly and alongside our pursuits in space, and that it should function as a reflective guide for these pursuits.

Let us move slowly, then—as slowly as our ignorance demands. Let us reflect on both the knowledge we already possess and the knowledge we gain as we direct ourselves increasingly toward extraterrestrial environments. When we have performed the proper analyses and come to well-justified conclusions about how we may permissibly treat not only one another but also other organisms, ecosystems, artificial entities, abiotic objects, and sterile environments in outer space, we will have set the stage for just and morally appropriate attitudes and assumptions regarding our potential effects on the cosmos of which we and our living planet are part.

Notes

1 It is worth noting that there is a philosophical debate about whether anything can be valuable in and of itself, i.e. intrinsically valuable, but the details of that debate are beyond the scope of this chapter, and we will proceed under the common ethical assumption that intrinsic value exists.
2 Immanuel Kant's (1724–1804) work in moral philosophy provides the most canonical account of ratiocentrism (see Kant 1999).
3 This position is well-established but certainly not uniform among animal ethicists.
4 See Taylor (1986).
5 This is not a strictly Western intuition, and many non-Western systems of thought include attitudes of respect toward all life. Perhaps most notable is the Indian spiritual tradition of Jainism, whose followers pursue *ahimsa*, which is the commitment not to harm other life.
6 I note that I am not through this quotation subscribing to the assumption of a sexual binary according to which one must either be a man or a woman.
7 See McShane (2004) for an exposition on how to define ecosystems and a viable response to the charge that assignations of ecosystem-status are a purely subjective matter.
8 There is, relatedly, a very interesting debate about whether, and the extent to which, robots are moral agents in addition to considering whether they qualify as moral subjects or patients—which is beyond the scope of this chapter.
9 While Rolston focused specifically on the possibility of the intrinsic value of planetary bodies, which is our focus here, others have also considered the extension of intrinsic value and respect to all natural things, both living and not—sometimes referred to as "deep ecology"—most notable of whom is Arne Næss (1973).

References

Bentham J (1789/1998) *An Introduction to the Principles of Morals and Legislation*. Amherst, NY: Prometheus Books.

Cockell C (2005) Planetary Protection – A Microbial Ethics Approach. *Space Policy* 21: 287–292.

Cockell C (2016) The Ethical Status of Microbial Life on Earth and Elsewhere: In Defense of Intrinsic Value. In JS Schwartz and T Milligan (eds) *The Ethics of Space Exploration*. New York: Springer, pp. 167–179.

Coeckelbergh M (2010) Robot Rights? Towards a Social-relational Justification of Moral Consideration. *Ethics and Information Technology* 12(3): 209–221.

Coeckelbergh M (2018) What Do We Mean by a Relational Ethics? Growing a Relational Approach to the Moral Standing of Plants, Robots and Other Non-Humans. In A Kallhoff, M Di Paola, and M Schörgenhumer (eds) *Plant Ethics: Concepts and Applications*. London: Routledge, pp. 110–121.

De Beauvoir S (1949/2010) *The Second Sex*. New York: Alfred A. Knopf.

Fox W (2007) *A Theory of General Ethics: Human Relationships, Nature and the Built Environment*. Cambridge, MA: MIT Press.

Gunkel DJ (2018) The Other Question: Can and Should Robots Have Rights? *Ethics and Information Technology* 20(2): 87–99.

Johnson-Schwartz JS (2016) *On the Methodology of Space Ethics*. In JS Schwartz and T Milligan (eds) *The Ethics of Space Exploration*. New York: Springer, pp. 93–107.

Johnson-Schwartz JS (2019) Where no Planetary Protection Policy Has Gone Before. *International Journal of Astrobiology* 18: 353–361.

Johnson-Schwartz JS (2020) *The Value of Science in Space Exploration.* New York: Oxford University Press.

Kant I (1999) *Practical Philosophy.* Translation, AW Wood. Cambridge, UK: Cambridge University Press.

Leopold A (1949) *The Sand County Almanac, and Sketches Here and There.* New York: Oxford University Press.

Lupisella M (1997) The Rights of Martians. *Space Policy* 13: 89–94.

McKay CP (2013) Astrobiology and Society. In C Impey, AH Spitz, W Stoeger (eds) *Encountering Life in the Universe.* Tucson, AZ: The University of Arizona Press.

McShane K (2004) Ecosystem Health. *Environmental Ethics* 26: 227–245.

Milligan T (2013) *Scratching the Surface: The Ethics of Helium-3 Extraction.* Paper presented at the 8th IAA Symposium on the Future of Space Exploration: Towards the Stars, Torino, Italy.

Milligan T (2015) *Nobody Owns the Moon: The Ethics of Space Exploitation.* Jefferson, NC: McFarland.

Næss A (1973) The Shallow and the Deep, Long-range Ecology Movement. *Inquiry* 16, reprinted in *Sessions* 1995: 151–155.

Regan T (1985) *The Case for Animal Rights.* In P Singer (ed.) *In Defense of Animals.* Oxford: Basil Blackwell, pp. 13–26.

Rolston H (1986) The Preservation of Natural Value in the Solar System. In EC Hargrove (ed.) *Beyond Spaceship Earth: Environmental Ethics and the Solar System.* San Francisco: Sierra Club Books.

Rolston H (2012) *A New Environmental Ethics: The Next Millennium for Life on Earth.* New York and London: Routledge.

Sagan C (1980) *Cosmos.* New York: Random House.

Shiva V (1992) Women's Indigenous Knowledge and Biodiversity Conservation. *India International Centre Quarterly* 19: 205–214.

Singer P (1975) *Animal Liberation.* London: The Bodley Head.

Smith K (2009) The Trouble with Intrinsic Value: An Ethical Primer. In C Bertka (ed.) *Exploring the Origin, Intent, and Future of Life.* New York: Cambridge University Press, pp. 261–280.

Sullivan WT (2013) Planetocentric Ethics: Principles for Exploring a Solar System That May Contain Extraterrestrial Microbial Life. In C Impey, AH Spitz, W Stoeger (eds) *Encountering Life in the Universe.* Tucson, AZ: The University of Arizona Press.

Taylor PW (1986) *Respect for Nature: A Theory of Environmental Ethics.* Princeton, NJ: Princeton University Press.

Turner J (2019) *Robot Rules: Regulating Artificial Intelligence.* Berlin: Springer.

Conclusion – A New Hope?

The Future of Researching Space Crime, Criminal Justice and Ethics

Yarin Eski and Jack Lampkin

Lessons Learnt Through Compiling this Book

Let us go back to Leonard Simon Nimoy's words in the introduction of this book: "*the more we share the more we have*". It is easier said than done though, which is what this book testifies to; bringing three different disciplines together is simply not that easy. However, in having done so, we have started to have more; the contributions with having in each and every way their own disciplinary backgrounds, enriched one another. There are five (interacting) lessons that emerged from this book that we shall highlight here: (1) epistemological cogitations on space criminology; (2) methodological innovations; (3) space criminal jurisdictional challenges; (4) addressing environmental space crimes; and (5) anticipating space-enabled sexual violence.

Epistemological Cogitations on Space Criminology

Space criminology can be and do many things (Lampkin and White 2023), as discussed by several authors in this book, it has a pivotal role to play, one that has an emancipating quality, acquiring and developing holistic perspectives and ideas. In fact, not only space criminology, but also other disciplines ought to maintain a holistic approach to space crime, criminal justice and ethics. This is in line with **Rob White**'s idea of ecological citizenship that advocates for Earth-centric policies, challenging nationalistic agendas and aligning with "Earth Rights" and emphasising collective responsibility for all species. Similar principles should guide space exploration, prioritising common interests over corporate profit; something space criminology should study thoroughly.

In making space criminology holistic, perhaps it could – and should – be joined by the development of an astro-victimology? **Jack Lampkin** showed how space law's infancy and lack of jurisdiction complicate the theorising of deviance in space activities. Therefore, potential victimisation in outer space suggests that the terms "space deviancy" or "space zemiology" might be more appropriate terms/areas of study, which could assist in holistically understanding crimes and harms in space, addressing these proactively.

DOI: 10.4324/9781003437178-24

Ultimately, it is vital to consider victimisation at the outside of *space criminology* in order to avoid the sideling of victims, which has been the bedrock of justification for the study of victimology. Criminology has been criticised as historically focussing too heavily on the offender, thus neglecting the victim who is the most impacted by criminal behaviour.

Furthermore, it is important to integrate a *space criminology* with other social scientific fields of inquiry that have started to consider space already, such as *space anthropology* (Masali et al. 2010; Messeri 2016) and *space political science* (Galloway 1986). Given how political scientists and criminologists have explored the intersection of politics and crime before, investigating how institutions shape crime policy, **Sarah Poss, Yarin Eski** and **Jack Lampkin** suggest to do so as well regarding specifically off-Earth spaceport crime and policing. They propose to move toward a politico-criminology of spaceports, focusing on how geopolitical tensions influence security ideologies, particularly at spaceports, that reflect global power dynamics. Such understanding of outer space's geopolitical geographies of off-planet spaceports is essential for a holistic space criminology.

Nevertheless, we ought to be careful in assuming "new stuff", or new criminologies, receive their birthright simply for being new (Rock 2005). Sure, emerging trends in criminology bring excitement and relevance to the field, necessitating ongoing adaptation to societal changes, as space exemplifies. However, our focus should remain on Earth due to the consequences of human activity, notably environmental degradation, as **Anita Lam, Nigel South** and **Avi Brisman** argued. Criminologists must still prioritise addressing anthropogenic environmental issues before considering extraplanetary concerns, because, how we collectively address and steward our planet, will also determine our ability to prevent catastrophic outcomes predicted by the elite, urging a shift towards planetary preservation, including in an interplanetary setting.

In fact, if we want to establish a holistic, *credible* space criminology, we need to go back to the dark origin of space exploration, where **Yarin Eski** took us in **Chapter 1**. While criminologists played an important role in Nazi ideology, rationalising Aryan superiority, Nazi aerospace scientists were advancing space exploration. They birthed the ethos of "one space for one race", influencing subsequent space endeavours amidst the US–Soviet rivalry and contemporary space races. Acknowledging this tainted legacy, especially within the space community, is crucial. A critical space criminology must confront these "roots of evil" and Nazi influence, when fostering an epistemological critique to address historical fallacies. Only then can it effectively tackle both unlawful and lawful space crimes, illuminating overlooked aspects of our past and present.

Methodological Innovations to Go "Out There"

Space is exciting, and space crime, criminal justice and ethics are perhaps even more exciting to do research on. But how are we supposed to do so with normal earthling research budgets, when a trip to space costs approximately US $55,000,000 (Carter 2021)? We need to get as close as immersively possible, using science fiction, space crime (control) scripts, virtual reality (VR) and analogue astronaut missions.

Science fiction authors and space planners often propose simplistic solutions like space prisons or execution-by-airlock for future space settlements, neglecting the complexities of criminal justice. However, space exploration and settlement offer a chance to develop innovative, non-carceral approaches beyond Earth's traditional systems, as **Erika Nesvold** has convincingly described in **Chapter 10**. Sci-fi and criminal justice systems in space settlements could enable experimentation with various models. They may present risks, but will also foster ethical considerations and societal advancement.

In the case of resource mining in outer space, there are also unprecedented risks, demanding knowledge on how potential environmental crimes are associated with astro-mining, which requires proactive research through the use of space crime scripts that draw on crime prevention techniques, as **Emilia Ziosi** has explained in **Chapter 4**. Such scripts may also form the inspiration, or foundation even, for anticipatory regulatory frameworks, she concludes, especially because there is a lack of international regulations on space crimes and harms. Space crime scripts may also enable VR space crime scenarios and can be used in analogue astronaut missions.

In **Chapter 3, Iris van Sintemaartensdijk** explains how, since the introduction of VR for criminological research, we have seen many fields use this method to further understand behaviour when it comes to burglary, bystander effect and aggression. Now, we should use this method to take the next step and study space crime. We can use the ability to create any scenario possible for normally hard-to-study scenarios and expose people to various new scenarios, not only to learn what people think of crimes in outer space but also to encourage people to be interested in learning about this type of crime. Moreover, VR can serve as a conversation starter and thus as a starting point for further discussion in the spacefaring community.

Traversing Earth's poles, space analogues or analogue astronaut mission may prove useful in studying crime, justice, and ethics for outer space, as **Rebecca Kaiser** has shown in **Chapter 8**. These analogues offer practical insights, shaping future policies and protocols for space exploration. Analogues serve as a testing ground, ensuring that as humanity ventures into space, it carries an ethical ethos reflective of its values, and thus, the analogue lessons learned could enhance safety and success in cosmic endeavours, fostering fair and ethical practices.

Space Criminal Jurisdiction Challenges

"Every country is effectively a coastal state of the universe" (Ward Munters said in van Heemstra 2022). Put differently, Earth, especially jurisdictionally speaking, is the coast and everything outside our atmosphere, the open sea. Maybe next to Earthly understanding of law and jurisdiction (perhaps in particular when it comes down to the hidden acts of space crime or regarding criminal justice and ethics, including policing and incarceration), space should be seen as a realm of continuous change, focussing on space objects and their connections to various stakeholders. An answer to this question may reside in several contributions in this book on satellite warfare, criminal justice jurisdiction, space policing and ethics of space (criminal) law.

The intersection of space, ethics and law, including jurisdictional challenges, entails a complex and dynamic process, which then requires extensive conversation, delving into the essence of human nature and the universe. This discussion requires courage and dedication, mirroring the challenges inherent in space exploration and rational inquiry, as **Brian Green** explains in **Chapter 20**. To navigate this domain effectively, we must delve deep into its intricacies and commit to a thorough examination and understanding. It is a task that demands serious engagement and effort if we aspire to achieve a comprehensive grasp of space-related issues.

In **Chapter 5, Bruno Reynaud de Sousa** has shown that the current international landscape is shifting towards multipolarity, prompting chaos according to the UN Secretary-General. Space commercialisation introduces an "age of transparency", enabling unprecedented scrutiny of state actions. This transparency, facilitated by US-led Earth Observation (EO) companies, affects everything from armed conflict documentation to human rights violations. However, it poses challenges, such as reliance on profit-driven entities and Europe's limited access to EO capabilities. While satellite transparency offers potential for societal good, ethical considerations and geopolitical implications must be addressed. Ultimately, responsible use of this power is crucial to avoid descending into chaos.

In particular regarding geopolitical conflict on Earth, there are challenges posed by satellite-enabled intelligence. In discussing traditional ethical theories, including the likes of Kantian deontology that may not suffice for guiding intelligence ethics in outer space, Act and/or Rule Consequentialism may offer a more progressive approach, as **Benjamin Segobaetso** describes in **Chapter 21**. A Sophisticated Rule Consequentialist approach emerges as worthy of exploration for outer space ethics, given its capacity to surpass other theories in addressing moral convictions and offers flexibility.

Lacking jurisdictional clarity in relation to environmental harm in outer space, or not having clear international provisions, unilateral steps by states to criminalise such harmful activities in space are unlikely, as **Alla Pozdnakova** explains in **Chapter 16**. Regulatory competition among space nations may hinder strict rule imposition, however, and although rapid harmonisation of space environmental criminal law seems improbable, ecocide prevention initiatives and increasing space sustainability regulations offer some hope. Applied Global Commons research could enhance international law's effectiveness by defining environmental harm in space and facilitating its legal enforcement.

To have a space criminal justice act, or anything like it, is already challenging to establish. But even if there is law, it does not mean space crime and harm stop. In fact, next to law, law enforcement is just as necessary. These thoughts are pondered by **Giulio Calcara** and **Mika Launiala**, in **Chapter 15**, who delved into space policing and, more specifically, the legal frameworks that would allow states to conduct policing in outer space and upon celestial bodies. They conclude it is worth considering that future human space exploration and settlement may prompt a re-evaluation of current legal frameworks, as space settlements may seek independence akin to historical events, as Elon Musk's company Starlink already indicates when it states it recognises "Mars as a free planet and that no Earth-based government has authority or sovereignty over Martian activities" (Starlink 2024).

Tackling Environmental Space Crimes

Powerful space corporate conglomerates, including Jeff Bezos' Blue Origin, Richard Branson's Virgin Galactic and Elon Musk's SpaceX, promise space exploration for all humankind. But at what cost? And is it really for all of us, or just a select (wealthy) minority? Either way, how space exploration is undertaken at this moment is anything but eco-friendly (Lampkin 2021).

Launch sites are pivotal in discussions concerning the ethics and justice of space endeavours. They underscore ongoing debates about land rights and the broader implications of outer space activities. It has been suggested that contemporary interpretations of rights, sustainability and access to space must evolve alongside the space industry's growth. By integrating discussions about outer space with earthly concerns, a more inclusive dialogue can emerge, challenging notions of space exploration solely for the common good. Launch sites highlight the intersection of power dynamics and space ethics, emphasising the necessity of their inclusion in broader debates. **Karlijn Korpershoek** facilitated and enriched these discussions by exploring the relationship between terrestrial power structures and potential governance models for space activities (in **Chapter 19**), drawing on interdisciplinary perspectives to advocate for a more holistic approach to space governance and ethics.

From launching platforms, into the atmosphere, it becomes clear that global space industry emissions worsen global warming, emitting pollutants during launch and flight, mostly reliant on fossil fuels. Despite progress in green criminology on terrestrial environmental issues, space industry impacts are underexplored, as **Jack Lampkin** and **José Carpio-Domínguez** point out in **Chapter 12**. They call for more research on space industry emissions and their atmospheric harm, because failing to address these emissions risks exacerbating climate injustices, including polar melting, sea-level rise, ecosystem loss, and environmental displacement, making Earth an increasingly uninhabitable place. Hence, we could very well not have a liveable Earth anymore to launch from if not done so sustainably, a valid juxtaposition.

International space agreements stress space's non-appropriation, yet geopolitics influences binding legal definitions, while public discourse shapes cooperation, defining just and reasonable behaviour through soft law consensus, which matters for sustainable space exploration. In **Chapter 11, Sabine Winters** argues that Rawls' veil of ignorance could assist in promoting objectivity and reasonableness, especially when justice in space debris management requires policies aligning with future space users' needs and sustainability practices. Rawlsian principles advocate for fairness and impartiality in defining space norms, enhancing political legitimacy, by which current space debris uncertainties can be addressed and equitable regulations for a sustainable space future promoted.

One way to move toward such sustainable space futures through equitable regulation, would be by moving the international criminal law's focus on protecting the environment through safeguarding human interests during conflict, indirectly safeguarding the environment, to more specific legislation on environmental space crime, as **Eman Aboelkhair** argues in **Chapter 14**. To enact a true environmental shift in international criminal law, incorporating a new international crime in the Rome Statute specifically addressing environmental damage's severity, complexity, in varied contexts, is imperative.

The difficulty with clear and sustainable ways of protecting space environments, may lie in the idea that space, from a human-centred perspective, is dead space, or, non-living. **Chelsea Haramia** problematises this perspective and emphasises the need for ethical considerations surrounding early-stage space expansion, particularly regarding sterile environments devoid of life (in **Chapter 22**). She advocates for thorough engagement with debates about the intrinsic value and moral consideration of these environments and artificial entities, stressing the importance of deliberating on ethical principles to guide actions in space exploration. Ultimately, ethical considerations must evolve alongside space exploration, shaping just and morally appropriate attitudes towards the cosmos.

Anticipating Space-Enabled Sexual Violence

Whereas environmental space crimes have happened and will most likely continue to do so when they remain legally unaddressed, there is a very wide range of other criminal and harmful activities space criminal justice and ethics must expect to be confronted with in the future. Perhaps one of the types of crime and harms that raise much concern, in relation to "sex in space" still being a taboo (Gorman 2023), is that of sexual violence in/because of space. One of them being, space-enabled image-based sexual violence (IBSV).

IBSV could be classified as a space-related crime due to the potential for widespread victimisation and perpetration facilitated by global satellite Internet coverage, as **Valérie Pijlman** shows in **Chapter 6**. Given the significant and potentially enduring impacts of IBSV victimisation on victims' welfare, as well as the risk of ongoing victimisation, measures must be implemented to minimise harm, provide victims with avenues for redress and justice, and prevent future abuse. Valérie advises to go on further empirical and legal exploration of this overlooked yet critical issue.

Space-enabled sexual violence could also manifest itself in outer space, given that human activity expands into outer space. As existing legal frameworks lack provisions for crimes among individuals in space, criminology may already offer insights into preventive actions against sexual harassment and assault in space. Drawing parallels with remote environments like Antarctica and COVID-19 lockdowns, **Kerry Clamp** explores sexual harassment and assault in space (in **Chapter 17**), and makes a strong case to shift from punishment to risk management, advocating restorative justice practices to address harmful space conduct.

In fact, relying solely on punitive measures to police and/or retaliate against sexual harassment, like incarceration and solitary confinement in space colonies, risks exacerbating existing problems. Instead, restorative practices such as mediation could be vital for maintaining safety and harmony in extraterrestrial habitats, as **Victoria Nagy** demonstrates in **Chapter 18**. In environments like on Mars where resources are limited, it is essential to prioritise solutions that prevent extended isolation and ensure the well-being of victims, witnesses, offenders and other space workers and civilians. Mediation can validate victims' experiences, including those of sexual harassment and assault, promoting safety, and restoring equilibrium within the community before offenders are returned to Earth for trial.

Future Directions for Researching Space Crime, Criminal Justice and Ethics

At the intersection of space crime, criminal justice and ethics, the five lessons that are drawn from the chapters within this book are concerning the contemplation of knowledge-based inquiries into space criminology, thinking of inventive methodologies, addressing jurisdictional hurdles in space crime and criminal justice, thoughts on tackling ecological space offences, and anticipating forms of sexual violence in and via outer space and how to address them.

This book is just the beginning though, having served as an introductory collection of works, to lay the groundwork for exploring the intricate intersection of space crime, criminal justice, and ethical considerations. There is far more to explore together, and with the inclusion of many more disciplines working on crime, criminal justice and ethics; perhaps especially the ones who do not realise yet that they should (and maybe will after having considered this book).

So, we marvel: What is next? What is further possible? Where shall we go to, and with whom? We do not have definitive answers, but here are a few topics and ideas to pick up on, and to move forward with at the intersection of space crime, criminal justice and ethics.

Volume Space Crimes and Harms

Next to paying attention to specific forms of space crime, such as Earth crimes in or via space, including organised crime in space, space piracy, hijacking and non-terrestrial terrorism (Emery 2013; Miller 2019), there is a need for analyses of dangers and harms related to human-missions by criminologists, criminal justice scholars and ethicists, that focus on spacefarers who are living in confined spaces and may suffer from (mental) health issues that can cause (violent) transgressions, crime and suicide (Axpe et al. 2020; Patel et al. 2020; Robinson 1974; Szocik 2019). Think also of (unintended) physical harms caused by microgravitational effects on the human body, such as osteoporosis (Szocik et al. 2019), most likely to become prevalent among those who have been tasked (unwillingly) to undertake astromining.

Escapist White Collar Space Crime

Given the public concern on escapism of neoliberal space leadership (Vdovychenko 2022), as embodied by Elon Musk, Jeff Bezos and Richard Branson's Billionaire Space Race (Kern 2021), it is recommendable that scholarship on space crime, criminal justice and ethics gears itself toward the powerful that try to escape the consequences of climate change on Earth and enjoying impunity, while continuing their harmful and (eco)criminal corporate crimes in space and on other planets.

The Quantum-Space Criminal Entanglement

Quantum technologies optimise spacecraft trajectories, enhance interplanetary communication through instant transmission and fortify cybersecurity with quantum cryptography. Moreover, quantum simulations offer insights into cosmic phenomena, aiding astrophysics. The European Commission (2024) has made a priority of using quantum technologies especially in space applications. What does such quantum technological development in space exploration entail for crime, criminal justice and ethics in space, though? Next to it being deployed in space warfare (Krelina 2021), having all sorts of harmful, war crime and human rights violations and implications, quantum technology could accelerate human exploration and settlement in space. As we may see the rise of a culture of galactic control (Eski 2023a) and given the knowledge–power nexus (i.e. knowledge creates power, and power creates knowledge) (Foucault 1990), such quantum technologies could be deployed to indeed gain faster and greater insight into the universe, allowing us to wield more power over celestial bodies and potential alien (sentient) life. Therefore, the quantum-space criminal entanglement is a problem to be highly aware of and one to start critical research into.

Interplanetary Contamination

Next to environmental crimes that affect pristine places in outer space milieus, and perhaps not necessarily to be defined as a space crime in the strictest sense, are extraterrestrial-biological and anthropogenic damages, such as interplanetary microbial, organic contamination due to retrieval missions or because of returning space crews. As much as we may contaminate Earth, we may also contaminate other planets to such an extent that we would no longer be able to discover alien life (Rummel 2001). Such interplanetary contamination, holding real risks in terms of affecting extraterrestrial life and even making it go extinct before discovery, implies a range of important (existential) questions at the intersection of space crime, criminal justice and ethics that must be discussed further.

Conclusion

Let us imagine, from a Baumanian perspective (Hviid Jacobsen 2016), that outer space is a utopia for all humankind, that it has its enduring allure, and a role as a counterbalance to social stagnation, that outer space as a utopia has the potential for progress as well as peril, and is thus inherently ambiguous, also regarding the envisioning of alternative futures. We would then like to conclude that at the intersection of space crime, criminal justice and ethics, we could find and resolve, and maybe also sustain such perils of the outer space utopia. We in fact have the luxury of developing a before-the-act criminology (Eski 2023b), criminal justice scholarship and ethical consideration of utopian perils that reside in space crimes and harms. It is there, and being privileged because of it, where we can find that new hope, exactly by embracing not only the "light" but also the "dark" sides of humankind in outer space. That is only possible, though, if we start to acknowledge the urgency of addressing space crime, criminal justice and ethics to begin with. We are confident that this book has become a push in the right direction of achieving this.

References

Axpe E, Chan D, Abegaz MF, Schreurs AS, Alwood JS, Globus RK & Appel EA (2020) A Human Mission to Mars: Predicting the Bone Mineral Density Loss of Astronauts. *PloS One* 15(1), e0226434.

Carter J (2021) How Much Is a Ticket to Space? $100,000 if you Can Wait a Decade: But Here's How To Pay Nothing. From: https://www.forbes.com/sites/jamiecartereurope/2021/07/19/how-much-is-a-ticket-to-space-100000-if-you-can-wait-a-decade-but-heres-how-to-pay-nothing/.

Emery AJ (2013) The Case for Spacecrime: The Rise of Crime and Piracy in the Space Domain. From: apps.dtic.Mil/sti/pdfs/ADA615000.pdf.

Eski Y (2023a) Space, the Final Frontier to Exploit and Annihilate: Space Exploration, Inhabitation and Settlement. In *A Criminology of the Human Species: Setting an Unsettling Tone* (pp. 71–98). Cham: Springer Nature Switzerland.

Eski Y (2023b) Crime Among the Stars: Space Criminology and the Domain of Limitlessness (transl. from Dutch: Criminaliteit Tussen de Sterren: Ruimtecriminologie en het Domein van het Grenzeloze). From: https://criminologie.nl/portfolio/de-criminoloog-28/.

European Commission (2024) Quantum Technologies. From: https://defence-industry-space.ec.europa.eu/eu-space-policy/eu-space-research/quantum-technologies_en.

Foucault M (1990) *The History of Sexuality: An Introduction* (Volume I). New York: Vintage.

Galloway E (1986) Government in Action: The Role of Political Science in Outer Space Activities. *Acta Astronautica*, 13(6–7),467–472.

Gorman A (2023) Contact Zones and Outer Space Environments. In JS Schwartz & L Billings (Eds) *Reclaiming Space: Progressive and Multicultural Visions of Space Exploration*. Oxford: Oxford University Press, pp. 215–231.

Hviid Jacobsen M (2016) Zygmunt Bauman–An Ambivalent Utopian. *Revue Internationale de Philosophie* (3), 347–364.

Kern S (2021) No, Billionaires Won't "Escape" to Space While the World Burns. From: https://salon.com/2021/07/07/no-billionaires-wont-escape-to-space-while-the-world-burns/.

Krelina M (2021) Quantum Technology for Military Applications. *EPJ Quantum Technology*, 8(24), 1–53.

Lampkin JA (2021) Mapping the Terrain of an Astro-Green Criminology: A Case for Extending the Green Criminological Lens Outside of Planet Earth. *Astropolitics: The International Journal of Space Politics & Policy*, 18(3), 238–259.

Lampkin JA & White R (2023) *Space Criminology: Analysing Human Relationships with Outer Space*. Cham: Springer Nature.

Masali M, Ferrino M, Argenta M & Micheletti Cremasco M (2010) Anthropology: Physical and Cultural Adaptation to Outer Space. In H Benaroya (Ed.) *Lunar Settlements*. Boca Raton, FL: CRC Press (Taylor & Francis), pp. 165–173.

Messeri L (2016) *Placing Outer Space: An Earthly Ethnography of Other Worlds*. Durham, NC: Duke University Press.

Miller GD (2019) Space Pirates, Geosynchronous Guerrillas, and Non-Terrestrial Terrorists: Non-State Threats in Space. *Air & Space Power Journal* 33(3), 33–51.

Patel ZS, Brunstetter TJ, Tarver WJ, Whitmire AM, Zwart SR, Smith SM & Huff JL (2020) Red Risks for a Journey to the Red Planet: The Highest Priority Human Health Risks for a Mission to Mars. *NPJ Microgravity* 6(1), 1–13.

Robinson GS (1974) Psychoanalytic Techniques Supporting BioJuridics in Space. *Journal of Space Law* 2(1), 95–106.

Rock P (2005) Chronocentrism and British Criminology. *The British Journal of Sociology* 56(3), 473–491.

Rummel JD (2001) Planetary Exploration in the Time of Astrobiology: Protecting Against Biological Contamination. *Proceedings of the National Academy of Sciences* 98(5), 2128–2131.

Starlink (2024) Terms of Service. From: https://www.starlink.com/legal/documents/DOC-1020-91087-64.

Szocik K (2019) *The Human Factor in a Mission to Mars: An Interdisciplinary Approach*. Cham: Springer.

Szocik K, Campa R, Rappaport MB & Corbally C (2019) Changing the Paradigm on Human Enhancements: The Special Case of Modifications to Counter Bone Loss for Manned Mars Missions. *Space Policy* 48, 68–75.

van Heemstra M (2022) In de Ruimte zijn Nauwelijks Regels: Maar des te Meer Belangen. From: https://decorrespondent.nl/13432/in-de-ruimte-zijn-nauwelijks-regels-maar-des-te-meer-belangen/.

Vdovychenko N (2022) Elon Musk's SpaceX: How the 'Space Race' to Mars Adopted the Californian Ideology. *Diggit Magazin*. From: diggitmagazine.com/articles/elon-musk-spacex.

Index

A page number in *italics* denotes an illustration. The letter n following a page number indicates a reference in the notes.

FSC
www.fsc.org
MIX
Papier | Fördert gute Waldnutzung
FSC® C083411

Kolibri 360 GmbH
Ferdinand-Jühlke-Straße 7
99095 Erfurt, Deutschland
produktsicherheit@kolibri360.de